Chinese Fishes

DEVELOPMENTS IN ENVIRONMENTAL BIOLOGY OF FISHES

28

http://springer.com/series/5826

David L.G. Noakes · Aldemaro Romero · Yahui Zhao · Yingqi Zhou
Editors

Chinese Fishes

Reprinted from *Environmental Biology of Fishes*, Vol 86-1, 2009

Springer

Editors
David L.G. Noakes
Oregon State University,
Dept. Fisheries & Wildlife,
Corvallis OR 97331-3803,
USA

Aldemaro Romero
Arkansas State University,
Dept. Biological Sciences,
PO Box 599
State University, AR 72467,
USA

Yahui Zhao
Chinese Academy of Sciences, Institute of Zoology,
Key Lab. Zoological Systematics,
Datun Road, 100101 Beijing,
People's Republic of China

Yingqi Zhou
Shanghai Ocean University,
334 Jun Gong Road,
200090 Shanghai,
People's Republic of China

ISBN: 978-90-481-3457-1

Springer Dordrecht Heidelberg London New York

Library of Congress Control Number: 2009939494

Printed on acid-free paper

Springer is part of Springer Science+Business Media (www.springer.com)

中國畫

The photo shows the later stage of fishing process on Big head carp by means of Ju net in Qiandao Lake (Thousands Islands lake) of Zhejiang Province, China.
Photographer: Yingqi Zhou

Environmental Biology of Fishes

Volume 86 · Number 1 · September 2009

Chinese Fishes
Guest Editors: David L.G. Noakes, Aldemaro Romero, Yahui Zhao and Yingqi Zhou

Springer

Abstracted/Indexed in *Academic OneFile, AGRICOLA, ASFA, Biological Abstracts, BIOSIS Previews, CAB Abstracts, CAB International, CSA/Proquest, Current Abstracts, Current Awareness in Biological Sciences (CABS), Current Contents/Agriculture, Biology & Environmental Sciences, Elsevier Biobase, EMBiology, Environment Index, Gale, Geobase, GeoRef, Global Health, Google Scholar, IBIDS, INIS Atomindex, Journal Citation Reports/Science Edition, OCLC, PASCAL, Science Citation Index, Science Citation Index Expanded (SciSearch), SCOPUS, Summon by Serial Solutions, TOC Premier, Zoological Record*

Instructions for Authors for *Environ Biol Fish* are available at http://www.springer.com/10641.

Environ Biol Fish (2009) 86:1–9
DOI 10.1007/s10641-009-9440-4

Introducing China

David L. G. Noakes

Received: 4 January 2009 / Accepted: 18 January 2009 / Published online: 28 January 2009

For many years I have told my students that the answer to any question is "China". There is, of course, more than a grain of truth in that assertion. But for a country with a recorded history of 4,000 years and a current population of more than one billion people this might seem little more than mere probability. I first visited China in 2005 and could see for myself the truth of that statement. Chinese colleagues I met at the World Fisheries Congress in Vancouver, British Columbia in 2004 arranged everything for me. I visited universities, research institutions, museums and aquaculture facilities. I gave lectures to groups of eager students and senior colleagues. I was hosted graciously everywhere I traveled, including Beijing (Figs. 1a, b, c), the Great Wall (Fig. 2), the Ming Tombs (Fig. 3), Wuhan (Fig. 4), the Chang Jiang River (Fig. 5) and the Three Gorges Project (Fig. 6). The kindness, generosity and courage of my hosts, the unbridled enthusiasm of the students and the enormity of the teaching and research activity on fishes and fish biology overwhelmed me. The many historical sites, and even the popular tourist locations, proved to be necessary and informative as part of the larger context.

One thing was clear to me. There had to be a special volume of Environmental Biology of Fishes devoted to papers from colleagues in China, to let them present their research to the international scientific community. Suzanne Mekking, our Publisher at Springer, had recently been to China for her own first visit. She enthusiastically supported my idea for the dedicated volume on China, and so it grew from there. I contacted those colleagues I had met in China and encouraged them to develop and deliver manuscripts. I had recently added Professor Yingqi Zhou to our Advisory Editorial Board. He agreed to join me as a Co-Editor of the China Volume and he immediately began contacting his colleagues, solicited manuscripts, and co-authored his own contributions.

As chance would have it, another of our Advisory Editors, Aldemero Romero, contacted me in 2006 and asked if I would be interested in a special issue of Environmental Biology of Fishes devoted to the incredible story of hypogean fishes in China, to be put together by him and his colleague, Yahui Zhao. I immediately informed him of the ongoing China volume and incorporated their hypogean proposal. We solicited manuscripts, reviewed and edited them and now we present the volume, like an elaborate scientific banquet (Fig. 7).

We have brought together manuscripts by authors from research institutions, universities and

D. L. G. Noakes (✉)
Fisheries and Wildlife Department,
Oregon Hatchery Research Center,
Oregon State University,
Corvallis, OR 97331-3803, USA
e-mail: David.Noakes@oregonstate.edu

Fig. 1 **a** My colleague and guide in China, Jianzhong Shen, in the Forbidden City, Beijing. **b** The entrance to Tian' anmen Square, Beijing. **c** The clock counting down to the 2008 Olympics in Beijing

government agencies. There is aquaculture, life history, genetics, marine and freshwater biology, conservation, physiology, new species descriptions, and of course the amazing hypogean fishes. There are papers dealing with some of the largest fishes, such as the Chinese sturgeon (*Acipenser sinensis*) (Figs. 8a, b), and some of the smallest cave species. We have papers dealing with some of the most traditional forms of aquaculture (Fig. 9) and others with the most modern molecular techniques (Fig. 10). We have papers on the most threatened fishes, the Chinese paddlefish (*Psephurus gladius*) (Fig. 11), and the most common food species, grass carp (*Ctenopharyngodon idella*).

We had several objectives in publishing this special volume. I have already stated my initial intention after my visit to China. Certainly the primary purpose was to provide Chinese colleagues the opportunity to publish their individual manuscripts in an English language journal in a highly competitive international scientific climate. We wanted to publish as broad a collection of papers as possible, to give a sense of the current state of fish biology in China. We wanted to publish information on rare and threatened species, and how China is dealing with that situation, particularly in light of our series on Threatened Fishes of the World in Environmental Biology of Fishes. The situation of Chinese hypogean fishes is of course remarkable. There are so many species, most described only recently, and much of that literature as yet unknown to those of us who cannot read the Chinese language. Those fishes clearly deserve as much attention as this publication can bring to them. We also invited as many Chinese colleagues as possible as external reviewers of manuscripts, to include their evaluations, opinions and perspectives. It is as important to have reviewers and editors, as it is to have authors.

So is the answer, China? There is perhaps an inevitable sense of paternalism that those of us with a scientific history of a few hundred years, and a peculiar western European linguist heritage should in any sense "discover" China or Chinese science. Of course we only discover it for ourselves, and those who share our particular circumstances. China and Chinese science

Fig. 2 The Great Wall, perhaps the most universally recognized international view of China

Fig. 3 The Ming Tombs, one of numerous sites presenting the extensive historical record of China

has developed, survived and adapted over millennia, and will continue to do so (Fu 2008). The monumental series by Joseph Needham (e.g., Bray 1986) documents much of that history. This most recent contact will benefit us all, and the expanded opportunities we can realize will enrich our science and our societies. Other western scientists have reported on fishes in China, a notable example being John Treadwell Nichols (1943). However, their tone has often been one of "discovery" of the fishes and not much about the fish biologists or their science. Western science "discovered" Chinese fishes only in the Linnaean sense. Not only had the Chinese discovered their own fishes, they were first domesticated in China, the goldfish (*Carassius auratus*) in the Yangtze region.[1]

We have a great deal to learn by continued contacts and collaborations with Chinese colleagues and

[1] http://www.af.czu.cz/en/?r=2233&dep=36&part=7&pub=1187683011&wp=katedry.publikace..info

Fig. 4 The entrance to Guazhong Agricultural University in Wuhan, home of the very active Faculty of Fisheries

Chinese fishes, from remarkable new species to solutions of invasive species and threatened fishes we share with them (Hwang et al. 1982; Zhu 1995). My own experience has included study of invasive Asian carp in North America (Fig. 12; Morrison et al. 2004), and discussions with colleagues in China about the potential threatened status of those same species in their native range. I have studied threatened North American sturgeon (Noakes et al. 1999) and discussed the status of the threatened Chinese sturgeon with colleagues in China (Fig. 13). I have studied the impacts of dams on North American

Fig. 5 Chang Jiang (Yangtze) River, in many ways central to Chinese culture and history, as well as fish research

Fig. 6 The Three Gorges Project, a monumental work in progress

fishes (McLaughlin et al. 2006) and have shared discussions of dams and their impacts on fishes with colleagues in China. I have taken my own steps along the course of continued contacts and collaborations by hosting Chinese colleagues at my research laboratory (Fig. 14).

I thank all my Chinese colleagues, especially Yingqi Zhou, Daiquin Chen and Jianzhong Shen, for everything they have done to assist me in bringing this special volume together. Aldemaro Romero and Yahui Zhao handled all the material on hypogean fishes. They also helped review and edit of a number

Fig. 7 Typical dishes in a Chinese banquet in Wuhan

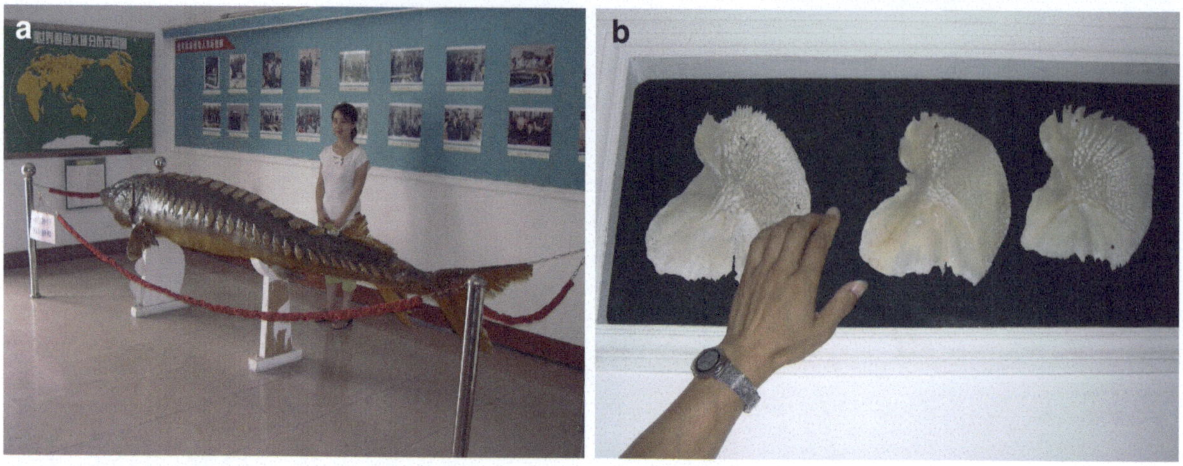

Fig. 8 **a** A Chinese sturgeon, *Acipenser sinensis*, on display at the Yangtze River Fisheries Research Institute. **b** Individual scutes from Chinese sturgeon, *Acipenser sinensis*, at the Yangtze River Fisheries Research Institute

of other manuscripts, and wrote their own major review of hypogean fishes for this volume. Suzanne Mekking, Martine van Bezooijen and Lynn Bouvier were as helpful as always in handling all the details. A special thanks to Jianzhong Shen and his many colleagues for arranging and supporting my travel in China, and to Yongwen Gao for his continued friendship and support, especially his personal translation of my name card to make my introduction to China so much more interesting (Fig. 15).

Fig. 9 A traditional farm pond near Iching

Fig. 10 The Key Lab of Freshwater Fish Germ Plasm and Biotechnology of the Ministry of Agriculture at the Yangtze Research Institute

Fig. 11 A display specimen of a Chinese paddlefish, *Psephurus gladius*, at the Yangtze River Fisheries Research Institute

Fig. 12 Students from my laboratory (Craig Blackie, Paul Vecsei and Dylan Weese, left to right) with a specimen of bighead carp, *Hypopthalmichthys nobilis*, I identified from Lake Erie

Fig. 13 Qiwei Wei, Key Laboratory of Freshwater Fish Germplasm Resources and Biotechnology, Ministry of Agriculture of China, Yangtze River Fisheries Research Institute, Chinese Academy of Fisheries Science

Fig. 14 Student group from Nanjing Agricultural University on a tour of the Oregon Hatchery Research Center

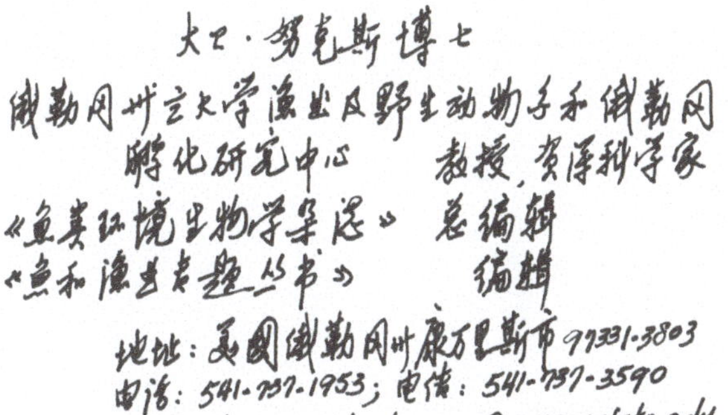

Fig. 15 My Chinese name card, courtesy of Yongwen Gao

References

Bray F (1986) Science and Civilization in China. Volume 6. Biology and Biological Technology. Cambridge University Press, Cambridge

Fu B (2008) Blue skies for China. Science 321:611

Hwang HC, Yueh PC, Yu SF (1982) The freshwater fishes of China in colored illustrations. Vol. 1, 2, 3. Shanghai Sciences and Technology Press, Shanghai, China

McLaughlin RL, Porto L, Noakes DLG, Baylis JR, Carl LM, Dodd HR, Goldstein JD, Hayes DB, Randall R (2006) Effects of low-head barriers on stream fishes: taxonomic affiliations and morphological correlates of sensitive species. Can J Fish Aquat Sci 63:766–779

Morrison BJ, Casselman J, Johnson TB, Noakes DLG (2004) New Asian carp genus (*Hypopthalmichthys*) in Lake Erie. Fish 28:1–3

Nichols, JT (1943) The freshwater fishes of China. Natural History of Central Asia: Volume IX. The American Museum of Natural History

Noakes DLG, Beamish FWH, Rossiter A (1999) Conservation implications of behaviour and growth of lake sturgeon, *Acipenser fulvescens*, in northern Ontario. Environ Biol Fish 55:135–144

Zhu S (1995) The synopsis of freshwater fishes of China. Jiangsu Science and Technology Publishing House, Nanjing

Environ Biol Fish (2009) 86:11–12
DOI 10.1007/s10641-008-9337-7

Threatened fishes of the world: *Brachymystax lenok tsinlingensis* Li, 1966 (Salmonidae)

Yahui Zhao · Chunguang Zhang

Received: 3 December 2006 / Accepted: 28 January 2008 / Published online: 4 March 2008

Keywords *Brachymystax lenok tsinlingensis* · Threatened fish · China

Common name: Hua Yu, Lu Yu (Chinese), Qinling lenok, five-coloured fish.

Conservation status: Second class state protected wild animal in China. Vulnerable—China Red Data Book of Endangered Animals, Pisces (Yue and Chen 1998). Vulnerable—China Species Red List, Vol. 1 Red List (Wang and Xie 2004). **Identification:** Dorsal fin iii–iv, 10–11; anal fin iii, 9; pectoral fin i, 15–16; pelvic fin i, 9. Lateral line scales 123–135. Gill rakers on the first gill arch 19–23. Vertebra 58–59. Head blunt, top broad, flat and slightly convex at center; mouth sub-inferior; jaws reaching to below eye level. Scales minute, lateral line complete and straight. Adipose fin opposite to anal fin. Auxiliary scales to pelvic fin present, long. Caudal fin forked. Pyloric caeca 65–75. Back of living fish dark brownish, fading on sides and turning to whitish on belly; long-oval black spots with pinkish edges scattered on back and sides, with largest spots at least half of eye diameter; also 4–5 round, blackish spots on dorsal fin base (Li 1966; Yue and Chen 1998). Standard length generally not longer than 300 mm (around 5 years old, Yang et al 2006), largest known specimen measured 450 mm (Song 1987). **Distribution:** *Branchymystax lenok tsinlingensis* is endemic to China, and only found in parts of the Taibaishan Moutains segment of the Qinling Moutains, especially in the Heihe, Shitouhe, Xushui and Taibaihe Rivers (Li 1966; Song 1987). **Abundance:** The *B. lenok tsinlingensis* population size has been little studied and thus is difficult to estimate. There were around 50,000 individuals in Xushui River based on a rough estimation in 1997 (Yang et al 1999). **Habitat and ecology:** *B. lenok tsinlingensis* inhabits cold-water mountain streams and rivers with elevations ranging from 900 to 2,300 m above sea level. It strongly prefers swift currents and clear water over large gravel substrates. This carnivorous fish mainly feeds on insects including fireflies, ladybugs, gadbees,

Y. Zhao · C. Zhang (✉)
Institute of Zoology, Chinese Academy of Sciences,
Beijing 100080, People's Republic of China
e-mail: fish@ioz.ac.cn

Y. Zhao
e-mail: zhaoyh@ioz.ac.cn

dayflies and ants. Small fish are also sometimes found among stomach contents (Li 1966). **Reproduction:** Maturity is reached after no less than 2 years. Egg diameter is generally from 1.0 to 1.3 mm (1.5 mm at most). In February and March, with water temperatures still below 10°C, mature fish spawn in shallow water along the bottom over a substratum of sand and gravel (Li 1966; Yue and Chen 1998). **Threats:** Overfishing is one of the main reasons for population decline, especially in conjunction with illegal fishing methods such as poisoning, electrification and the use of explosives. A shrinking geographical distribution due to environmental change by dam construction and mining also enhances the vulnerability to extinction. **Conservation recommendations:** A protected area has been established for protecting water resources in the Heihe River region. Therefore, Heihe River population enjoys some protection from local officials. There is an artificial breeding station in Zhanghe River, preliminary feeding and breeding has been successfully made. Local governments should enforce related regulations prohibiting harmful fishing practices in other places where *B. lenok tsinlingensis* has been recorded. A detailed survey of the stocks of this fish is also urgently needed. **Remarks:** *B. lenok tsinlingensis* is markedly different from the nominal species *B. lenok lenok* by having fewer pyloric caeca (65–75 vs. 91–111), lateral line scales (123–135 vs. 132–175) and gill rakers on the first gill arch (19–23 vs. 24–26). Spots on the body and maturity are also different between two sub-species (Li 1966). This sub-species probably should be promoted to full species status. Our recent genetic work also supports it.

References

Li SC (1966) On a new subspecies of fresh-water trout, *Branchymystax lenok tsinlingensis*, from Taipaishan, Shensi, China. Acta Zootaxon Sin 3(1):92–94

Song SL (1987) Salmonidae. In: Shaanxi Institute of Zoology, Institute of Hydrobiology, Chinese Academy of Sciences, and Department of Biology, Lanzhou University (eds) Fishes in Qinling Mountain Area. Science Press, Beijing, pp 14–15

Wang S, Xie Y (2004) China species red list, vol. 1. Red list. Higher Education Press, Beijing, p 160

Yang DG, Wei QW, Li XX et al (1999) The distributing actuality and protecting countermeasure of rare aquatic animals in Xushui River of Qinling Mountains. J Fish Sci Chin 6(3):123–125

Yang JY, Zhong L, Wang KF et al (2006) Biological studies and protection proposes of *Brachymystax lenok* in Heihe River Shaanxi Province. J. Shaanxi Normal Univ. 34(Sup.): 67–69

Yue PQ, Chen YY (1998) China red data book of endangered animals, Pisces. Science Press, Beijing, pp 107–109

Environ Biol Fish (2009) 86:13–22
DOI 10.1007/s10641-009-9498-z

Changes in abundance of larvae of the four domestic Chinese carps in the middle reach of the Yangtze River, China, before and after closing of the Three Gorges Dam

Xinbin Duan · Shaoping Liu · Mugui Huang · Shunling Qiu · Zhihua Li · Ke Wang · Daqing Chen

Received: 14 August 2008 / Accepted: 2 June 2009 / Published online: 16 June 2009

Abstract The four domestic Chinese carps, i.e. grass carp (*Ctenopharyngodon idella*), black carp (*Mylopharyngodon piceus), silver carp (Hypophthalmichthys molitrix)* and bighead carp (*Hypophthalmichthys nobilis*), cast their drifting eggs in the running water of the Yangtze River. In 1997~2005, a systematic study on the dynamics of the abundance the larvae of the four domestic Chinese carps was performed in the middle reach of the Yangtze River, close to Sanzhou city. It was calculated that in breeding seasons from early May to late June every year the abundance of fry drifting through the section of Sanzhou was 3.59 billion in 1997, 2.75 billion in 1998, 2.15 billion in 1999, 2.85 billion in 2000, 1.90 billion in 2001, 1.90 billion in 2002, 406 million in 2003, 339 million in 2004 and 105 million in 2005 respectively. Several environmental parameters related to the dynamics of larval abundance were also examined simultaneously. The construction of the Three Gorges Dam near Yichang between the upper and middle reaches of the Yangtze River has apparently had a drastic influence on the abundance of fish larvae.

Keywords Fish larvae · Asian carps · Dynamic analysis · Three Gorges Dam · Yangtze River

Introduction

Grass carp (*Ctenopharyngodon idella*), black carp (*Mylopharyngodon piceus*), silver carp (*Hypophthalmichthys molitrix) and bighead carp (Hypophthalmichthys nobilis),* referred to in China as the four domestic fishes, are commercially important fish species in China with high abundance in the catch. In the 20th century, 60% of the entire catch of freshwater fishes consisted of the four domestic fishes (Wu et al. 1992).

Investigations on the reproductive ecology of the four domestic fishes in the Yangtze River have been carried out since 1960's (Yih and Liang 1964; Yu 1985; Yi et al. 1988a; Zeng 1990; Qiu et al. 2002). It was reported by Yih and Liang (1964) that the spawning grounds ranged from Chongqing to Pengze with a range of 1,695 km. There are 36 spawning grounds distributed in the mainstream of the Yangtze River, and the mean egg production each year was over 100 billion in the 1960's. The breeding season of the four species was from May to July every year in the Yangtze River (Yih and Liang 1964). Water temperature and hydrograph were the most important

X. Duan · S. Liu · M. Huang · S. Qiu · Z. Li · K. Wang · D. Chen (✉)
Key Laboratory of Freshwater Fish Germplasm & Biotechnology of Ministry of Agriculture, Yangtze River Fisheries Research Institute,
Chinese Academy of Fishery Sciences,
Jingzhou 434000, China
e-mail: chdq@yfi.ac.cn

factors controlling spawning (Stanley 1978; Liu et al. 1992). The lowest water temperature for spawning is 18 . If the water temperature is below 18 during the breeding season, spawning will stop even if there is flooding (Cao and Yu 1987). The triggering stimulus for spawning is an increase in discharge. Spawning occurs when the water level rises, but it stops immediately as soon as the water level starts to recede (Yi et al. 1988b).

After the construction of the Gezhouba Water Control Project (completed in the 1980's), six spawning areas were inundated by the reservoir, 30 spawning grounds remained, with yearly egg production of 20 billion or more in the 1980's (Yu 1988). The effect of the Three Gorges Project on the reproduction of the four species became a highly sensitive issue in the early 1990's in China (Sun 1990). This investigation was initiated to determine the influence of the Three Gorges Project on the abundance of larvae of the four domestic Chinese carps, and to make countermeasures required for protection of the four domestic Chinese carps.

Materials and methods

Sampling area

The investigation was in the Sanzhou section of the Yangtze River (E112°55′ N29°32′), close to the inlet of Dongting Lake (Fig. 1). The width of the main channel was about 1200 m between revetments. The maximum water depth measured at the sampling sites was 23 m.

Sampling design and collecting methods

Every year from 1997 to 2005, sampling was conducted daily between May 1 and June 30. Samples were collected twice daily, once between 9:00 and 11:00, and once between 15:00 and 17:00. The gears used for capture of larvae were a ring net and a trap net based on the methods of Pavlov (Pavlov and Sbikin 1978; Pavlov et al. 1983; Pavlov 1994; Pavlov et al. 1995 Pavlov et al. 2000), (Yi et al. 1988a) and Oesmann (2003). A ring net was used to gather larvae and eggs from different water depths. The ring net consisted of a round mouth formed by a steel loop (diameter=50 cm; area=0.1963 m^2), a conical net and a collection bucket (Fig. 2). The netting was made of fine silk with the grid size of 0.5 mm. A steel frame supported the whole body. A flow meter was fixed in the mouth of the net. A large weight was attached to the bottom of the frame to assist sinking the net into the flowing water.

Trap nets were used to collect eggs and larvae from near water surface in littoral areas (Fig. 3). The trap nets had a semi-circular mouth (area=0.5384 m^2). It is usually made up of fine flax cloth suspended below a frame in a half-cone configuration terminating in a silk tube which runs to the collection box. The cloth and frame funnel the larvae to the collection box at the cod end. Below the main body is a heavy metal weight so as to open the net to its full size. The frame is made from two long bamboos. Two short bamboos were situated at the front and the back. The collection box was made of the same thin silk material as the trap net. The collection box was 40 cm long and had width and height of 30 cm. The upper part of the case is wide open for gathering eggs and larvae. The buoyancy of the collection box keeps its open top above the water surface.

The trap net was anchored in place, and allowed to float on the surface of the water, keeping parallel with the bank. The trap nets were deployed both day and night. The anchor positions of the three sampling stations, one on the right side of the main channel (right), one on the left side (left), and one in the middle (midstream), were marked using bearing stakes on the riverbank. Two stakes marked the position of the straight cross-section in the longitudinal direction of the main channel. The intersections of this line with three other straight lines marked the distinct anchoring positions.

Samples were collected at three depths: 1 m below the water surface (surface), in the middle of the water column (mid layer), and 1 m above the substrate, referring to the middle of the net's mouth (Fig. 4).

Sample preservation

After sampling, the larvae were taken out from the trap net and the ring net and fixed with 4% formaldehyde and 70% ethanol mixed solution. Species identification was carried out in the laboratory according to Yi et al. (1988b).

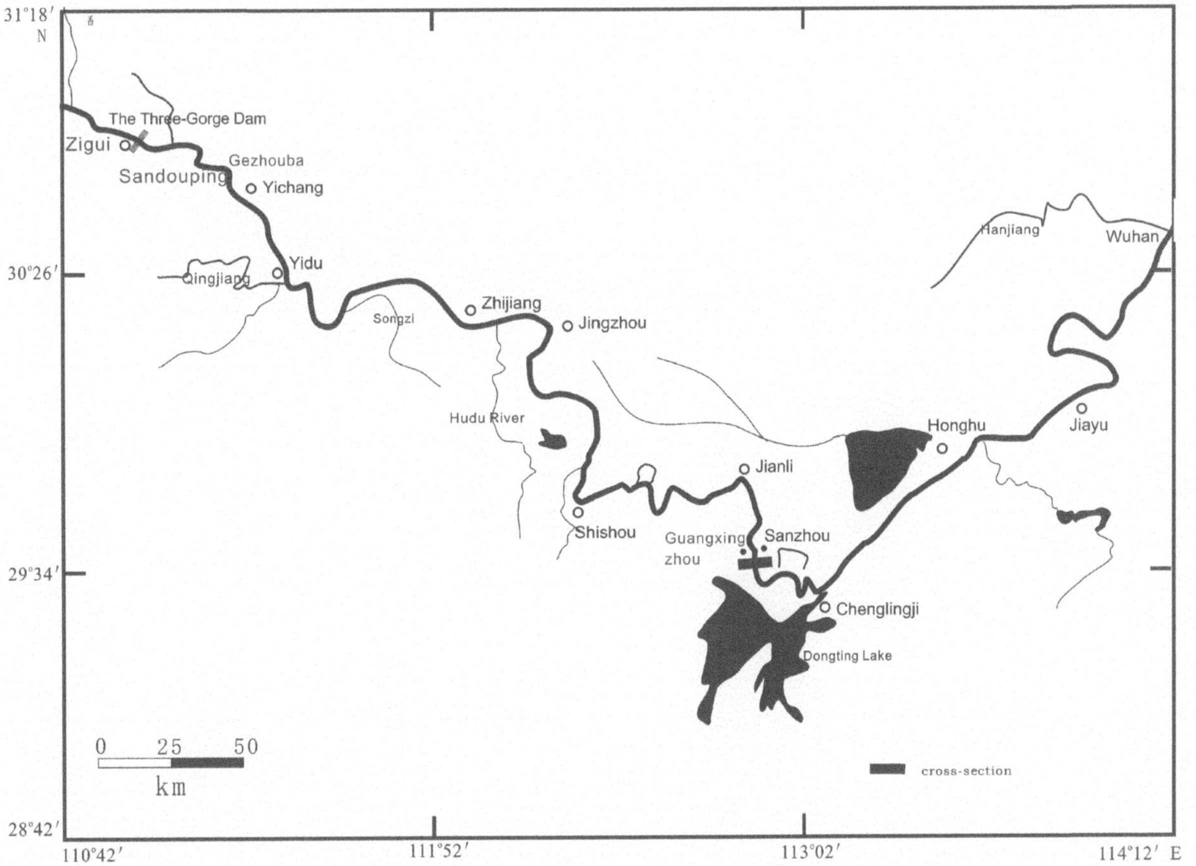

Fig. 1 The distribution of the four domestic fishes' spawning grounds in the middle of Yangtze River

Environmental parameters

Several abiotic parameters were measured at the different sampling sites to determine their relationships to the abundance of eggs and larvae. The fishing depth was measured with a Biosonics DT-X sonar system (Biosonics Sound Navigation and Ranging Inc.). The distance to the shoreline was measured (precision ±0.1 m) using a mobile global positioning system (Model 315 Magellan). The current velocity was measured with a Current Velocity Meter (Model LJD 3, Chongqing Current Velocity Meters Inc.). The water temperature was measured in Celsius scale using a hand-held thermometer.

Fig. 2 Ring net (used to gather larvae and eggs from different water depths)

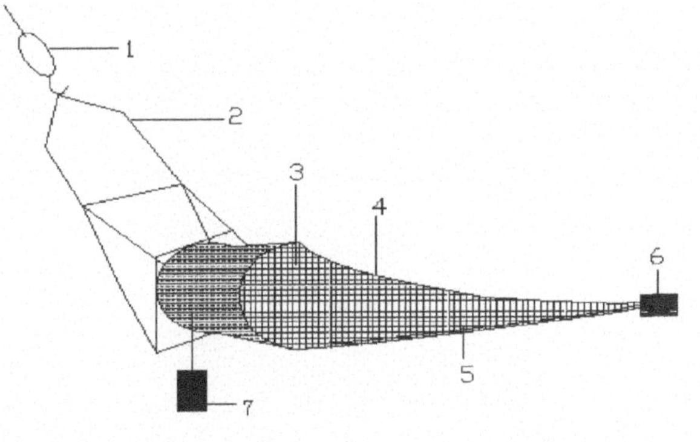

(1 windlass pothook; 2 frame; 3 current velocity meter; 4 conical net; 5 net-covering strength strip; 6 collection bucket; 7 weight)

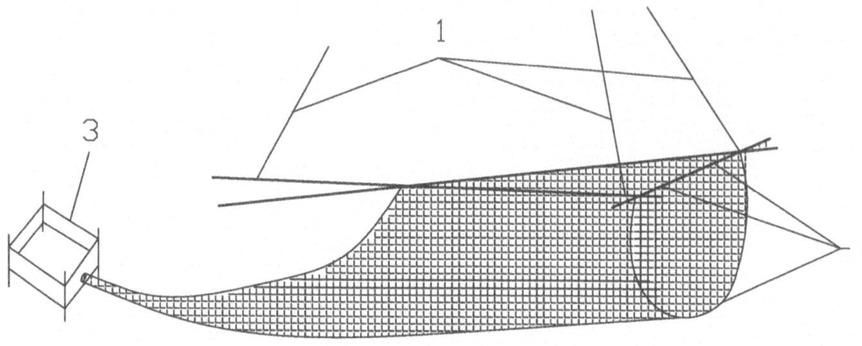

Fig. 3 Trap net (used to collect eggs and larvae from near water surface in littoral areas)

(1 security rope; 2 draught rope; 3 collection box)

Calculation of spawning ground and larvae amount

The location and size of the spawning ground was determined according to (Yi et al. 1988a). The abundance of the larvae Bm was calculated as

$$Bm = QMC/SV$$

Where *Bm* is the number of eggs and larvae at the cross-section; Q is the average water flow of the collecting point at the cross-section in $m^3 \cdot s^{-1}$; M is the number of larvae collected at the fixed point; C represents the runoff, emendation coefficient of the eggs or larvae at the cross-section; S is the area of the net mouth in m^2 and V is the current velocity at the net mouth in $m \cdot s^{-1}$.

Results

Reproductive habit of the four domestic fishes

Silver carp

In the Yangtze River, silver carp usually reproduce in late April to early July, and middle May to middle June is the best time for breeding (Yi et al.1988a). Spawning grounds are in the mainstream and tributaries of the Middle and Upper Yangtze River. The 396 km-long Jingjiang section (Fig. 1) is an especially important spawning area. This part of the river has sharp rock outcropping and there are sandbars in middle of the river, which create conditions favorable to silver carp spawning. The reproductive behavior of silver carp is in many ways similar to the other domestic fishes. Silver carp spawn when the water temperature is above 18 and the river begins to rise. After the discharge peaks, the spawn will stop. Spawning often occurred near the surface. Sometimes we could see the female fish swimming upside-down near the surface, the abdomen quivering with the discharge of eggs.

Bighead carp

The breeding activity of bighead carp is usually from May to July, slightly later than that of silver carp. Bighead carp have stricter requirements for hydrologic conditions than the other domestic fishes. Bighead carp spawn mostly at sharp increases in the hydrograph. Spawning will stop at once after the hydro-

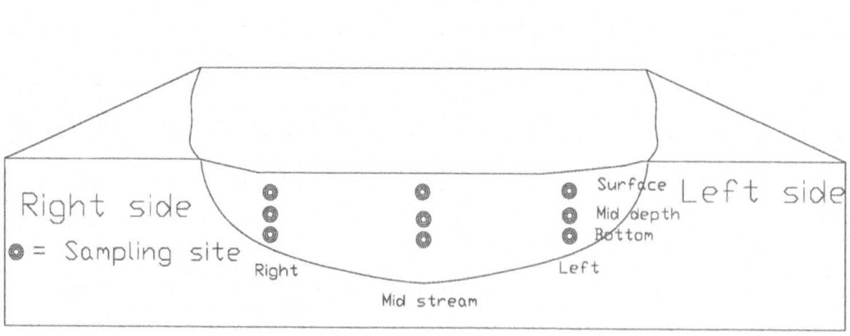

Fig. 4 Collection points for larval ring nets

graph peaks. When spawning, female and male fish also chase each other but not as violently as silver carp.

Grass carp

The reproductive period of grass carp is usually earlier than that of black carp. Generally spawning begins in late April or early May and ends in late June or early July. The reproductive requirements of grass carp are similar to the other domestic fishes. When the water temperature is above 18 and discharge increases, grass carp will be stimulated to spawn. During the peak spawning season, grass carp can also sometimes spawn during a level or receding hydrograph. The amount of larvae reproduced is little related with the beginning water level, but has great relation with the intensity of water rise or decline.

Black carp (Mylopharyngodon piceus)

The spawning period of black carp is from May to July, a little later than that of grass carp and silver carp. The spawning requirements are not as strict as for the other three species. Spawning will begin with only a slight increase in discharge. Compared to the other three domestic fishes, black carp spawning is relatively dispersed; the breeding season begins somewhat later and the total length of the spawning season is longer. In addition, black carp usually spawns deeper in the river and the quantity of eggs is highly variable.

Spawning grounds

From 1997 to 2005, there were ten spawning grounds for the four domestic fishes from Yichang to Chenglingji below the dam, a distance of 400 km (Table 1). Among these sites, Yichang, Zhijiang, Jiangkou, Hexue, Shishou and Tiaoguan were most important. Compared with the conditions in 1980's (Yu 1985), the spawning locations were almost the same, but larval abundance was lower.

Larval abundance

From May to June in 1997~2002, before the closing of the Three Gorges Dam on June 1 2003, the larval abundance of the four domestic fishes at Sanzhou were 3.6 billion, 2.7 billion, 2.2 billion, 2.9 billion, 1.9 billion and 1.9 billion respectively (Fig. 5). The average larval abundance before the closing of the Three Gorges Dam was 2.5 billion.

Table 1 The distribution of the four domestic fishes' spawning grounds from Yichang to Chenglingji

No.	Name	Range	Distance(km)
1	Yichang	Shilihong ~ Gulaobei	24
2	Yidu	Yunchi ~ Yidu	10
3	Zhijiang	Yangxi ~ Zhijiang	29
4	Jiangkou	Jiangkou ~ Yuanshi	25
5	Shashi	Huduhe ~ Guanyinsi	27
6	Haoxue	Majiazai ~ Xinchang	28
7	Shishou	Ouchihekou ~ Shishou	15
8	Tiaoguan	Laijiazhai ~ Tiaoguan	34
9	Jianli	Tasiyi ~ Shajiabian	25
10	Fanzui	Yanchuantao ~ Jingjiangmen	6

The closing of the Three Gorges Dam drastically changed the hydrology of the Yangtze River. After the dam was completed, larval abundance decreased sharply. Total larval abundance in May and June of 2004 was 0.339 billion, 13.4% of the average value before closing of the dam. By the year 2005, larval abundance dropped to 0.105 billion, 4.2% of the average value of the past few years (Fig. 5) and the lowest level in history.

Relative abundances of fish larvae

Before the peak in abundance of the larvae of the four domestic fishes, there are other peaks in the abundance of larvae of other fishes. However, these other species will spawn in the absence of a substantial increase in discharge. The investigation of 1997~2005 indicated that the larvae of the four domestic fishes accounted for 9.77%; 8.59%; 5.91%; 5.43%; 3.65%;3.12%; 2.72%; 0.80% and 0.85% of the total larval abundance, respectively (Table 2).

The investigation of 1997~2002 (before the Three Gorges Dam was closed) indicated that grass carp and black carp were the most abundant larvae among the four domestic fishes, and grass carp especially dominated (60~85%). Silver carp and bighead carp accounted for 26 000 000 (21.81%) and 213 000 000

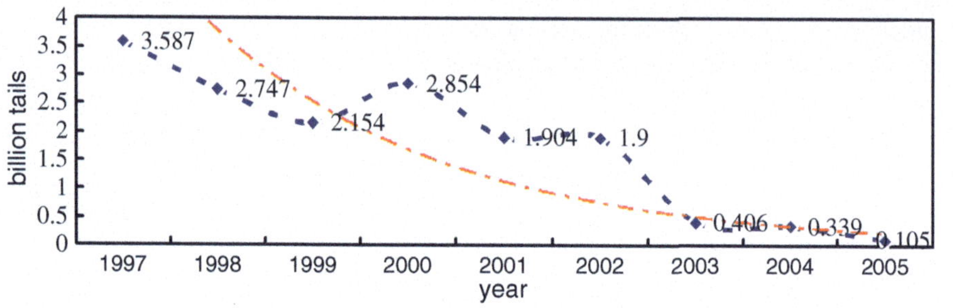

Fig. 5 Production of larvae of the four domestic fishes at Sanzhou cross-section (1997~2005)

(8.40%) individuals respectively. Since 2003 (after the Three Gorges Project was closed), the composition of larvae has undergone great changes. The proportion of silver carp increased; this value reached 66.11% in 2005. The composition of the four domestic fishes in each year can be seen in Table 2.

Peaks in larval abundance and hydrology

In the breeding season of the four domestic fishes, whether the adult fishes spawn or not is generally determined by external conditions.

Water temperature

The lower limit of water temperature for the four fishes' breeding is 18 (Yi et al. 1988b). If it is below 18 the breeding activity will stop. The best temperature for breeding is 21–24 . The water temperature in May and June of 1997–2005 at the Sanzhou cross-section ranged from 18.0 to 26.1 , which is appropriate for spawning.

Table 2 The composition of larvae at Sanzhou cross-section

Year	Proportion (%)			
	Silver carp	Bighead carp	Grass carp	Black carp
1964	13.5	8.0	43.0	35.5
1965	31.0	7.3	33.0	28.7
1981	2.6	6.8	47.9	42.7
1997	2.6	8.5	67.5	21.4
1998	1.3	7.1	67.7	23.9
1999	0.3	2.2	85.3	12.3
2000	3.0	5.0	84.9	7.1
2001	2.4	3.5	72.7	21.4
2002	5.62	2.13	74.35	17.90
2003	21.81	6.13	60.21	11.86
2004	36.84	1.94	52.24	8.98
2005	66.11	5.12	24.52	4.25

Explanation: the data of 1964, 1965, 1981 (Yu1985)

Hydrograph and larval flood

The daily hydrograph at Sanzhou cross-section in 1997~2005 is shown in Figs. 6 and 7. From 1997 to 2005, 15 larval floods (spikes in production of larvae are known in China as "larval floods") had been detected at Sanzhou cross-section (Table 3).

The correlation of water level, flux and larvae flood at Jianli cross-section in 1997~2005 can be seen in Table 3. The results indicated that the larvae flood has great relationship with the water level and the flux. In the 2 months of reproducing period, two or three large-scale larvae floods formed the laying activity for domestic fishes, and each larvae flood lasted for 3~7 days. The lasting time of a large-scale larvae flood was usually ascending simultaneously with the increase of the flux of water. Twelve large-scale larvae floods of the four famous domestic fishes had been detected before the operation of the Three Gorges Project. The maximum amount of them was 2.006614 billion individuals, which lasted for 5 days, accounting for 72.39% of the year and the minimum amount was 0.31084 billion individuals, which lasted for 9 days, accounting for 10.89% of the year.

Discussion

Relationship between hydrograph and larval flood

A large number of eggs were spawned during increases in water velocity and hydrograph. Spawning activity was positively correlated with the amount of hydrograph increase. Commonly, when the increase

Fig. 6 Daily discharge at Sanzhou cross-section from 1997 to 2002

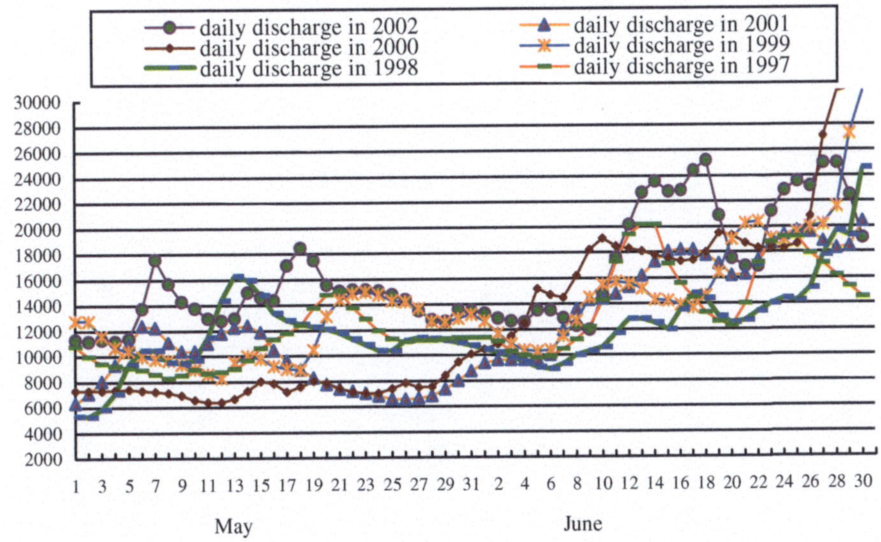

in discharge is greatest, the larval flood is bigger, most of the fishes spawned during increases in discharge.

Effect of the three gorges project on larval abundance and composition

The Yangtze River is the most important living and breeding location for the four domestic fishes. Historically, the catch of the four domestic fishes in the Yangtze River accounts for 63% of the total natural fry production in the whole country (Wu et al. 1992). Due to the building of this large-scale hydraulic project and other anthropogenic changes including water pollution, overfishing, the creation of farmland by filling lakes and the loss of connectivity between floodplain lakes and the Yangtze River, the four domestic fishes are declining rapidly. From 1997 to 2002, the mean number of larvae produced was 2.5 billion (range 1.9–3.6 billion), an obvious decline compared to the 1980's, before the construction of the Gezhouba and Three Gorges dams.

In May and June of 2003, the larval production was 0.406 billion, 16.1% of the production before

Fig. 7 Daily discharge at Sanzhou cross-section from 2003 to 2005

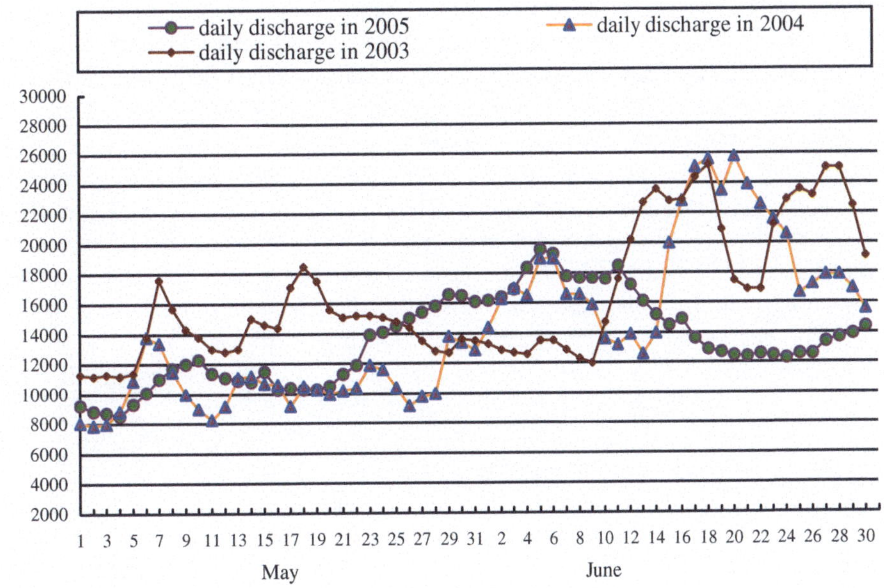

Table 3 The relationship between discharge and larval flood of the four domestic fishes at the Sanzhou cross-section

Year	Date of discharge		Water level (m)			Current (m^3/s)			Date of larval flood		Larvae amount(million) individuals)
	From	To	From	To	Range	From	To	Range	From	To	
1997	5.09	5.21	27.34	30.51	3.17	8550	15100	6550	5.20	5.24	156674
	6.06	6.15	28.13	31.60	3.47	9900	20200	10300	6.11	6.14	118
1998	5.01	5.15	25.58	30.86	5.28	5320	14800	9480	5.14	5.31	722
	6.06	6.30	28.69	35.61	6.92	9300	24600	15300	6.11	6.27	1,915
1999	5.16	5.26	27.67	30.78	3.11	9200	14200	5000	5.18	5.31	731
	6.06	6.11	29.99	30.72	0.73	10400	15700	5300	6.10	6.20	348
	6.17	6.30	30.38	34.43	4.05	13700	30600	16900	6.23	6.30	851
2000	5.12	5.15	26.63	27.09	0.46	6390	8010	1620	5.19	5.27	311
	6.07	6.10	30.06	31.49	1.43	14500	19100	4600	6.10	6.14	2,066
2001	5.01	5.06	27.79	29.59	1.80	6370	12400	6030	5.09	5.13	584
	6.05	6.09	28.16	29.75	1.59	9650	14300	4650	6.11	6.14	684
2002	5.16	5.18	32.65	33.46	0.85	14400	18500	4100	5.20	5.24	540
	6.09	6.14	30.50	32.21	1.71	12000	23600	11600	6.14	6.17	687
2003	5.8	5.9	28.15	30.24	2.09	7240	10900	3660	5.16	5.18	89
	6.11	6.14	26.91	31.13	4.22	4960	21100	16140	6.17	6.20	38
2004	6.13	6.18	30.12	32.21	2.09	12600	25600	13000	6.24	7.02	286

the closing of the Three Gorges Dam; from May to June in 2004 and 2005 larval production was 0.339 and 0.105 billion, which accounted for 13.4% and 4.2% respectively of the amount before the closing of the Three Gorges Dam. Larval abundance in 2005 was the lowest in history.

From 1997 to 2002 the most abundant fry at the section from Yichang to Chenglingji were grass carp and black carp, with grass carp as a dominating species (range 60%~85%). After the closing of the Three Gorges Dam, the composition of the fry underwent enormous changes. The proportion of silver carp rose to 66.1% in 2005. Based on the larval abundance and composition, the production of silver carp and grass carp was estimated at 0.07 and 0.026 billion respectively. Comparing with the value before the closing of the Three Gorges Dam, when the mean production of silver carp and grass carp was 0.08 and 1.9 billion individuals respectively, it is obvious that the change in silver carp production was small but the change in production of grass carp was very large. This indicates that the Three Gorges Project had little impact on silver carp but a large impact on grass carp.

Effect of the three gorges project on spawning ground and spawning

The construction of the Three Gorges Dam has inundated the existing spawning grounds above the dam. It has also caused a depression of spawning activities below the dam, because the regulated water flow does not perform similarly to the natural hydrograph. The peaks of floods were leveled off (Cao and Yu 1987; Zeng 1990). Our investigation during 1997 to 2005 indicated that from Yichang to Chenglingji below the Three Gorges Dam, a distance of 400 km, the distribution of the existing ten spawning grounds including Yichang, Zhijiang, Jiangkou, Hexue, Shishou and Tiaoguan etc. changed little compared to the 1980's, but the amount of fry was reduced. Before the closing of the Three Gorges Dam, many fry came from the upper reach of the Yangtze River. After its closing, all fry came from the reach below the dam.

The drastic alteration of the natural hydrograph has resulted in depression of spawning. From June 1 to 11, 2003, the Three Gorges Dam began storing water, no spawning activities and larvae were observed. After 11 June 2003, normal discharge was resumed, and a larval flood of the four domestic fishes occurred from 17–20 June 2003. In June 2004, discharge at Sanzhou increased from 12 600 $m^3 \cdot s^{-1}$ to 25 600 $m^3 \cdot s^{-1}$ with an increase of 13 000 $m^3 \cdot s^{-1}$. From 24 June to 2 July 2004 a larval flood occurred with a production of 0.286 billion, accounting for 95% and 84% of the larvae amount in June and May of 2004 respectively. These results indicate that sharp increases of discharge are important in the stimulation of spawning activity in the four domestic carps. It is possible that water releases from the Three Gorges Dam might be designed to simulate these natural discharge peaks, to the benefit of the four domestic carps.

Protection of the four domestic fishes

Silver carp, bighead carp, grass carp and black carp are the main cultured fishes in China. Their yield always occupies the chief position in Chinese freshwater production. From 1997 to 2003, the number of fry at the Sanzhou cross-section declined rapidly. Furthermore, the adult resources of the four domestic fishes descended very quickly; the proportion of the four fishes in the catch also dropped from 20–30% in the mid-1960s to about 5% at present, which brought great attention of government agencies charged with the protection of fisheries.

The causes of the declines of the four domestic fishes are likely the dams and gates which obstruct both longitudinal and lateral migrations of the four domestic fishes. The larvae cannot pass from the river into floodplain lakes, which are thought be feeding and nursery areas (Fig. 8) and the adult fish cannot move from floodplain lakes into the river to spawn. In addition, excessive land reclamations around the waters, water pollution and overfishing have caused declines in the beeeding community (Cao and Yu 1987).

Artificial reproduction and stocking is a common method for mitigation of negative impacts. Russia, America and Japan have benefited much from this method (Michel 2001). We recommend a closed fishing season from May to June and the release of fry of the four domestic fishes into the river at different points. This should assist in the protection and recovery of the four domestic fishes. In addition, improving lateral connectivity, reducing fishing mortality (possibly including constraints on boats, horsepower, tonnage and the size of the nets etc.), should positively affect this important fishery. Meanwhile, corresponding basic research such as the ecological effect of the Three Gorge Project on the four domestic fishes, the recovery and reconstruction of the spawning environment, the changes of the biological resource and the continuing usage of the fishery resources should be carried out. The above will help to realize a change in the prevailing trend in the

Fig. 8 Connectivity of floodplain lake and the Yangtze River

abundance of the four domestic fishes after the Three Gorge Project's operation.

Acknowledgements Many thanks to my colleagues at the Yangtze River Fisheries Research Institute of CAFS, who helped in measuring the current velocity of main channel and sorting the samples. Thanks to all participant student assistants. This study was financed by the National Science Foundation of China (No. 30490234 and No.30371105) and the Three Gorges Project Company.

References

Cao W, Yu Z (1987) Preliminary assessment of impacts of the Three Gorges Project on fish resources of the Changjiang River and approaches to the resource proliferation. Pages 2–20 *in* Symposium on Impacts of the Three Gorges Project On Ecology and Environment and Possible Countermeasures. Science Press, Beijing (in Chinese)

Liu J, Wang Z, Deng Z (1992) Reproduction of the farm fishes. Pages 65–103 *in* Liu, J. and He, B., editors. Cultivation of the Chinese Freshwater fishes. Science Press, Beijing (in Chinese)

Michel L (2001) Environmental issues, dams and fish migration. Pages 33–61 in Marmulla G., editors. Dams, Fish and Fisheries Opportunities, Challenges and Conflict Resolution. FAP Fisheries Technical Paper 419, Rome

Oesmann S (2003) Vertical, lateral and diurnal drift patterns of fish larvae in a larger lowland river, the Elbe. Appl. Icthyol 19:284–293

Pavlov DS, Sbikin YN (1978) The number of visual reference points and their influence on the manifestation of rheoreaction in fishes. Icthyol 18:1022–1026

Pavlov DS, Skorobogatov MA, Shtaf LG (1983) Threshold rheoreaction rates of roach in flows with different degree of turbulence. Dokl. An. S. S. S. R 268:510–512

Pavlov DS (1994) The downstream migration of young fishes in rivers: mechanism and distribution. Folia Zool 43:193–208

Pavlov DS, Nezdoliy VK, Urteaga AK, Sanches OR (1995) Downstream migration of juvenile fishes in the river of Amazonian Peru. Icthyol 35:227–248

Pavlov DS, Sadkovskii RV, Kostein VV, Lupandin AI (2000) Experimental study of young fish distribution and behaviour under combined influence of bacro-, photo-and thermo-gradients. Fish. Biol 57:69–81

Qiu S, Liu S, Huang M (2002) The resources of four Chinese farm fishes in the middle reach of Yangtze River. ACTA Hydrobiologica Sinica 26(6):716–719 in Chinese with English abstrac

Stanley JG (1978) Reproductive requirements and likelihood for naturalization of escaped grass carp in the United States. Trans. Am. Fish. Soc. 107(1):119–128

Sun J (1990) Study on effects of the Three Gorges Project on fish resource of the Yangtze River and lakes adjacent to the river and on approaches to the resource proliferation. Research Report of Chinese National Significant Science and Technology Project in "7.5" period. No. 75-16-06-02-01 (in Chinese)

Wu X, Rao J, He B (1992) The histoiy of the Chinese freshwater fisheries. In: Liu J, He B (eds) Cultivation of the Chinese Freshwater Fishes. Science Press, Beijing, China, pp 5–29

Yi B, Yu Z, Liang S (1988a) he distribution, natural conditions and breeding production of the spawning grounds of four domestic freshwater fishes on the mainstream on the Yangtze River. In: Yi B, Yu Z, Liang Z (eds) oject and four domestic fishes in Yangtze River. Hubei Science and Technology Press, Wuhan, China, pp 1–46 In Chinese with English abstract

Yi B, Liang Z, Yu Z, Lin R, He M (1988b) A comparative study of the early development of grass carp, black carp, silver carp and big head of the Yangtze River. In: Yi B, Yu Z, Liang Z (eds) Gezhouba Water Control Project and four domestic fishes in Yangtze River. Hubei Science and Technology Press, Wuhan, China, pp 69–135 In Chinese with English abstract

Yih P, Liang T (1964) Natural conditions of the spawning grounds of the domestic fishes in Yangtze River and essential external factor for spawning. ACTA Hydrobiologica Sinica 5(1):1–15 in Chinese with English abstract

Yu Z (1985) On spawning grounds of four Chinese farm fishes in the River Changjiang after damming of Gezhouba area. Transaction of the Chinese Ichthyological Society 4:1–12 (in Chinese with English abstract

Yu Z (1988) The present situation of the spawning grounds of the four Chinese domestic fishes in the Yangtze River after the construction of the Gezhouba water control project. In: Yi B, Yu Z, Liang Z (eds) Gezhouba Water Control Project and Four Domestic Fishes in Yangtze River. Hubei Science and Technology Press, Wuhan in Chinese with English abstract

Zeng X (1990) Study on the promotion of natural reproduction of domestic fish by regulating water rising. Ecology 9 (4):20–23 in Chinese with English abstract)

Environ Biol Fish (2009) 86:23–27
DOI 10.1007/s10641-009-9496-1

The fish fauna of mountain streams in the Guanshan National Nature Reserve, Jiangxi, China

Maolin Hu · Zhi-Qiang Wu · Yinlan Liu

Received: 2 December 2007 / Accepted: 2 June 2009 / Published online: 11 June 2009

Abstract Guanshan National Nature Reserve (GNNR) is located in the middle reach of the Jiuling Mountain Range in the southern China and the first national nature reserve within Jiangxi Province. The fish fauna of mountain streams in the GNNR was investigated from September 2006 to August 2007. A total of 379 fish were collected and classified into two families and seven species. *Acrossocheilus parallens* [relative abundance (RA) 44.3%] was the most abundant fish species. Subdominant species were *Onychostoma barbata* (RA, 20.3%), *Zacco platypus* (RA, 14.2%) and *Opsariichthys bidens* (RA, 12.4%).A greater number of species (7 vs. 2) and individuals (295 vs. 84) were collected from the West Stream compared to the East Stream. Among the seven fish species collected in the GNNR, *Acrossocheilus parallens*, *Onychostoma barbata* and *Vanmanenia stenosoma* are endemic to the Chinese mountain streams. This study suggests that mountain streams in the nation reserves including GNNR are very important for freshwater fish diversity and conservation in Jiangxi Province.

M. Hu · Z.-Q. Wu
School of Environment Science and Engineering,
Nanchang University,
Nanchang, Jiangxi 330031, People's Republic of China

Z.-Q. Wu (✉) · Y. Liu
School of Life Sciences, Nanchang University,
Nanchang, China
e-mail: wuzhiqiang108@hotmail.com

Keywords Mountain streams · Fish community · Conservation · Guanshan National Nature Reserve · Jiangxi · China

Introduction

Guanshan National Nature Reserve (GNNR) is located in the middle reach of the Jiuling Mountain Range in the northwestern Jiangxi Province between latitudes 28°30′ and 28°40′ N and longitudes 114°29′ and 114°45′ E. GNNR has an area of 115 km^2 and ranges in altitude from 200 m to 1,480 m. The climate in the GNNR belongs to the subtropics warm humid climate. The annual temperature is 16.2°C, annual precipitation ranges from 1,950 mm to 2,100 mm, and forest coverage rate is up to 93.8%. Most mountain streams in the GNNR flow into the East and West Streams which drain into the Poyang Lake. The GNNR was designated as a Chinese National Nature Reserve in April 2007 and the first national nature reserve within Jiangxi Province primary for the conservation of wildlife resources (Liu et al. 2003; Liu and Wu 2005).

There have been several notable surveys of the flora and fauna within the GNNR. These studies include surveys of the avifauna (Dai et al. 1997; Xu et al. 2007), insect fauna (Ding et al. 2005; Guo and Ding 2005) and flora (He et al. 2001; Chen 2002; He and Ji 2002; Li and Tang 2003; Ji et al. 2005). However, until our survey there have been no studies on the distribution and abundance of fish species. There is a need to document the diversity and consequent conservation value of the GNNR for fishes, as well as environmental factors impacting the fish communities. Therefore, the purpose of this paper is to present the results of our fish surveys in the GNNR.

Methods

Site selection

We surveyed a total of nine sites on East Stream and West Stream within the GNNR (Fig. 1). All study sites were located between an altitude of 460 m and 760 m. Sampling site selections were based on the representative habitat types present and accessibility during the study period. The habitat types were recorded for all sites. Water temperature, dissolved oxygen, pH, and stream width and water depth were measured at each site.

Fish sampling

Fish populations were surveyed seasonally from September 2006 to August 2007. At each site, samples were collected using an electrofishing device consisting of two copper electrodes on wooden handles, powered by a 500-watt portable AC generator. Stunned fish were collected using dip nets or caught by hand. A cast net (mesh 5×5 mm; $\pi\times0.6^2$ m=1.13 m^2) was also used within shallow pools of the stream system to collect fish. These fish sampling devices were most suitable since all the stations were characterized by shallow water depths, narrow channel widths, and relatively fast water currents flowing over boulder substrate. Collected specimens were preserved in 10% formalin solution until counting, after which they were stored in 5% formalin solution. All specimens were identified according to Zhu (1995), Chen (1998), Chu et al. (1999) and Yue (2000).

Data analysis

All the fish collected were identified to species. The relative abundance of each species at each sampling site was estimated by: $P_{jk} = N_{jk}/T_k$, where N_{jk} = the number of species j collected in site k; T_k = the total number of all fish collected in site k. The Margalef index (D) and Shannon-Wiener index (H) were used to calculate fish species richness for each site (Peet 1974; Magurran 1988): $D_k = (S_k-1)/\ln T_k$ and $H_k = -\sum P_{jk} \ln P_{jk}$, where S_k = the total number of species collected in site k.

Results

Stream characteristics and physicochemical parameters

The physical characteristics of each site are described in Table 1. The substrata of both East Stream and West Stream are mainly composed of sand, gravel,

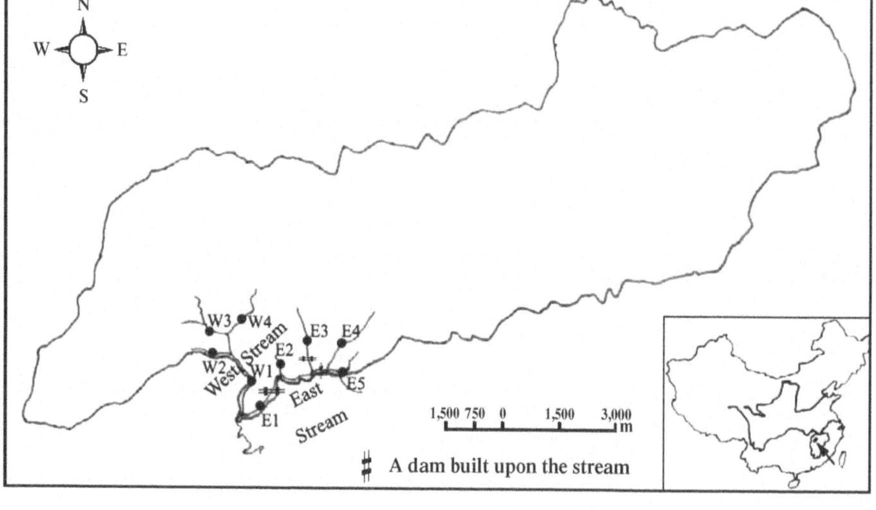

Fig. 1 Map showing fish sampling sites on East Stream and West Stream, GNNR, China

Table 1 Characteristics of survey sites within East Stream and West Stream, GNNR, China

Sites	Depth (m)	Width (m)	Distance sampling (m)	Habitat description
E1	0.5–0.8	5–7	150	Slow flowing and clear water, gravel and boulder substrate, stream shaded by riparian vegetation
E2	0.3–0.6	2–4	100	Fast flowing and clear water, boulder substrate, shaded by riparian vegetation
E3	0.8–1.0	3–5	100	Slow flowing and clear water, gravel and sandy substrate, shaded by riparian vegetation
E4	0.3–0.5	2–3	100	Fast flowing and clear water, with boulder and gravel substrate, shaded by riparian vegetation
E5	0.3–0.5	2–4	100	Fast flowing and clear water, with boulder substrate, shaded by riparian vegetation
W1	1.2–1.5	6–8	150	Slow flowing and clear water, gravel and sandy substrate, shaded by riparian vegetation
W2	0.5–1.0	2–4	100	Fast flowing and clear water, gravel and sandy substrate, shaded by riparian vegetation
W3	0.5–0.8	4–5	100	Fast flowing and clear water, with gravel and boulder substrate, shaded by riparian vegetation
W4	0.3–0.5	2–3	100	Fast flowing and clear water, boulder substrate, shaded by riparian vegetation

and boulders. Most of the study sites surveyed were in high gradient sections of streams with pools and riffles. All sites had clear water and were shaded by riparian vegetation.

Physicochemical characteristics were similar among all study sites in the streams. The mean (± SE) water temperature was 15.6±4.3°C, ranging from 9.4°C to 20.8°C at all sites. All sites in the streams were fully saturated with dissolved oxygen (mean ± SE, 9.7±1.5 mg·L^{-1}). The high DO could be attributed to low water temperature and high water velocity. Site average pHs were between 5.8 to 7.6 (mean ± SE, 6.5±1.3).

Fish fauna of the GNNR

A total of 379 fish were collected and classified into two families and seven species. The dominant family, Cyprinidae, accounted for 85.7% (six species) of the total number of fish species collected. Fish species richness from the nine sites ranged from two to six species. *Acrossocheilus parallens* and *Onychostoma barbata* were the most common species, being collected at all sites sampled from September 2006 to August 2007 (Table 2). *Acrossocheilus parallens* [relative abundance (RA) 44.3%] was the most abundant fish species collected (Fig. 2), followed by

Table 2 Fish species and individuals sampled from East Stream and West Stream, GNNR, China

Species name	East Stream					West Stream				Total
	E1	E2	E3	E4	E5	W1	W2	W3	W4	
Cyprinidae										
Acrossocheilus parallens	14	10	9	11	12	26	30	35	21	168
Onychostoma barbata	9	4	2	6	7	16	11	12	10	77
Zacco platypus						16	13	13	12	54
Opsariichthys bidens						15	13	11	8	47
Hemibarbus maculatus						20				20
Pseudorasbora parva						8				8
Homalopteridae										
Vanmanenia stenosoma									5	5
Number of individuals	23	14	11	17	19	101	67	71	56	379

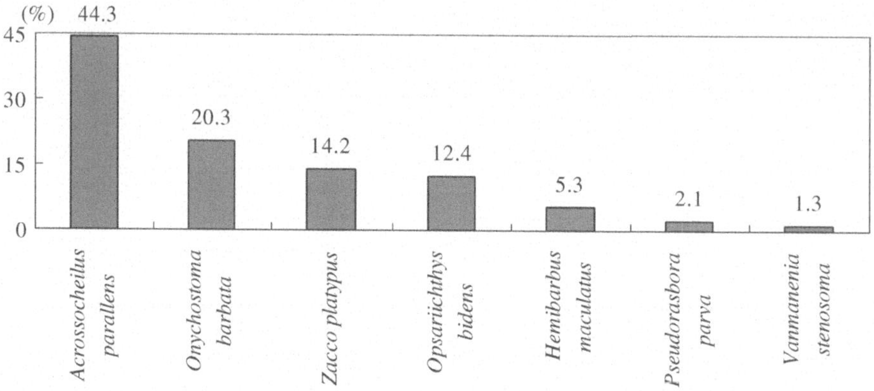

Fig. 2 Relative abundance of each fish species identified from sampling sites in East Stream and West Stream, GNNR, China

Onychostoma barbata (RA, 20.3%), *Zacco platypus* (RA, 14.2%) and *Opsariichthys bidens* (RA, 12.4%).

We recorded a greater number of species (*S*) and individuals (*N*) in West Stream compared to East Stream. In addition, the Margalef diversity index (*D*) and Shannon-Wiener diversity index (*H*) were lower in the East Stream as compared with the West Stream (Table 3). This indicates that higher fish diversity in West Stream compared to East Stream may be because the East Stream watershed has been modified by small dams (Fig. 1). The dams do not have effective fish passage facilities, and this may impede fish dispersal and migration throughout the watershed. Our observations suggest that there is a need to enhance the connectivity of the East Stream for fish dispersal and migration through modification of the dams to allow passage.

Discussion

Jiangxi Province is located in the southern part of China. A total of 205 fish species (more than 20% of all Chinese freshwater species) have been reported within the province (Guo and Liu 1995; Fu et al. 2003). This indicates that freshwater fish biodiversity in the Jiangxi Province plays an important role in the Chinese freshwater fish species bank.

None of the seven fish species collected in the GNNR is exotic species. Among them, *Acrossocheilus parallens*, *Onychostoma barbata* and *Vanmanenia stenosoma* are common in the mountain streams within Jiangxi Province and endemic to the Chinese mountain streams (Guo and Liu 1995; Huang et al. 2008; Li et al. 2008). They are anatomically well adapted to live in fast flowing current with clear water and relatively higher dissolved oxygen concentration. This suggests that mountain streams in the nation reserves including GNNR are very important for freshwater fish diversity and conservation in Jiangxi Province.

During recent decades, streams and rivers in China have been drastically modified for agricultural usage and drinking water supplies with the construction of multi-purpose dams, artificial reservoirs, levees, and weirs. These physical alterations and other human influences, such as road construction and deforestation, have accelerated eutrophication (Fu et al. 2003). This pattern is reflected throughout much of East Asia, and the resultant degradation of aquatic habitats for fish is of great concern (Inoue and Nakano 2001). Small and fast-flowing streams have often been changed to large, slow-flowing streams, which would cause organisms adapted to rapid streams to become restricted to mountainous areas, and to be replaced by organisms adapted to slow-flowing streams (Yuma et al. 1998).

Table 3 Comparison of fish species diversity and abundance between East Stream and West Stream, GNNR, China

Mountain streams	Total number of species (S)	Total number of individuals (N)	Margalef diversity index (D)	Shannon-Wiener diversity index (H)
East Stream	2	84	0.23	0.64
West Stream	7	295	1.06	1.62

The conservation status of nature reserves has greatly limited the extent of such influences within its boundaries and so conditions have remained relatively suitable for the maintenance of native fish communities.

Worldwide, freshwater fishes are the most diverse of all vertebrate groups, but are also the most highly threatened (Duncan and Lockwood 2001). Protected areas such as the GNNR play an important role in conservation of freshwater fish diversity within Jiangxi Province. This study allows consideration of a long-term conservation strategy for fish diversity in the GNNR. Long-term management and conservation of the fish fauna of the GNNR will require development of reference sampling sites as part of the development of a long-term monitoring protocol.

Acknowledgments We are grateful to the staff of Guanshan National Nature Reserve Management Station (GNNRMS) for their help provided during the survey. We thank W.X. Cao and E Zhang (Institute of Hydrobiology, the Chinese Academy of Science) for their help with fish identification.

References

Chen YY (1998) Fauna sinica: Osteichthyes cypriniformes II. Science, Beijing, China. in Chinese

Chen LS (2002) A survey of flora in the Guanshan Nature Reserve. Jiangxi Forestry Science and Technology 1:13–15. (in Chinese)

Chu XL, Zheng BS, Dai DY (1999) Fauna sinica: Osteichthyes siluriformes. Science, Beijing, China. in Chinese

Dai NH, Liu W, Cai RL (1997) A survey of avifauna **in the Guanshan Nature Reserve**. Jiangxi Science 15:243–246. (in Chinese)

Ding DS, Qiu NF, Chen CF, Lin YJ (2005) Geographical distribution of insects in the Guanshan Nature Reserve. Jiangxi Forestry Science and Technology 3:6–7. (in Chinese)

Duncan JR, Lockwood JL (2001) Extinction in a field of bullets: a search for the causes in the decline of the world's freshwater fishes. Biological Conservation 102:97–105

Fu CZ, Wu JH, Chen JK, Wu QH, Lei GC (2003) Freshwater fish biodiversity in the Yangtze River basin of China: patterns, threats and conservation. Biodiversity and Conservation 12:1649–1685

Guo ZZ, Liu RL (1995) The fishes of Jiangxi Province. Journal of Nanchang University 19:222–232. (in Chinese)

Guo ZF, Ding DS (2005) Analysis of butterfly fauna in the Guanshan Nature Reserve. Entomological Journal of East China 14:100–104. (in Chinese)

He YG, Ji XX (2002) A survey on wild ornamental plants in the Guanshan Nature Reserve. Jiangxi Forestry Science and Technology 5:7–9. (in Chinese)

He F, Zheng QY, Liu KW (2001) The community characteristics of *Amentotaxu argotaenia* (Hance) Pilger in the Guanshan Nature Reserve. Journal of Central South Forestry University 21:74–77. (in Chinese)

Huang LL, Wu ZQ, Hu ML, Li Q, Zong DS, Wan ZQ, Zhao WQ (2008) Fish species diversity in the Lushan Nature Reserve, Jiangxi, China. Journal of Nanchang University 32:161–164. (in Chinese)

Inoue M, Nakano S (2001) Fish abundance and habitat relationships in forest and grassland streams, northern Hokkaido, Japan. Ecological Research 16:233–247

Ji MC, Zheng G, Xie Y, Wu HP, Qiang S (2005) Distribution of bryophyte **in the Guanshan Nature Reserve**. Journal of Zhejiang Forestry College 22:370–374. (in Chinese)

Li KM, Tang XZ (2003) Community characteristics and diversity of *disanthus cercidifolius* var. *longipes* shrubland in the Guanshan Nature Reserve. Journal of Nanjing Forestry University 27:1–3. in Chinese

Li Q, Wu ZQ, Huang LL, Chen HM, Lu HJ (2008) Fish resources of Qiyunshan Nature Reserve, Jiangxi, China. Acta Zootaxonomica Sinica 33:324–329. (in Chinese)

Liu XZ, Wu HP (2005) Scientific survey and study on the Guanshan Nature Reserve in Jiangxi Province. Chinese Forestry, Beijing, China. in Chinese

Liu LZ, Li XY, Yang F (2003) Function and significance of the **Guanshan Nature Reserve**. Central South Forest Inventory and Planning 22:37–39. (in Chinese)

Magurran AE (1988) Ecological diversity and its measurement. Cambridge University Press, London, London

Peet RK (1974) Measurement of species diversity. Annual Review of Ecology and Systematics 5:285–307

Xu YP, Zheng JW, Ding P, Jiang PP, Cai LJ, Huang XF, Yao XH, Xu XR, Yu ZP (2007) Seasonal change in ranging of Elliot's pheasant and its determining factors in the Guanshan National Nature Reserve. Biodiversity Science 15:337–343. (in Chinese)

Yue PQ (2000) Fauna sinica: Osteichthyes cypriniformes Ш. Science, Beijing, China. in Chinese

Yuma M, Hosoya K, Nagata Y (1998) Distribution of the freshwater fishes of Japan: an historical overview. Environmental Biology of Fishes 52:97–124

Zhu SQ (1995) The synopsis of freshwater fishes of China. Jiangsu Science and Technology, Nanjing, China. in Chinese

Environ Biol Fish (2009) 86:29–30
DOI 10.1007/s10641-007-9302-x

Threatened fishes of the world: *Schizothorax taliensis* Regan, 1907 (Cyprinidae)

Youji Wang · Menghong Hu ·
Ling Cao · Weimin Wang

Received: 9 April 2007 / Accepted: 10 August 2007 / Published online: 16 November 2007

Keywords *Schizothorax taliensis* · Dali schizothoracin · Threatened fishes · Erhai Lake

Common name: Dali schizothoracin. **Conservation status**: Endangered in China Red Data Book of Endangered Animals – Pisces (Yue and Chen 1998). Identification: Dorsal fin iii-7–8; anal fin iii-5; pectoral fin i-16–19; pelvic fin i-8–10; lateral line scales 95–109, pharyngeal teeth in three rows 2,3, 4–4,3,2 or 2,3,5–5,3,2; mouth terminal; lower jaw without sharp horny edge; lower lip smooth, in two lobes, without median lobe; barbel two pairs, less than half of eye diameter in length; thorax between isthmus and fore-abdomen naked; lateral line complete; last unbranched ray of dorsal fin hard, with posterior serrations (Yue 2000). **Distribution**: Endemic to Erhai Lake (25°36′–25°58′N, 100°05′–100°17′E, 1,974 m above sea level, 249.34 km^2), Lancang River, Yunnan Province. **Abundance**: Accounted for 30% of total harvest in Erhai Lake before the mid-20th century (Wu and Wang 1999), but now is rarely found (Huang 1996). **Habitat and ecology**: Usually inhabits quiet, open areas of lakes, mainly feeding on zooplankton. Migrates upstream or assembles around spring-fed caves during spring for spawning (Yue and Chen 1998). **Reproduction**: Matures at 3–4 years (Leng et al. 2006). Spawns over sandy and gravel bottom or upwelling groundwater in lakes (Huang 1996). **Threats**: It lives and spawns mainly in ditches and caves of shallow lakes and creeks entering the lake, many of which have dried up since 1970s from human activity. Introduced exotic fish species are serious egg predators (Yang 1996). Overfishing contributed to the drastic population depletion (Yue and Chen 1998). **Conservation action**: Cangshan Mountain and Erhai Lake National Nature Reserve was established in 1981 and became national level in 1994. Local government enacted the "Erhai Lake Administrative Regulation" to protect this species (Yue and Chen 1998). **Conservation recommendation**: Research for population preservation and restoration is required. Fishing regulations need to be enhanced. Exotic fishes should be removed. Natural habitats should be protected.

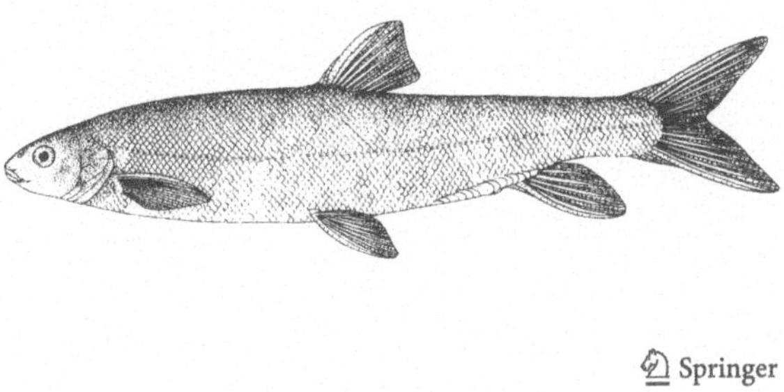

Y. Wang · M. Hu · L. Cao · W. Wang (✉)
College of Fisheries, Huazhong Agricultural University,
Wuhan 430070, China
e-mail: wangwm@mail.hzau.edu.cn

References

Huang KY (1996) Progenitive protection of *Schizothorax taliensis* Regan. Sci Fish Farming 1:16(in Chinese)

Leng Y, Xu WY, Liu YT, Bao JH, Yang GQ, Wang DW, Yang ZX (2006) Experiment on artificial propagation of *Schizothorax taliensis*. Reserv Fisheries 26:26–27 (in Chinese)

Wu QL, Wang YF (1999) On the succession of aquatic communities in Erhai Lake. J Lake Sci 11:267–273 (in Chinese with English abstract)

Yang JX (1996) The exotic and indigenous fishes in Yunnan: effective mode, extent and other issues. In: Wang S et al (ed) Protection on the biodiversity in China (2). Chinese Environmental Science Press, Beijing, p 129–138(in Chinese)

Yue PQ, Chen YY (1998) Endangered in China red data book of endangered animals – Pisces. Science Press, Beijing, pp 150–152

Yue PQ (2000) Fauna sinica-cypriniformes III. Science Press, Beijing, pp 320–321

Environ Biol Fish (2009) 86:31–44
DOI 10.1007/s10641-008-9421-z

A review of the genus *Distoechodon* (Teleostei: Cyprinidae), and description of a new species

Yahui Zhao · Fang Kullander · Sven O Kullander · Chunguang Zhang

Received: 18 September 2007 / Accepted: 22 September 2008 / Published online: 21 October 2008

Abstract *Distoechodon* species are widely distributed in rivers of southern China. Previous studies recognized only two species in the genus, viz., *D. tumirostris* and *D. hupeinensis*. A morphological analysis of 164 specimens from different localities permits recognition of four species. *Distoechodon tumirostris* is mainly restricted to small coastal rivers in the Zhejiang Province. *Distoechodon multispinnis* is widely distributed in the Chang Jiang and Zhu Jiang basins. *Distoechodon compressus* is known only from the Fujian and Jiangxi Provinces, and Taiwan. *Distoechodon macrophthalmus*, the new species, is known only from the Chenghai Lake, Chang Jiang basin in Yunnan Province.

Keywords China · Jiang river · Zhu Jiang river · Distoechodon · Systematics

Y. Zhao (✉) · C. Zhang (✉)
Institute of Zoology, Chinese Academy of Science,
Beijing 100101, China
e-mail: zhaoyh@ioz.ac.cn

F. Kullander · S. O. Kullander
Department of Vertebrate Zoology, Swedish Museum of Natural History,
POB 50007, SE-104 05 Stockholm, Sweden

Introduction

The endemic Chinese fish genus *Distoechodon* (Cypriniformes: Cyprinidae) was established by Peters (1881) with the type species, *D. tumirostris,* type locality Ningpo (now known as Ningbo), located on the Qiantang Jiang River, Zhejiang Province, China. It can be distinguished from the most similar genus *Xenocypris* by having a transverse instead of a curved mouth and possessing only two instead of three rows of pharyngeal teeth (Peters 1881). The status of *Distoechodon* is ambiguous and confusing at both the genus and species levels.

Nichols (1928) synonymized *Distoechodon* with *Xenocypris* and treated it as a subgenus without providing any specific explanation. Yang (1964) revalidated *Distoechodon* on the basis of possessing only two instead of three rows of pharyngeal teeth. Bănărescu (1970) disagreed with Yang's (1964) synonymy because reduction of the number of rows of pharyngeal teeth is common among cyprinid fishes. He thus synonymized *Distoechodon* back as a subgenus of *Xenocypris*. But almost all contemporary Chinese systematic work on xenocyprinid genera follows Yang (1964), recognizing *Distoechodon* as a valid genus having two rather than three rows of pharyngeal teeth (Lian 1984; Chen and Li 1989; Mao 1991; Ding 1994; Liu and He 1998).

Nichols (1925) described *Xenocypris compressus*, which has a deeper body and longer pectoral fins, from Yenping (now known as Nanping), Min Jiang River, Fujian Province. Later, he put the species into the subgenus *Distoechodon* with *X. tumirostris* (Nichols 1928). Yang (1964) recognized both of them as species of *Distoechodon*. Bănărescu (1970) synonymized *X. compressus* with *X. tumirostris*, but retained it as a subspecies. Lian (1984) followed Yang (1964), recognizing *D. compressus* as a valid species. Liu and He (1998) synonymized *D. compressus* with *D. tumirostris*, based on the fact that the number of lateral-line scales and gill rakers among populations of *Distoechodon* are overlapping. Shan (1998) described differences between *D. compressus* and *D. tumirostris* in lateral-line scale count, shapes of the dentary and basioccipital masticatory plates, and revalidated *D. compressus*. Molecular analysis of nucleotides of the mitochondrial D-loop region and the complete cytochrome *b* gene supported *D. compressus* as a distinct species (Liu 2002).

Yih (1964) added another species to the genus, *Distoechodon hupeinensis*, from the Liangzihu Lake, Chang Jiang River basin, in Hubei Province. Compared to all other *Distoechodon* species, *D. hupeinensis* has fewer lateral-line scales, fewer gill rakers, and lacks the dark horizontal stripes along the sides of the body. Liu and He (1998) accepted this description in their review. In a phylogenetic study based on morphological characters, Shan (1998) placed *D. hupeinensis* under *Xenocypris*. A molecular study of the cytochrome *b* gene from a number of xenocyprinid species showed that *D. hupeiensis* is indeed more closely related to *X. argentea* and *X. davidi* than to other *Distoechodon* species (Xiao et al. 2001).

With synonymising *Distoechodon* to *Xenocypris*, Bănărescu (1970) also recognized three subspecies of *X. (D.) tumirostris*, based mainly on lateral-line scale counts and number of gill rakers: *X. (D.) t. tumirostris* (Zhejiang Province, 71–77 lateral-line scales, 80–85 gill rakers), *X. (D.) t. compressus* (Fujian Province; 74–75 lateral-line scales, 93–98 gill rakers), and *X. (D.) t. multispinnis*, a new subspecies from the Chang Jiang basin, Sichuan Province, with 77–84 lateral-line scales and 91–108 gill rakers. Ding (1994) tentatively synonymized *X. tumirostris multispinnis* with *D. tumirostris* without providing any explanation.

In the course of reviewing the genus *Distoechodon* as well as clarifying the species diversity of the genus, we examined 164 specimens of *Distoechodon* from different localities in China, allowing the recognition of the genus, redescription of the type species, *D. tumirostris*, revalidation of both *D. compressus* and *D. multispinnis*, as well as a new species, which is herein described.

Materials and methods

We examined 164 specimens of *Distoechodon* and 57 specimens of *Xenocypris*. Comparative material of *Xenocypris* is listed separately. Measurements were taken point to point with a digital caliper (Mitutoyo CD-20C) to 0.1 mm. Morphometric and meristic characters were selected according to methods described by Chen (1998) and Fang (1997). The anterior-most measuring point is at the tip of the snout instead of the upper jaw symphysis. Post-predorsal length was measured from the posterior end of supraoccipital to the dorsal-fin insertion. Measurements and counts were taken on the left side of the body whenever possible. Gill rakers were counted only in the lower part of the first gill arch, on the ceratobranchial and hypobranchial bones.

Specimens were radiographed with Philips MG 150 low voltage X-ray unit with Kodak X-Omat V films. Dorsal, anal and caudal-fin rays and vertebrae were counted on radiographs. Abdominal vertebrae include the Weberian apparatus. Predorsal vertebrae were counted from the first free vertebra to the vertebra before the first dorsal-fin pterygiophore. Caudal vertebrae were counted from the first vertebra with haemal spine to the last half centrum.

Morphometric and meristic information was analyzed and organized using SYSTAT version 10. Most of the coordinates provided in the material examined list were researched by the authors. Institutional abbreviations follow Leviton et al. (1985) and Leviton and Gibbs (1988). The local Chinese toponymy is employed for all localities. The Chinese names for the Yangtze and Pearl Rivers are Chang Jiang and Zhu Jiang, respectively.

Results

Key to Xenocyprinid genera

Previous keys were mainly based on the number of pharyngeal tooth rows and thus impractical as they require extensive dissection (Liu and He 1998). An alternative key is proposed here, using external characters only, which is more user-friendly and practical.

Key to genera of the Xenoxyprininae

1. Abdominal keel longer than distance from pelvic-fin base to vent 2
1. Abdominal keel absent or rather short 3
2. Lateral line scales fewer than 50; gill rakers on first gill arch more than 100............................ *Pseudobrama*
3. Lateral line scales more than 70; gill rakers on first gill arch fewer than 50 .... *Plagiognathops*
4. Mouth transverse; pectoral-fin axillary fold present ....*Distoechodon*
5. Mouth curved; pectoral-fin axillary fold absent ....*Xenocypris*

Key to species of Distoechodon

Distoechodon Peters 1881

Distoechodon Peters, 1881, Mber. Akad. Wiss., Berl. 45: 912 (type species: *Distoechodon tumirostris* Peters)

Body elongated and compressed; abdomen round without keel; mouth inferior, transverse (slightly curved), with cuticular edge (Fig. 1A); snout round (Fig. 1B); pectoral-fin axillary fold present (Fig. 1B); pharyngeal teeth (Fig. 1C) in 2 rows, outer row well-developed, deeply compressed; lateral line curved, possessing scales more than 60. The genus includes four species, all distributed in China (Chang Jiang River basin, Zhu Jiang River basin and some coastal rivers in south-east China; historical records in Taiwan Island).

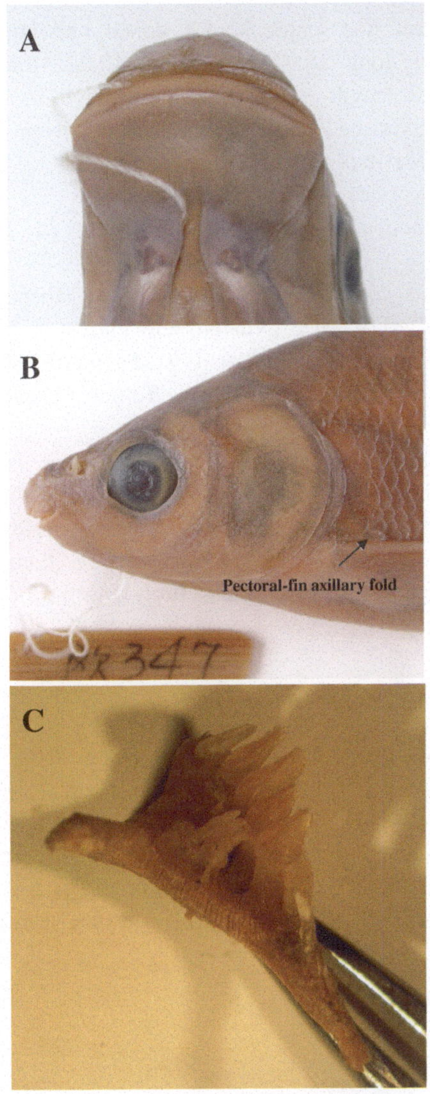

Fig. 1 Some typical characteristics of the *Distoechodon* species. **a** Mouth shape. **b** Snout and pectoral-fin axillary. **c** Pharyngeal teeth

Key to species of *Distoechodon*

1. Eye relatively large among the same-sized individuals, 5.97–10.82% of standard length; distributed in Fujian Province and Chenghai Lake in Yunnan Province 2
1. Eye relatively small among the same-sized individuals, 4.58–9.08% of standard length; distributed Chang Jiang and Zhu Jiang River basins and coastal rivers in Zhejiang Province 3
2. Lateral line scales 67–74, predorsal scales 27–33....................... *D. compressus*
3. Lateral line scales 78–85, predorsal scales 34–39................ *D. macrophthalmus*
4. Lateral line scales 68–77; dorsal and anal fins relatively short, 16.49–21.73% and 11.47–14.73% of standard length respectively, distributed in costal rivers in eastern China .... *D. tumirostris*

Lateral line scales 72–82, dorsal and anal fins relatively long, 18.97–24.13% and 12.22–16.58% of standard length respectively, distributed in Chang Jiang River and Zhu Jiang River basins................. *D. multispinnis* sp.nov.

Distoechodon tumirostris Peters 1881 *(Fig. 2A, Table 1)*

Distoechodon tumirostris Peters 1881, Mber. Akad. Wiss., Berl. 45: 925 (Ningpo, now known as Ningbo in Zhejiang Province); Mao 1991, Fauna of Zhejiang, Fresh Water Fishes: 76 (Anji, Tonglu, Tiantai and Yunhe in Zhejiang Province)

Xenocypris (*Distoechodon*) *tumirostris* Nichols 1928, Bull. Am. Mus. Nat. Hist. 58(1): 24 (Ningpo); Nichols 1943, Nat. Hist. Central Asia 9: 124 (Ningpo)

Xenocypris (*Distoechodon*) *tumirostris tumirostris* Bănărescu 1970, Rev. Roum. Biol. (zool.) 15(6): 400 (Ningpo)

Specimens examined

All from China. Zhejiang Province: Ou Jiang River basin: ASIZB 64104–64105, 2, 154.4–156.2 mm SL, Lishui (28°27′ N, 119°54′ E), 27 Apr. 1972; ASIZB 61771–61773, 3, 111.5–126.2 mm SL, Qingtian (28°09′ N, 120°18′E), 28 Apr. 1972; IHB 0002420, 4, 159.2–318.0 mm SL, Lishui and Jinyun (28°39′N, 120°04′E). Ling Jiang River basin: ASIZB 73089–73092, 4, 99.1–156.8 mm SL, Xianju (28°51′N, 120°44′E) and Linhai (28°51′N, 121°07′E), 29 May 1972; ASIZB 62312–62316, 5, 172.3–243 mm SL, Xianju and Linhai, 31 May 1972; IHB 0002422, 15, 91.3–242.0 mm SL, Linhai, Xianju and Tiantai (29°08′N, 121°02′E), Sep. 1974. Yong Jiang River basin: IHB 0002421, 5, 146.0–245.0 mm SL, Ningbo (29°52′N, 121°33′E) and Fenghua (29°39′N, 121°24′E), Sep. 1974 and May 1976. Xin'an Jiang River basin: IHB 0002426, 1, 298.3 mm SL, Chun'an (29°37′N, 119°02′E), June 1977. Cao'e Jiang River basin: IHB 0002423, 2, 103.3–149.5 mm SL, Shengzhou (29°36'N, 120°49'E), Sep. 1974. River unknown: ASIZB 62137, 1, 75.5 mm SL, May 1972; ASIZB 62139, 1, 86.6 mm SL, May 1972; IHB 0002425, 3, 146.1–203.1 mm SL.

Diagnosis Distoechodon tumirostris can be distinguished from all other congeners by having a combination of the following characters: lateral-line scales 68–77, modally 73; predorsal scales 30–36, mostly 33 (Fig. 3); relatively smaller eye (5.07–9.08% in SL). Similar to *D. compressus* in lateral-line scale count, but is different from it by having more predorsal scales (30–36 vs. 27–33, Fig. 3) and always smaller eyes at the same body size (5.07–9.08% vs. 5.97–10.82% in SL, Table 1 and Fig. 4). Similar to *D. multispinnis* in predorsal scale count and eye diame-

Fig. 2 Four species of *Distoechodon*. **a** *Distoechodon tumirostris*, ASIZB 73089, 201.9 mm SL. **b** *Distoechodon compressus*, ASIZB 66087, 116.0 mm SL. **c** *Distoechodon multispinnis*, NRM T3486, 141.9 mm SL. **d** *Distoechodon macrophthalmus* sp. nov.; KIZ 797297, holotype, 155.4 mm SL

Table 1 Morphometry of four species of the genus *Distoechodon*. Measurements are expressed as percents of SL

	Distoechodon tumirostris					*Distoechodon compressus*				
	N	Min	max	mean	SD	N	Min	max	mean	SD
SL (mm)	46	75.5	318.0	171.2	56.3	60	81.2	247.7	145.8	37.6
Body depth	46	22.9	29.7	25.6	1.6	60	22.2	32.0	26.9	2.0
Predorsal length	46	50.0	56.3	53.3	1.4	60	50.6	59.6	53.8	1.7
Preanal length	46	73.6	83.1	77.3	2.2	60	73.1	82.9	77.9	1.9
Prepectoral length	46	20.8	26.1	23.8	1.3	60	21.2	28.2	25.2	1.5
Prepelvic length	46	50.3	56.2	53.0	1.5	60	51.0	57.6	53.7	1.5
Caudal-peduncle length	46	13.1	17.1	14.9	1.0	60	12.3	17.2	14.6	0.9
Caudal-peduncle depth	46	9.4	11.8	10.8	0.6	60	9.6	12.6	11.2	0.6
Pectoral-fin length	46	14.3	19.0	16.7	1.0	60	16.2	19.9	18.1	0.9
Pelvic-fin length	46	12.1	16.0	14.2	0.8	60	13.8	17.5	15.5	0.8
Dorsal-fin length	46	16.5	21.7	19.4	1.2	59	15.1	24.3	21.6	1.5
Anal-fin length	46	11.5	14.7	13.2	0.8	59	12.8	16.2	14.4	0.8
Head length	46	20.2	26.4	22.9	1.4	60	21.4	26.8	24.3	1.3
Head depth	46	15.4	19.7	17.7	0.8	60	15.9	20.7	18.0	1.0
Snout length	46	7.1	10.0	8.2	0.6	60	6.9	10.1	8.6	0.8
Eye diameter	46	5.1	9.1	7.0	1.1	60	6.0	10.8	8.2	1.0
Interorbital width	46	7.7	10.6	8.9	0.6	60	7.6	10.8	9.0	0.7
Mouth width	46	5.8	9.3	7.7	0.7	60	6.4	9.5	8.1	0.7
	Distoechodon multispinnis					*Distoechodon macrophtalmus* sp.nov.				
	N	min	max	mean	SD	*N*	min	max	mean	SD
SL (mm)	52	111.6	326.8	191.7	43.8	15	104.2	200.1	153.0	31.1
Body depth	52	21.7	32.4	26.0	2.4	15	22.9	26.6	24.8	0.9
Predorsal length	52	49.7	56.4	52.5	1.5	15	51.6	55.8	53.5	1.2
Preanal length	52	73.7	81.7	76.6	1.7	15	74.7	78.8	76.7	1.3
Prepectoral length	52	21.6	26.4	23.5	1.1	15	21.0	25.6	23.7	1.5
Prepelvic length	52	49.9	55.4	52.2	1.3	15	49.2	56.6	52.4	1.8
Caudal-peduncle length	52	12.9	17.5	15.4	1.1	15	12.1	16.4	14.4	1.1
Caudal-peduncle depth	52	9.3	13.2	11.4	0.8	15	9.3	11.3	10.4	0.6
Pectoral-fin length	52	15.4	20.9	16.8	1.0	15	16.7	18.9	17.8	0.8
Pelvic-fin length	52	13.6	17.6	15.1	0.9	15	14.5	15.9	15.3	0.5
Dorsal-fin length	42	19.0	24.1	20.5	1.2	15	18.1	22.8	20.2	1.1
Anal-fin length	42	12.2	16.6	14.4	1.1	15	12.4	15.4	14.3	0.7
Head length	52	20.3	25.0	22.3	1.1	15	21.4	24.5	22.8	1.1
Head depth	52	16.0	19.7	17.4	0.9	15	16.2	17.8	17.1	0.5
Snout length	52	7.1	9.2	8.2	0.5	15	7.3	8.4	7.8	0.4
Eye diameter	52	4.6	8.3	6.4	0.8	15	6.5	9.4	8.0	1.0
Interorbital width	52	8.0	10.2	9.1	0.4	15	7.8	9.3	8.4	0.5
Mouth width	52	6.3	8.6	7.7	0.5	15	6.6	7.8	7.2	0.4

ter, but differs by having fewer lateral-line scales (68–77 vs. 72–82, Fig. 3), shorter dorsal-fin and anal-fin length (16.49–21.73% vs. 18.97–24.13% and 11.47–14.73% vs. 12.22–16.58% in SL respectively).

Description Body elongate and moderately compressed; dorsal profile straight or slightly convex; ventral profile concave; abdomen round; abdominal keel absent. Head rather small and compressed; snout round, anterior tip truncated; snout length longer than eye diameter; mouth inferior, transverse, with cuticular edge; mouth wide, as wide as head width at that point; no barbel; nostrils closer to eyes than to tip of snout; eyes relatively small.

Scales moderately large. Lateral line complete, scales 68(5), 71(5), 72(7), 73(6), 74(7), 75(6), 76(6), 77(4).

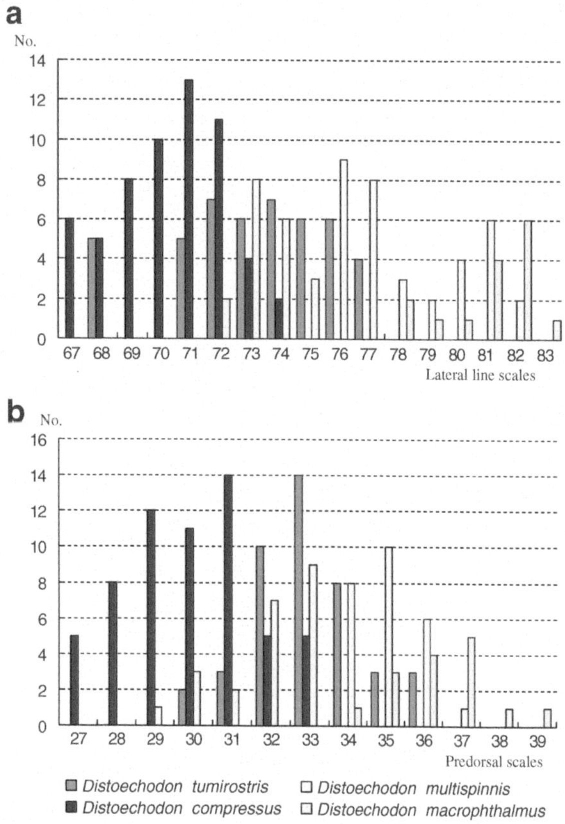

Fig. 3 Differences in lateral-line scale (above, A) and predorsal scale counts (below, B) among four species of *Distoechodon*

Lateral line descending from top of gill opening to above pectoral-fin ray, then running parallel to horizontal scale row and ascending to body midline above anal-fin base and extending to end of caudal peduncle. Scales above lateral line 12–13, rarely 14; scales below lateral line 7–8, rarely 9; circumpeduncular scales 26; predorsal scales irregularly arranged, 30–36, mostly 33; scales cover abdomen. Pectoral-fin axillary fold present; pelvic-fin axillary scale well-developed. Gill rakers tiny, numerous and triangular, 51(1), 52(3), 53(7), 54(3), 55 (1), 56(5), 57(1), 58(3), 59(6), 61(1), 62(3), 63(2), 64(1) in lower part of first gill arch.

Pectoral-fin insertion slightly anterior to vertical through posterior margin of opercle. Pectoral fin short, reaching to midway between pectoral and pelvic-fin insertions. Pelvic fin short, reaching to midway between pelvic-fin insertion and vent. Dorsal-fin origin midway between snout tip and caudal-fin base, and slightly posterior to pelvic-fin insertion. Anal-fin insertion midway between pelvic-fin origin and caudal-fin base. Dorsal-fin rays iii, 7 (11), last unbranched dorsal-fin ray ossified, smooth. Anal-fin rays iii, 9 (9), 8 (2). Pectoral-fin rays i, 14 (6), 15 (3), 16 (2). Pelvic-fin rays i, 8 (11). Caudal fin forked, lobes almost equal in length; procurrent caudal-fin rays 9 (2), 10 (7), 11 (2) dorsally, 8 (2), 9 (9) ventrally.

Total vertebrae 25+20=45 (3), 26+19=45 (5), 26+20=46 (2), 27+19=46 (1); predorsal vertebrae 9(3), 10 (7), 11 (1). Pharyngeal teeth 2 in rows, 6, 3; 3, 6. Outer row of pharyngeal teeth strong or well developed, narrow and compressed, with curved and pointed tip. Inner row of pharyngeal teeth short, not compressed, with very sharp tips.

Coloration in preservative Alcohol preserved specimens generally grayish to brownish, without markings. All fins light brownish.

Geographical distribution Coastal rivers (Ou Jiang, Ling Jiang, Yong Jiang, Qiantang Jiang and Cao'e Jiang River basins) in Zhejiang Province (Fig. 5).

Remarks Characters agree with data from the original description (Peters 1881), except for body depth, which was stated as 4.75 in the original description, but comes out as 3.4–4.3 in our study. The original description was

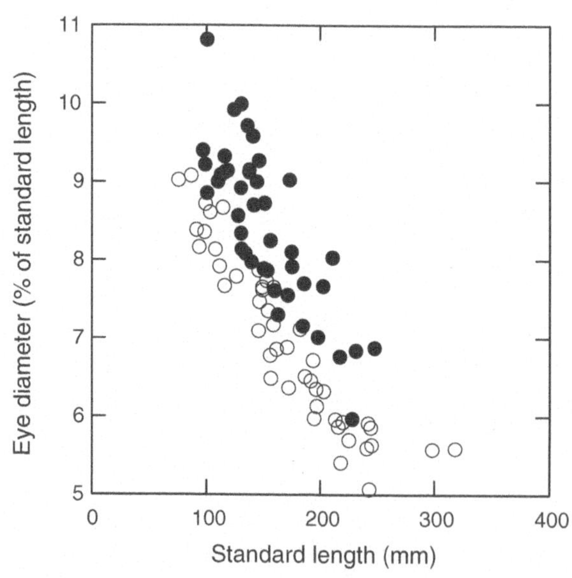

Fig. 4 Eye diameter (% SL) plotted against standard length (mm) for *Distoechodon tumirostris* and *D. compressus*. ○ *Distoechodon tumirostris*. ● *Distoechodon compressus*

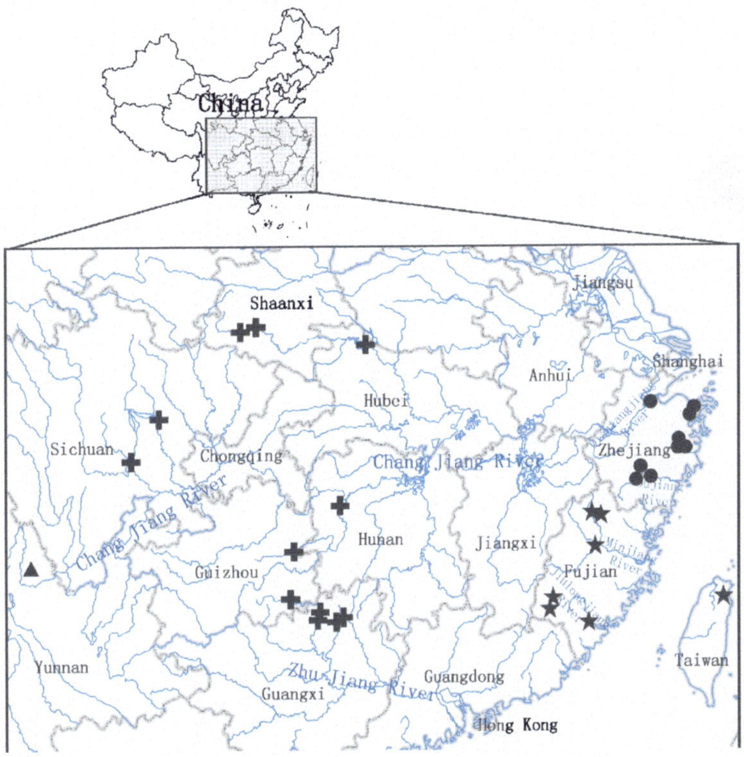

Fig. 5 Distribution of four *Distoechodon species*. ● *Distoechodon tumirostris*. ★ *Distoechodon compressus*. *Distoechodon compressus*. ✚ *Distoechodon multispinnis*. ▲ *Distoechodon macrophthalmus* sp. nov

based on one specimen only. The difference in body depth may reflect a different way of measuring standard length. The holotype was not re-examined. Some of our material is from Ningbo, the type locality of the species, and serve as topotypes.

Distoechodon compressus (Nichols), 1925
(Fig. 2B, Table 1)

Xenocypris compressus Nichols 1925, Am. Mus. Novit. (185):6 (Yenping, now known as Nanping in Fujian Province);

Xenocypris (*Distoechodon*) *compressus* Nichols 1928, Bull. Am. Mus. Nat. Hist. 58(1): 24 (Fukien, i.e. Fujian Province); Nichols 1943, Nat. Hist. Central Asia 9: 124 (Kienning, now known as Jianning and Yenping in Fujian Province).

Distoechodon compressus Yang 1964, The Cyprinid fishes of China, 131 (Nanping); Lian 1984, The fishes of Fujian Province: 254 (Nanping, Jianyang, Jian'ou etc. in Fujian Province).

Xenocypris (*Distoechodon*) *tumirostris compressus* Bănărescu 1970, Rev. Roum. Biol. (zool.) 15(6): 400 (Yenping, Fukien)

Distoechodon tumirostris Lian 1984, The fishes of Fujian Province: 253 (Nanping, Jianyang, Chong'an etc. in Fujian Province); Shen and Tzeng 1993, Fishes of Taiwan: 141 (Jilong River, Yilan in Taiwan Island).

Specimens examined All from China. Fujian Province: Min Jiang River basin: ASIZB 67298–67305, 8, 134.2–160.6 mm SL, Jianyang (27°20' N, 118°07'E), Jianxi River, Jun. 15, 1977; ASIZB 66084–66089, 6, 81.2–115.0 mm SL, Jian'ou (27°03'N, 118° 19'E), Jianxi River, Apr. and May, 1977; IHB 0002431, 3, 112.7–153.6 mm SL, Nanping (26°38'N, 118°10'E), Apr., 1957. Jiulong Jiang River basin: IHB 0002432, 4, 98.8–134.2 mm SL, Zhangzhou (24°31'N, 117°39'E), May, 1974; SHFU A004217, 1, SHFU 6164–6169, 6, 130.8–159.9 mm SL, Jimei (24°34'N, 118°07'E). Ting Jiang River basin: IHB 0002423, 1, 96.9 mm SL, Shanghang (25°03'N, 116°24'E), Jun., 1974; IHB 0002434, 7, 171.5–247.7 mm SL, Liancheng (25°42'N, 116°44'E), Jun., 1974 and 1976; SHFU 6185, 1, SHFU 6212, 2, SHFU uncat., 1, SHFU 6181, 1, SHFU 6184, 1, SHFU 6219, 1, SHFU 6199, 1, SHFU 6217, 1, SHFU 6204, 1, SHFU 6201, 1, SHFU 6206, 1, 100.7–210.8 mm SL, Liancheng, Jun., 1975. Ao Jiang River basin: IHB

0002437, 5, 115.9–139.8 mm SL, Lanjiang (26°12'N, 119°31'E), Jun. and Jul., 1974.

Diagnosis Distinguished from other *Distoechodon* species by a combination of the following characters: fewer lateral–line scales (67–74, mostly 70, Fig. 3), fewer predorsal scales (27–33, mostly 30, Fig. 3), and relatively larger eye, 8.28% vs. 6.37–7.77% in SL. The most similar species is *Distoechodon tumirostris*, which can be distinguished by having more predorsal scales (Fig. 3) and smaller eyes (Fig. 4).

Description Body elongate and moderately compressed; dorsal profile straight or slightly concave; ventral profile convex; abdomen round; abdominal keel absent. Head relatively large and compressed; snout round, anterior tip truncated; snout length longer than eye diameter; mouth inferior, transverse, with cuticular edge; mouth wide, as wide as head width at that point; no barbel; nostrils closer to eyes than to tip of snout; eye diameter relatively greater than that of *D. tumiostris* and *D. multispinnis* (Fig 4. Table 1.)

Scales moderately large. Lateral-line complete, scales 67(6), 68(5), 69(8), 70(10), 71(13), 72(11), 73 (4), 74(2). Lateral line descending from top of gill opening to above pectoral-fin ray, then running parallel to the horizontal scale row and ascending to body midline above anal-fin base, and running until end of caudal peduncle. Scales above lateral line 12–13; scales below lateral line 7–8; circumpeduncular scales 26; predorsal scales irregularly arranged, 27–33, mostly 30; scales cover abdomen. Pectoral-fin axillary fold present; pelvic-fin axillary scales well-developed. Gill rakers tiny, numerous and triangular, 51(1), 52(1), 53(3), 54(2), 55(5), 56(6), 57(4), 58(6), 59(4), 60(3), 61(4), 62(2), 63(9), 64(2), 65(1), 66(1) in lower part of first gill arch.

Pectoral-fin insertion slightly anterior to vertical through posterior margin of opercle. Pectoral fin short, reaching to midway between pectoral and pelvic-fin insertions. Pelvic fin short, reaching to midway between pelvic-fin insertion and vent. Dorsal-fin origin midway between snout tip and caudal-fin base, and slightly posterior to pelvic-fin insertion. Anal-fin insertion midway between pelvic-fin origin and caudal-fin base. Dorsal-fin rays iii, 7 (8), last unbranched dorsal-fin ray ossified but smooth. Anal-fin rays iii, 9 (8). Pectoral-fin rays i, 14 (3), 15 (2), 16 (3). Pelvic-fin rays i, 8 (8). Caudal fin forked, lobes almost equal in length; procurrent caudal-fin rays 9 (1), 10 (6), 11 (1) dorsally, 8 (3), 9 (5) ventrally.

Total vertebrae 25+19=44 (1), 26+18=44 (1), 26+19=45 (5), 26+20=46 (1); predorsal vertebrae 9(2), 10 (5), 11 (1). Pharyngeal teeth in 2 rows, 3, 6; 6, 3. Outer row of pharyngeal teeth strong, deeply compressed, with curved and pointed tips; inner row of pharyngeal teeth not well-developed, with very sharp tips.

Coloration in preservative Alcohol preserved specimens generally grayish to brownish, all fins light yellowish.

Geographical distribution Coastal rivers (Min Jiang, Jiulong Jiang, Ting Jiang and Ao Jiang River basins) in Fujian Province, and northern Taiwan Island (Yilan, Jilong River, Shen and Tzeng 1993, Chen and Chang 2005) (Fig. 5)

Remarks The characters observed for this study generally agree with the original description (Nichols 1925). The holotype was not re-examined but topotypes are included in the material. It was difficult to check the specimens from Taiwan. Based on the data from the Fish Database of Taiwan (http://fishdb.sinica.edu.tw), nine specimens have 66–74 lateral line scales, 30–32 predorsal scales. The rest of the characters are also consistent with our description. Therefore, we put them under *Distoechodon compressus*. It still needs to be further studied in the future.

Distoechodon multispinnis (Bănărescu), 1970
(Fig. 2C, Table 1)

Distoechodon tumironstris Tchang 1933, Zool. Sinica (B) 2 (1): 110 (Kaiting, should be Kiating); He 1989, Fishes of Zhujiang River: 115 (Longsheng and Rong'an counties, Guangxi); Ding 1994, The fishes of Sichuan, China: 163 (Huanglongxi, Leshan, Yibin, Luzhou and Shengzhong in Sichuan Prvince)

Xenocypris compressus Tchang 1933, Zool. Sinica (B) 2 (1): 117 (Szechuan i.e. Sichuan Province)

Xenocypris (Distoechodon) tumirostris multispinnis Bănărescu 1970, Rev. Roum. Biol. (zool) 15 (6): 401 (Suifu, Szechwan, now known as Yibin City in Sichuan Province).

Specimens examined All from China. Chang Jiang River basin: Sichuan Province: NRM T3483, 1; NRM

T3486, 1; NRM T3489–3495, 7, 141.9–187.6 mm SL, Leshan (29°34'00"N, 103°43'59"E), 7 Aug 2002, S. O. Kullander et al.; ASIZB 39779, 1, 183.0 mm SL; Min Jiang River, Jun 1957; ASIZB 55822, 1, 214.0 mm SL, Tuo Jiang River; IHB 0002417, 6, 111.6–224.1 mm SL, Tuo Jiang River, 1958. IHB 0002578, 3, 120.3–200.8 mm SL, Leshan, 1978, 1982 and 1987. Shaanxi Province: ASIZB 61555, 2, 135.0–146.7 mm SL, Hanzhong (33°04'00"N, 107°02'00"E), Hanshui River, Apr 1990, C. G. Zhang and W. L. Jia; ASIZB 73551–73554, 4, 146.6–199.6 mm SL, Yangxian (33°13'00"N, 107°33'00"E), Hanshui River, 12 Jul 2003 T. Q. Xu; IHB 0002439, 2, 210.9–326.8 mm SL, Hanzhong, Hanshui River, 1966 and May 1980. Hubei Province: IHB 0002444, 7, 198.4–260.2 mm SL, Danjiangkou (32°33'00N, 108°30'00E), Hanshui River, Jun 1976. Guizhou Province: IHB 0002418, 1, 201.35 mm SL, Tongren (27°43'N, 109°12'E), Yuan Jiang River, Apr 1988. Hunan Province: IHB 0002436, 1, 145.2 mm SL, Luxi (28°17'N, 110°09'E), Yuan Jiang River, Jul 1987. Zhu Jiang River basin: Guangxi Zhuang Autonomous Region: ASIZB 63456–63457, 2, 188.4–237 mm SL, Sanjiang (25°47'N, 109°36'E), Rong Jiang River, 14 May 1975. IHB 0002441, 2, 193.1–214.5 mm SL, Longsheng (25°47'N, 110°02'E), Rong Jiang River, Apr 1975; IHB 0002440, 3, 173.1–319.7 mm SL, Guilin (25°17'N, 110°17'E) and Yangshuo (24°46'N, 110°28'E), Li Jiang River, 1958, 1975 and 1981; IHB 0002442, 5, 125.9–179.9 mm SL, Rong'an (25°13'N, 109°22'), Siweihe reservoir, Jul 1973 and Nov 1974. Guizhou Province: ASIZB 65694–65697, 4, 146.6–224.0 mm SL, Duliujiang River, 1973.

Diagnosis Distinguished from other *Distoechodon* species by a combination of characters: lateral-line scales 72–82, mostly 77; predorsal scales 29–37, mostly 34 (Fig. 3); relatively small eye (6.37% vs. 6.98–8.28% in SL). *Distoechodon multispinnis* can be distinguished from another similar species, *D. tumirostris,* by having more lateral-line scales (Fig. 3), and longer dorsal and anal fins. (Table 1).

Description Body elongate and moderately compressed; dorsal profile straight or slightly convex; ventral profile concave; abdomen round; abdominal keel absent. Head rather small and compressed; snout round, anterior tip truncated; snout length longer than eye diameter; mouth inferior, transverse, with cuticular edge; mouth wide, as wide as head width at that point; no barbel; nostrils closer to eyes than to tip of snout; eyes relatively small.

Scales moderately small. Lateral-line complete, scales 72(2), 73(8), 74(6), 75(3), 76(9), 77(8), 78(3), 79(2), 80(4), 81(6), 82(2). Lateral line descending from top of gill opening to above pectoral-fin ray, then running parallel to horizontal scale row and ascending to body midline above anal-fin base and extending to end of caudal peduncle. Scales above lateral line 12–13; scales below lateral line 6–8; circumpeduncular scales 26; predorsal scales irregularly arranged, 29–37, mostly 34; scales cover abdomen. Pectoral-fin axillary fold present; pelvic-fin axillary scales well-developed. Gill rakers tiny, numerous and triangular, 52(2), 53(1), 54(5), 55(5), 56(3), 57(8), 58(4), 59(8), 60(2), 61(1), 62(6), 63 (2), 64(2), 65(2), 66(1) in lower part of first gill arch.

Pectoral-fin insertion slightly anterior to vertical through posterior margin of opercle. Pectoral fin short, reaching to midway between pectoral and pelvic-fin insertions. Pelvic fin short, reaching to midway between pelvic-fin insertion and vent. Dorsal-fin origin midway between snout tip and caudal-fin base, and slightly posterior to pelvic-fin insertion. Anal-fin insertion midway between pelvic-fin origin and caudal-fin base. Dorsal-fin rays iii, 7 (17), last unbranched dorsal-fin ray ossified but smooth. Anal-fin rays iii, 8, (1), 9 (14), 10 (2). Pectoral-fin rays i, 14 (3), 15 (10), 16 (1), 17 (3). Pelvic-fin rays i, 8 (17). Caudal fin forked, lobes almost equal in length; procurrent caudal-fin rays 8 (1), 9 (2), 10 (14) dorsally, 7 (1), 8 (7), 9 (8), 10 (1) ventrally.

Total vertebrae 26+19=45 (3), 26+20=46 (7), 27+19=46 (2); 26+21=47 (2), 27+20=47 (3); predorsal vertebrae 9 (1), 10 (13), 11 (3). Pharyngeal teeth in 2 rows, 3, 7; 7, 3. Outer row of pharyngeal teeth strong, compressed, with curved and pointed tips; inner row of pharyngeal teeth not well-developed, with very sharp tips.

Coloration in preservative Alcohol preserved specimens generally grayish to brownish; a dark spot usually present on each side of gill cover; Horizontal scale-rows above the lateral line are marked by dark spots along the posterior margin of each scale, forming 10 to 12 dark stripes. In poorly preserved specimens, this character is not distinct.

Geographical distribution Upper Chang Jiang basin (Min Jiang, Tuo Jiang in Sichuan Province, Hanshui

Rivers in Shanxi and Hubei Provinces, Yuan Jiang River in Guizhou and Hunan Provinces) and Zhu Jiang basin (Rong Jiang and Li Jiang Rivers in Guangxi Zhuang Autonomous Region and Duliu Jiang River in Guizhou Province) (Fig. 5).

Remarks Characters observed generally agree with the original description (Bănărescu 1970), except for caudal peduncle length and preanal length (19.5–21.9% and 69.8–72.5%, respectively in the original description). Differences in these two characters may be explained by different measuring methods. Eight specimens were included in the original description but were not re-examined by us. Several specimens from near the type locality are included in our material.

Nucleotide sequences of the mitochondrial D-loop region and complete cytochrome *b* gene show that the p-distances between *Distoechodon* specimens from the upper Chang Jiang basin in Sichuan and the Zhu Jiang basin in Guangxi Zhuang Autonomous Region, and *D. compressus* from costal rivers in Fujian Province are higher than the common intraspecific difference (Liu 2002), which supports the current recognition of *D. multispinnis* (including the specimens both from upper Chang Jiang and Zhu Jiang basins) as distinct from *D. compressus*. *D. multispinnis* is most similar to *D. tumirostris* in body shape and scales count. Principal Component Analysis clearly separated the two species on PCII, characterized by high loadings of pectoral fin-base length, dorsal-fin length as well as anal-fin length (Table 2, Fig. 6).

Table 2 Character Loadings on Principal Components I–III for Measurement from specimens of *Distoechodon tumirostris* (46 specimens) and *D. multispinnis* (36 specimens)

	PC I	PC II	PC III
Standard length	0.3172	0.0157	0.0135
Body depth	0.3353	−0.0315	0.0212
Predorsal length	0.3042	0.0141	0.0124
Post-predorsal length	0.3226	0.0119	0.0252
Prepectoral length	0.2787	0.0149	−0.0054
Prepelvic length	0.3005	0.0161	0.0074
Preanal length	0.3102	0.0151	0.0131
Caudal peduncle length	0.3347	0.0015	0.0297
Caudal peduncle depth	0.3461	−0.0328	0.0123
Dorsal-fin base length	0.3210	0.0036	0.0027
Anal-fin base length	0.3122	−0.0061	−0.0003
Pectoral-fin base length	0.3240	**0.0384**	−0.0103
Pelvic-fin base length	0.3458	−0.0123	0.0095
Dorsal-fin length	0.3181	**−0.0375**	−0.0113
Anal-fin length	0.3542	**−0.0452**	−0.0140
Pectoral-fin length	0.3032	−0.0087	−0.0151
Pelvic-fin length	0.3193	−0.0327	−0.0086
Head length	0.2703	0.0134	−0.0063
Head dorsum length	0.2774	0.0241	−0.0142
Head depth	0.3042	0.0061	0.0108
Head width	0.3308	0.0183	0.0158
Snout length	0.3316	0.0069	−0.0368
Eye diameter	0.1735	0.0176	−0.0069
Interorbital width	0.3546	−0.0026	0.0144
Upper jaw length	0.2963	−0.0165	0.0050
Mouth width	0.3554	0.0286	−0.0087
Prenostril length	0.3323	0.0008	−0.0605
Variance explained by components	2.6935	0.0122	0.0095
Percent of total variance explained	97.4509	0.4421	0.3446

Three highest loadings on PC II indicated by boldface

Distoechodon macrophthalmus sp. nov. *(Fig. 2D, Table 1)*

Distoechodon tumirostris Chen and Li, 1989, The fishes of Yunnan, China, Part I Cyprinidae: 96 (Changhai Lake, Yunnan Province).

Holotype KIZ 797297, 155.4 mm SL. Chenghai Lake, Chang Jiang basin in Yongsheng County (26°42’N, 100°44’E), Yunnan Province, P. R. China, 1979.

Paratypes KIZ 797298, 1; KIZ 799738, 1; KIZ 797295, 1; KIZ 818413, 1; 104.2−155.4 mm SL, same data as holotype; IHB 0002446, 10, 115.2 −200.2 mm SL, same locality as holotype.

Diagnosis Distinguished from other species of *Distoechodon* in possessing the highest number of lateral line scales (78–85, mostly 82), more pre-dorsal scales (34–39, mostly 36) (Fig. 3) and relatively bigger eye, being 7.97% vs. 6.37–6.98% in SL. *Distoechodon macrophthalmus* is similar to *D. multispinnis* in having higher lateral line and predorsal scale counts, and both occur in the upper Chang Jiang basin. However, *D. macrophthalmus* is easily distinguished by having larger eyes at the same body size (Fig. 7, Table 1).

Description Body elongate and moderately compressed; dorsal profile straight or slightly convex;

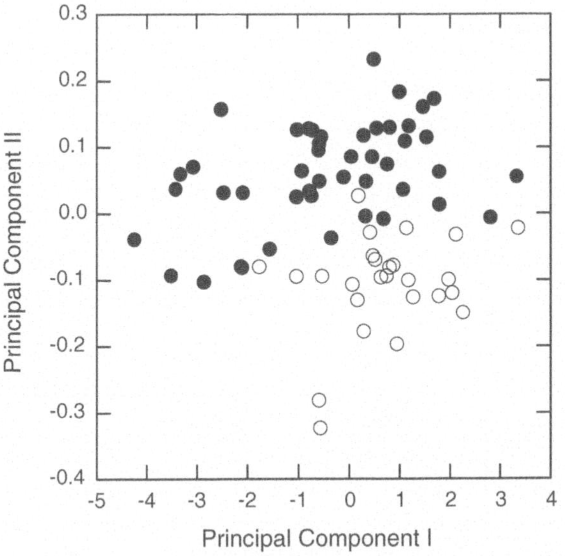

Fig. 6 *Distoechodon tumirostris* and *D. multipinnis*. Plot of scores of principal component II on principal component I. ○ *Distoechodon multispinnis*. ● *Distoechodon tumirostris*

ventral profile concave; abdomen round; abdominal keel absent. Head rather small and compressed; snout round, anterior tip truncated; snout length longer than eye diameter; mouth inferior, transverse, with cuticular edge; mouth wide, as wide as head width at that point; no barbel; nostrils closer to eyes than to tip of snout, first pair of nostrils small and round, second pair of nostrils transverse, after first pair immediately with membranes; eye relatively bigger than that of *D. tumirostris* and *D. multispinnis* (Table 1, Fig. 7).

Scales moderately large. Lateral-line complete, scales 78(2), 79(1), 81(1), 82(4), 84(6), 85(1). Lateral line descending from top of gill opening to above pectoral-fin ray, then running parallel to horizontal scale row and ascending to body midline above anal-fin base and extending to end of caudal peduncle. Scales above lateral line 12–13; scales below lateral line 7–8; circumpeduncular scales 26 (14), rarely 24 (1); predorsal scales irregularly arranged, 34–39, mostly 36; scales cover abdomen. Pectoral-fin axillary fold present; pelvic-fin axillary scales well-developed. Gill rakers tiny, numerous and triangular, 53(3), 55 (2), 56(6), 57(2), 59(1) in lower part of first gill arch.

Pectoral-fin insertion slightly anterior to vertical through posterior margin of opercle. Pectoral fin short, reaching to midway between pectoral and pelvic-fin insertions. Pelvic fin short, reaching to midway between pelvic-fin insertion and vent. Dorsal-fin origin midway between snout tip and caudal-fin base, and slightly posterior to pelvic-fin insertion. Anal-fin insertion midway between pelvic-fin origin and caudal-fin base. Dorsal-fin rays iii, 7 (15), last unbranched dorsal-fin ray ossified but smooth. Anal-fin rays iii, 9 (15). Pectoral-fin rays i, 14 (6), 15 (9). Pelvic-fin rays i, 8 (18). Caudal fin forked, lobes almost equal in length.

Pharyngeal teeth in 2 rows, 2, 7; 7, 2. Outer row of pharyngeal teeth strong, deeply compressed, with curved and pointed tips; inner row of pharyngeal teeth not well-developed, with very sharp tips.

Color in life. Dorsum dark grayish; abdomen silvery whitish; side with 10 to 12 horizontal stripes composed of black dots; dorsal and caudal fin light yellowish; pectoral, pelvic and anal fin light reddish (Chen and Li 1989).

Coloration in preservative Alcohol-preserved specimens generally brownish, dorsum blackish, no specific color pattern observed.

Geographical distribution *Distoechodon macrophthalmus* is known only from Chenghai Lake, upper Chang Jiang basin in Yunnan Province (Fig. 5). Chenghai Lake is situated about 1,503 m above sea level, with an average depth at 25.7 m (Tao et al., 1999). The lake was isolated from the Jinsha Jiang

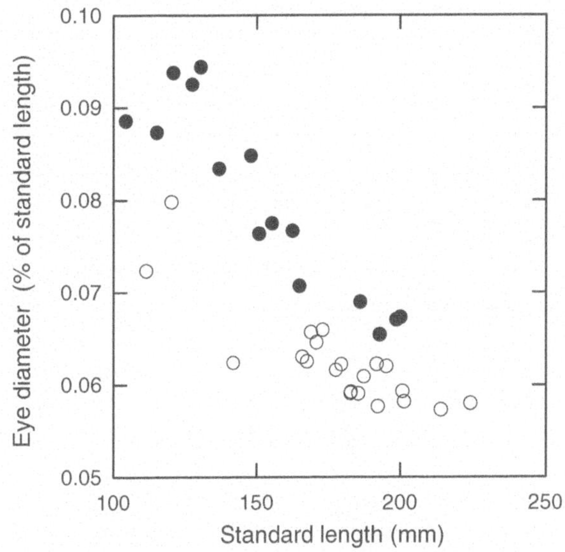

Fig. 7 Eye diameter (% SL) plotted against standard length (mm) for *Distoechodon multispinnis* and *D. macrophthalmus* sp. nov. ○ *Distoechodon multispinnis*. ● *Distoechodon macrophthalmus* sp. nov.

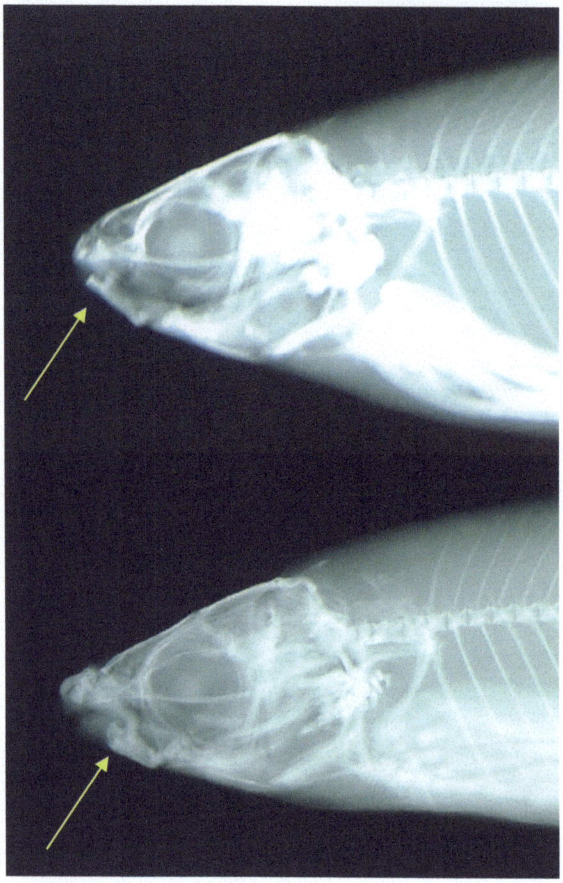

Fig. 8 Radiograph of *Xenocypris argentea* (ASIZB 63153, above) and *Distoechodon compressus* (ASIZB 67303, below), showing the difference in the dentary

River (belonging to Chang Jiang drainage) at least 400 years ago (Wang and Dou, 1998).

Etymology The specific name *macrophthalmus* is a noun composed of the Greek words *macros,* meaning big or long, and *ophthalmus* meaning eye, referring to its large eyes.

Local name Hongchiyu (Red-wing fish) in Yongsheng County, obviously referring to the light reddish pectoral, pelvic and anal fins in live specimens.

Remarks The population is shrinking quickly. It was a very important component of the local fishery before the 1990s. But during our field survey in 2004, we could not find any adult specimen. The primary threat might be the introduction of icefishes (family Salangidae), based on our survey.

Discussion

Compared to *Xenocypris*, *Distoechodon* has a transverse instead of concave mouth, a shorter dentary (Fig. 8), a well-developed (vs. weak or not developed) cutting edge on the lower jaw, distinct pectoral-fin axillary folds, more gill rakers (85–122 vs. 39–61), and more predorsal vertebrae (Fig. 9). *Distoechodon* and *Xenocypris* are both diagnosible and obviously valid genera.

Distoechodon, and the other three genera, *Xenocypris*, *Pseudobrama*, and *Xenocyprioides*, was classified into the subfamily Xenocypridinae (Liu and He, 1998). The number of rows of pharyngeal teeth has been considered to be a main diagnostic character in the subfamily (sensu Liu and He 1998): species from *Xenocyprioides* and *Xenocypris* possess three rows of pharyngeal teeth; *Distoechodon*, two rows; and *Pseudobrama* one row. Recent phylogenetic studies based on both morphological and molecular characters have shown that the number of pharyngeal teeth rows is not a phylogenetically informative character (Shan 1998; Xiao et al. 2001).

The genus *Xenocyprioides* was established and designated into the subfamily Xenocypridinae by Chen in 1982, mainly because of the similar pharyngeal teeth pattern to other members of the subfamily. *Xenocyprioides* includes 2 min species (21–30 mm

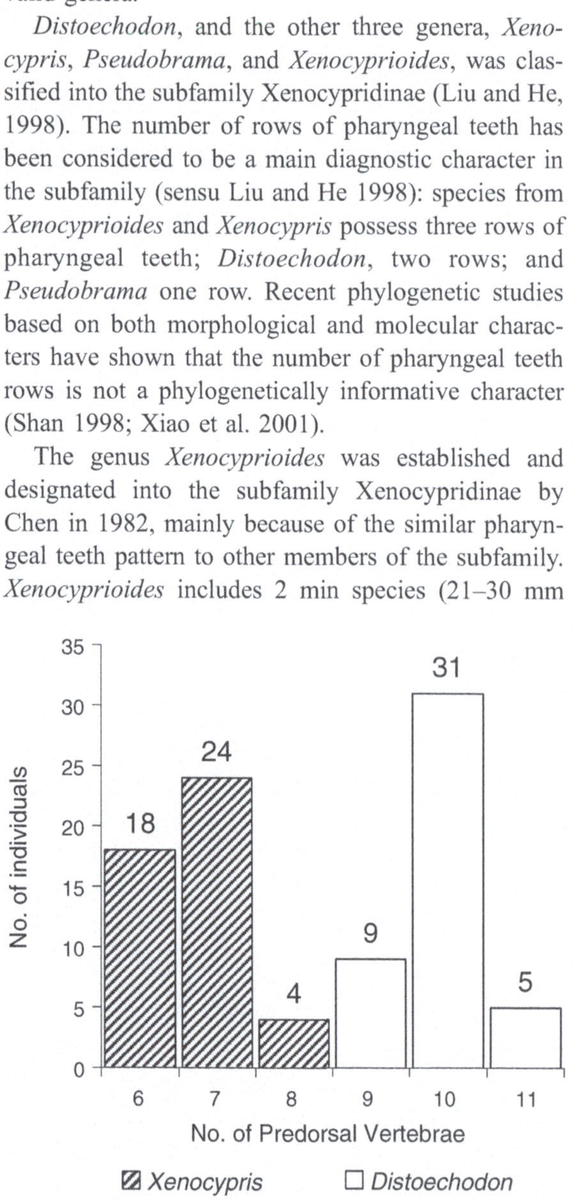

Fig. 9 Differences in the number of predorsal vertebrae between *Xenocypris argentea* and *Distoechodon* species (including *D. tumirostris, D. compressus,* and *D. multispinnis)*

SL) from southern China. They possess a flexible last unbranched dorsal-fin ray, and fewer vertebrae (33–35 vs. 40–48) and scales in a lateral row (31–35 vs. 57–84,) and lack lateral line, in contrast with all other xenocypridines. Shan's phylogenetic work based on morphology demonstrated that this genus falls outside the true Xenocypridinae, although she thought it might be a primary clade (Shan, 1998). The phylogenetic tree of Xenocypridinae based on combined Cyt *b*, ND4L, ND4, $tRNA^{Ser}$, and $tRNA^{His}$ data sets also had a similar result (Xiao et al, 2001). The position of the genus *Xenocyprioides* in the family Cyprinidae still need to be further discussed. It might be closer to *Danio* or *Rasbora*. Therefore, four genera were included in the subfamily Xenocypridinae by this study. Except for *Distoechodon* and *Xenocypris*, *Pseudobrama* is diagnosed by an abdominal keel from the pelvic-fin base to the vent. It also has the lowest number of lateral-line scales (41–47) and the highest number of gill rakers (128–139) among the xenocypridin genera. In addition, based on the development degree of the abdominal keel and cutting edge on the lower jaw, we are apt to admit *Plagiognathops* as a valid genus.

In a recent review (Liu and He 1998), the genus *Distoechodon* included two species, *D. hupeinensis* and *D. tumirostris*. *D. hupeinensis* has a curved mouth and fewer lateral line scales and gill rakers, distinguished from *D. tumirostris*, but close to the species of *Xenocypris*. *D. hupeinensis* was arranged under the genus *Distoechodon* by Yang (1964) mainly because of two rows of pharyngeal teeth, the same as the species from *Distoechodon*. In fact, it has many characters in common with the species of *Xenocypris* including mouth shape, premaxilla and dentary morphology, and the number of lateral-line scales. Therefore, Shan (1998) thought two rows of pharyngeal teeth might be the result of parallel evolution and suggested placing *D. hupeinensis* within *Xenocypris*. We accepted her opinion in the present study. The evidence from molecular systematics work shows that *D. hupeinensis* is closely related to *X. argentea* and *X. davidi*, but is distant from the species of *Distoechodon* (Xiao et al 2001). Even the study of the related parasite fauna also supports this viewpoint (Xiao & Wang 2000).

After removing *Distoechodon hupeinensis*, four species can now be recognized in the genus *Distoechodon*. Bănărescu (1970) used the number of gill rakers and lateral-line scales to distinguish them. Our study, however, shows no particular variation in the gill-raker count among the different species. The number of lateral-line scales, on the other hand, does show great variation among the different species. In addition, the four species occupy different areas. *D. multispinnis* has the widest distribution among them. The new species is the westernmost record of the genus (Fig. 5).

Comparative material examined All from China. *Xenocypris argentea*: Hunan Province: Yuan Jiang River: NRM-EC2002-2 (4), 12.9–24.5 mm SL, Taoyuan (28°53′59″N,′ 111°28′59″E), 18–19 Mar, 2002, F. Fang et al. NRM-EC2002-3 (13), 63.0–123.0 mm SL, Taoyuan (28°53′59″N, 111°28′59″E), 18–19 Mar, 2002, F. Fang. Hebei Province: IZB 15758–15763 (6), 74.5–131.5 mm SL, Tangshan. *Xenocypris davidi*: Hunan Province: Lishui River: NRM-EC2002-18 (2), 111–112 mm SL, Shimen (29° 35′21″N, 111°17′43″E), 25 Mar, 2002, F. Fang et al. Guangxi Zhuang Autonomous Region: IZB 63153–63156 (4), 123–185 mm SL. *Xenocypris* sp.: Hunan Province: Yuan Jiang River: NRM-EC2002-5 (1), 50.0 mm SL, Chehuyuan (28°59′46″N, 111°31′35″E), Taoyuan, 18 Mar, 2002, A. Malzahn et al. NRM-EC2002-9 (25), 50–78 mm SL, Xinlongjie (28°47′54″ N, 111°7′25″E), Taoyuan, 20 Mar, 2002, C. G. Zhang. NRM-EC2002-41 (2), 134.0–146.0 mm SL, Hengdong (27°4′59″N, 112°56′59″E), 9 Apr, 2002, F. Fang et al.

Acknowledgments The work was carried out within the ECOCARP project (New native fish species for Asian aquaculture: conserving natural genetic reserves and increasing options for sustainable use of aquatic resources), European Commission INCO-DEV contract number ICA4-CT-2001-10024, and also supported by Swedish Biodiversity Centre, Swedish University of Agricultural Sciences and Chinese National Science Fund for Fostering Talents in Basic Research (NSFC-J0030092). Many thanks go to Liu H. Z. for constructive discussions and suggestions, to Tang W. Q., Cai M. J., Zhang E., Xu T. Q., Cui G. H. and Yang J. X. for allowing to access specimens under their care, to M. Höggren and V. Areskoug for their guidance, and kind concern and help. A. Silfvergrip helped to take fish photos, and E. Åhlander helped to take all the radiographs.

References

Bănărescu P (1970) Remarks on the genus *Xenocypris* (Pisces, Cyprinidae) with description of a new subspecies. Rev Roum Biol Zool 15:395–402

Chen IS, Chang YC (2005) A photographic guide to the inland-water fishes of Taiwan, vol 1 Cypriniformes. Sueichan, Keelung, pp 84–89

Chen YR, Li ZY (1989) Xenocyprininae. In: Chu, XL (eds) The fishes of Yunnan, China, Part I Cyprinidae. Science, Bejing, pp 93–98

Chen YY (1982) Description of a new genus and species of Cyprinid fish. Acta Zootax Sin 7:425–427

Chen YY (1998) General Introduction to Cyprinidae. In: Chen, YY (eds) Fauna Sinica. OsteichthyesCypriniformes II. Science, Beijing, China, pp p1–18

Ding RH (1994) The fishes of Sichuan, China. Sichuan Publishing House of Science and Technology, Chengdu, pp 163–165

Fang F (1997) Redescription of *Danio kakhienensis*, a poorly known cyprinid fish from the Irrawaddy basin. Ichthyol Explor Freshwat 7:289–298

He MJ (1989) Xenocyprininae. In: Zheng, CY (eds) Fishes of Zhujiang River. Science, Beijing, pp 112–115

Leviton AE, Gibbs RH, Heal E (1985) Standards in herpetology and ichthyology: part I. Standard symbolic codes for institutional resource collections in herpetology and ichthyology. Copeia 1985:802–832

Leviton AE, Gibbs RH (1988) Standards in Herpetology and ichthyology, standard symbolic codes for institution resource collections in herpetology and ichthyology, supplement No. 1: additions and corrections. Copeia 1988:280–282

Lian ZS (1984) Xenocyprininae. In: "Fishes of Fujian province" editorial subcommittee (ed) The fishes of Fujian province (Part I). Fujian Science and Technology Press, Fuzhou, pp 252–255

Liu HZ (2002) Mitochondrial DNA sequence variations and systematics of the genus *Distoechodon* (Teleostei: Cyprinidae). J Appl Ichthyology 18:181–184 doi:10.1046/j.1439-0426.2002.00325.x

Liu HZ, He MJ (1998) Xenocyprinae. In: Chen, YY (eds) Fauna Sinica, OsteichthyesCypriniformes II. Science, Beijing, pp 208–223

Mao JR (1991) Xenocyprinae. In: Mao, JR (eds) Fauna of Zhejiang, Fresh Water Fishes. Zhejiang Science and Technology Publishing House, Hangzhou, pp 73–78

Nichols JT (1925) Some Chinese fresh-water fishes. 7. New carps of the genera *Varicorhinus* and *Xenocypris*. Am Mus Novit 182:1–8

Nichols JT (1928) Chinese fresh-water fishes in the American Museum of Natural History's collections. A provisional check-list of the fresh-water fishes of China. Bull Am Mus Nat Hist 58:1–62

Nichols JT (1943) The fresh-water fishes of China. Nat Hist Central Asia 9.

Peters WCH (1881) Über die von der chinesischen Regierung zu der internationalen Fischerei-Ausstellung gesandte Fischsammlung aus Ningpo. Monatsb Akad, Wiss Berl, pp 921–927

Shan XH (1998) Phylogeny and dispersal-variance explanation of the extant fishes of Xenocyprinae (Teleostei: Cyprinidae). Acta Zootax Sin 23(Supplement):56–66

Shen SC, Tzeng ZC (1993) Cypriniformes. In: Shen SC (ed) Fishes of Taiwan. Zoology Department, Taiwan University, Taibei, p 141

Tao WD, Xia F, Xing CY (1999) On environmental issues of Lake Chenghai and its management strategy. Resour Environ Yangtze Basin 8:210–214

Tchang TL (1933) The study of Chinese Cyprinoid Fishes. Part I. Zool. Sin B 2(1):1–247

Wang SM, Dou HS (eds) (1998) Lakes of China. Science, Beijing

Xiao WH, Zhang YP, Liu HZ (2001) Molecular systematics of Xenocyprinae (Teleostei: Cyprinidae): taxonomy, biogeography and coevolution of a special group restricted in East Asia. Mol Phylogenet Evol 8:163–173 doi:10.1006/mpev.2000.0879

Xiao WH, Wang JG (2000) Phylogeny of *Hexamita* parasitic in Xenocyprinae based on cladistic analysis of morphological characteristics with consideration of coevolution between hosts and host-parasites. Acta Hydrobiol Sin 24:122–127

Yang GR (1964) Xenocyprininae. In: Wu, XW (eds) The Cyprinid fishes of China. Science, Beijing, pp 128–132

Yih BL (1964) Distoechodon hupeinensis. In: Wu, W (eds) The Cyprinid fishes of China. Science, Beijing, pp 129–130

Environ Biol Fish (2009) 86:45–52
DOI 10.1007/s10641-008-9334-x

Otolith microstructure of *Oxygymnocypris stewartii* (Cypriniformes, Cyprinidae, Schizothoracinae) in the Lhasa River in Tibet, China

Yintao Jia · Yifeng Chen

Received: 4 June 2007 / Accepted: 20 November 2007 / Published online: 26 February 2008

Abstract Otolith microstructure of *Oxygymnocypris stewartii* collected from the Lhasa River was examined and described with regards to the early life history events. The monthly changes in the number of microincrements on the margin of the otolith were examined to validate the approximately daily periodicity of otolith increment formation. The microstructure of otoliths was used to detect changes in microincrement deposition patterns corresponding with events during early life. The annuli, microincrements and checks including the hatch check, yolk absorption check and several recurrent patterns in the otolith were described. Periodicity of the recurrent patterns was weekly, fortnightly and monthly. Through counting the number of the microincrements, it was confirmed that the primary growth of *O. stewartii* was in a period of 7 or 8 months from late March to October; it was estimated that *O. stewartii* might hatch between April and May.

Y. Jia · Y. Chen (✉)
Laboratory of biological invasion and adaptive evolution, Institute of Hydrobiology, Chinese Academy of Sciences, Wuhan 430072, China
e-mail: chenyf@ihb.ac.cn

Y. Jia
e-mail: jiayintao@163.com

Y. Jia
Graduate School of Chinese Academy of Sciences, Beijing 100039, China

Keywords Tibetan Plateau · Endemic fish · Annulus formation · Environment

Introduction

Otolith microstructure research has developed rapidly since Pannella (1971) showed the presence of daily growth increments in fish otoliths (Campana and Neilson 1985). Aging techniques based on daily increments are commonly used in ecological studies of various fish species. In addition to age and growth, fish otoliths may also record life history events, such as the movement from one habitat to another (Neilson and Geen 1982), metamorphosis (Campana 1984b) and settlement (Victor 1982). Therefore, otolith microstructure analysis has become a useful tool for studying the biological characteristics of fish (Campana and Neilson 1985). A prerequisite for all the otolith microstructure work is to prove that the increments are formed daily (Campana and Neilson 1985). Most validation works focus on the larval or juvenile period under controlled condition in laboratory (Geffen 1987; Meisfjord et al. 2006). Studies of wild populations usually adopt method of introducing a mark onto the otolith (Campana and Neilson 1982), and such studies lack a direct and simple method to validate the periodicity of otolith increment formation. Due to clear increments on its otolith, the schizothoracinae is a suitable and reasonable fish to validate the periodicity of otolith increment formation, and otolith microstructure is used to further study the life-history of wild populations. Chen et al. (2002a) reported that *Gymnocypris selincuoensis* otoliths exhibited clear microincrements. Song et al. (2006) described otolith microstructure of larval *Gymnocypris potanini* from the Minjiang River in China. However, they did not give more information about the method to validate the daily period of the microincrements.

The Lhasa River is the longest tributary of the Yarlung Tsangpo River Systems (82° E–97° E, 28° N–31°15′ N) that is the highest river in the world (Team of Chinese Academy of Science of the Scientific Expedition to the Qinghai-Xizang Plateau 1983). There are abundant species of fish in the Yarlung Tsangpo River, and several of them are endemic to this river, for example, *Oxygymnocypris stewartii* (Chen and Cao 2000). This fish inhabits in stretches at an altitude of above 3,600 m and adapts to the clear and cool water. Limited distribution, slow growth and late maturity hinder the survival of this species. Recently wild survey found that *O. stewartii*'s quantity declined rapidly, and it has been listed to the China red data book of endangered animals as the endangered fishes (Yue and Chen 1998). The purposes of the present study are: (1) to validate the daily periodicity of microincrement deposition; (2) to describe the otolith microstructure of *O. stewartii* in relation to the occurrence of early life history events; (3) to study the relationship of early life history characteristics with plateau environmental conditions. It is hope that the study could provide useful information for the reasonable protection of species, *O. stewartii*.

Materials and methods

Sampling

A total of 430 specimens were used in the study, most of them were collected from the Lhasa River, a tributary of the Yarlung Tsangpo River system of the Tibet, China. Sampling was carried out monthly from April in 2004 to September in 2006 by purchasing at local fish markets and catching by gill net. The collection time was centralized during the last fifteen days of every month. Conventional biological measurements for specimens were carried out on fresh fish. The range of specimen's standard length was from 46.0 to 546.0 mm. Both lapilli were extracted from the utricle, cleaned with distilled water and stored in a polythene zip packet. In the text lapillus was referred to as otolith. Right otoliths were mainly used, and if the right otolith was broken then the left one was taken.

Otolith preparation

Among the three pairs of otoliths found in fish, the use of the lapillus was highly recommended because it had moderate size and was not easy to break in the process of preparing section. To get a transverse section, otoliths were mounted on a glass slide using thermoplastic cement, ground (1,500 grid wet sand paper) and polished (with 0.25-mm alumina paste) until the core region was reached. The section was then turned upside down and attached back to the slide. New grinding and polishing was carried out until the core was reached. The preparation was accomplished when the tightly packed microincrements around the core were exposed for reading (Chen et al 2002a).

Counting the number of microincrements

The longest growth axis in the otolith was from the nucleus to the posterior edge. Microincrement counts began at the nucleus and along the longest radius. Microincrements in a sectioned otolith were counted by two readers independently using a microscope with transmitted light at ×400 magnification. If the two counts differed by less than 2.5% of the lower count, they were averaged. If the difference exceeded 2.5%, both readers counted the section again. The second count usually resolved any between-reader difference, but if the counts still differed by more than 2.5%, and the section was discarded (Pannella 1971).

Marginal increment ratio analysis

To validate the deposition of the most recent translucent zone in otoliths, the authors calculated the index of completion for marginal increment ratio (MIR) of the otolith using the following formula: MIR = [(CE-CLA)/(CLA-CSA)]×100, where CE was the distance from centre of origin to the outer edge, CLA was the distance from centre of origin to the last formed annulus and CSA was the distance from centre of origin to the second last formed annulus (Hass et al. 1995; Fig. 1). To identify more precise age of the edge in the sectioned otoliths, edge type analysis was used: from the MIR value the last-formed annulus was judged as 'C' (close), 'Z'(other) or 'N'(now due) (Vilizzi and Walker 1999). In sectioned otoliths it was not possible to detect a newly formed opaque zone when its width was less than 10–30% of the previous increment, so the score were used to divide the type as follows: 'C' (if MIR≤ 40%), 'Z' (if 40%<MIR< 75%), and 'N' (if MIR≥75%; Chen et al. 2002b). In edge type analysis, an appearance of low percentage of type 'C' values suggested that in some specimens new rings were recently completed, and an appearance of high percentage of type 'N' suggested that most specimens were about to form new rings. The period when a low percentage of type 'C' occurred with high percentage of type 'N' (greater than 50.0%) and the highest percentage of type 'N' occurrence were considered the ring formation period.

The otolith radius was defined as a distance from the core to the maximum margin of the distal edge. The ring radius was measured from the core to the outer edge of the ring and followed the same axis of the otolith radius. Both measurements were done to the nearest 0.0001 mm by using a computer aided image analysis system designed by Zhu et al. (2002).

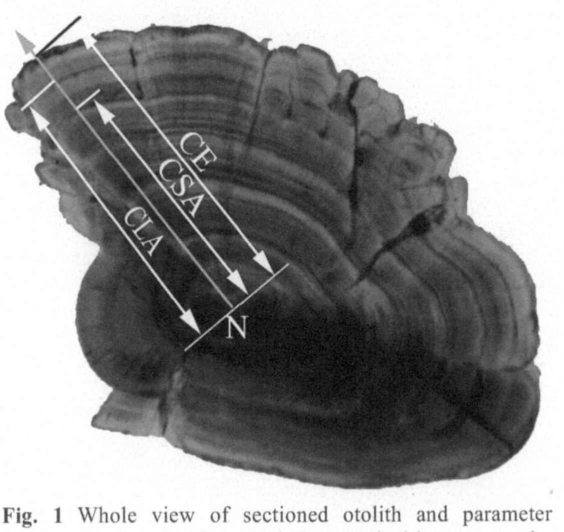

Fig. 1 Whole view of sectioned otolith and parameter measurement location for calculating marginal increment ratio. *CE* the distance from centre of origin to the outer edge, *CLA* the distance from centre of origin to the last formed annulus, and *CSA* the distance from centre of origin to the second last formed annulus

Analytical software

Statistical analyses were carried out using SPSS13.0 software and Microsoft Office Excel 2003.

Results

Otolith microstructure

In all of 430 specimens, 256 otolith sections exhibited fine microstructure and 174 otoliths were unreadable because of over grinding. Right and left otoliths of 53 specimens were examined, and the same results of the microstructure were gained, so both right and left otolith was suitable for studying and the results had no difference. Under a light microscope with transmitted light, the otolith section exhibited an irregular oval shape, and the color from nucleus to the distal edge thinned gradually. Otoliths usually had one nucleus (Fig. 2a). There were three types of rings on the otolith: microincrements, annuli and checks.

Validation of the daily periodicity of microincrements deposition

To determine if the periodicity of otolith microincrement formation was under an approximately daily rhythm, we counted the monthly changes in the number of microincrements in the margin of the otolith from the last annuli to the utmost edge on the otolith. The 2 (n=18), 3 (n=28) and 4 (n=12) years old samples which were collected from May to September were selected for study, because these samples' microincrements in the margin were clear and easy for counting. The mean number of microincrements were 37 [n=11, 36.67± 11.38 (mean ± SD)] in May, 62 (n=10, 61.67±12.31) in June, 87(n=12, 86.91±16.04) in July, 112 (n=13, 112.2±10.60) in August, 133 (n=12, 133.3±7.448) in September, respectively. The monthly increased in microincrements from May to June was 25, from June to July was 25, from July to August was 25 and from August to September was 21 (Table 1).

Characteristics of the microincrements

The microincrements of *O. stewartii* otoliths had two characteristics: The width of microincrements was approximate equal at the beginning of opaque zones,

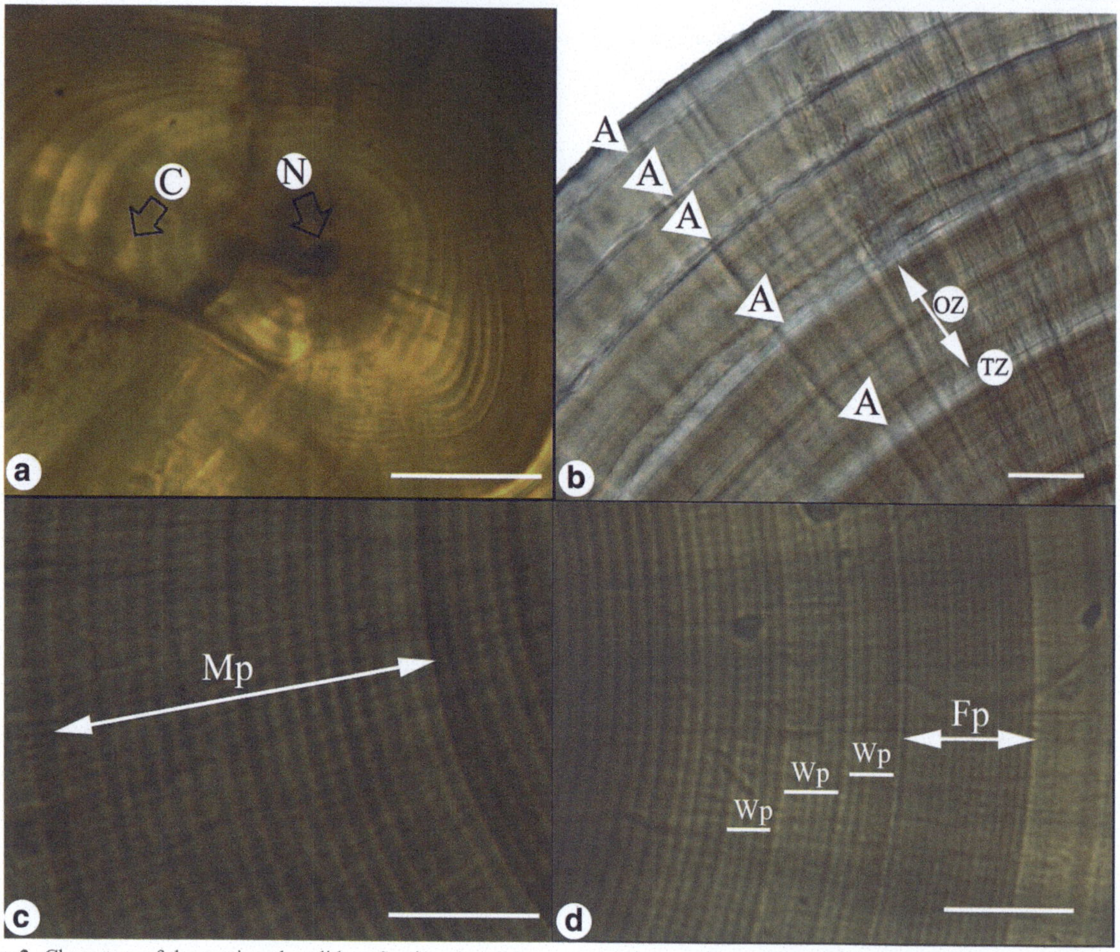

Fig. 2 Characters of the sectioned otolith. **a** Sectioned otolith, showing nucleus and check, scale bar = 20 μm; **b** sectioned otolith, showing annuli, translucent zone and opaque zone, scale bar = 100 μm; **c** sectioned otolith, showing monthly recurrent pattern, scale bar = 20 μm; **d** sectioned otolith, showing weekly and fortnightly recurrent pattern, scale bar = 20 μm. Explanation to abbreviation: *N* nucleus, *C* hatch check, *A* annulus, *TZ* translucent zone, *OZ* opaque zone, *Mp* monthly recurrent pattern, *Wp* weekly recurrent pattern, *Fp* fortnightly recurrent pattern

and gradually decreased close to the translucent zones within annulus; and the width of microincrements at different ages became thinner when fish's age increased.

To determine the main growth period *O. stewartii*, the number of the microincrements within the first annulus and between the first and the second annulus were counted, and the numbers were 121–184 (n=60, 153.4± 16.85) and 160–187 (n=58, 171.4±11.10), respectively.

Validation of the time of the annulus formation

In sectioned otoliths, the core was clearly discernible under a light microscope with transmitted light, and wide translucent zones and narrow opaque zones were alternately arranged (Fig. 2b). It was generally inter-

Table 1 Monthly change in the number of microincrements and microincrement mean number from May to September on the outer margin of the otolith

Month	Sample size	Microincrement mean increase per month	Microincrement mean number (SD)
May	11		37 (11.38)
		25	
June	10		62 (12.31)
		25	
July	12		87 (16.04)
		25	
August	13		112 (10.60)
		21	
September	12		133 (7.448)

preted that the translucent band formed during the slow or no growth period in the fall and winter and the opaque one formed during spring and summer. Opaque zones showed a slow transition to translucent zones, however, translucent zones changed rapidly to an opaque zone at the time of the formation of new annuli. The transition from translucent zone to opaque zone was defined as the new annuli formed.

As shown in Fig. 3, in March the percentage of type 'N' occurrence was maximal (66.7%), subsequently, the percentage decreased gradually to 33.4% in May and over the next several months, type 'N' edges did not occur at all until October. In April, type 'C' edges occurred for the first time in the year and, subsequently their percentage increased gradually until all specimens showed the type 'C' in June. Because percentage of type 'N' occurrence was the highest in March (the value exceeded 60%) and in April, percentage of type 'N' occurrence exceeded 50% with the lowest percentage of type 'C' occurrence, it was confirmed that the annulus of the otolith was produced chiefly from March to April. The occurrence once again of type 'N' in October indicated that *O. stewartii*'s growth became slow in October.

Checks in otolith sections

Checks were regarded as discontinuous and opaque rings under light microscope, which were greatly different from daily microincrements, and they were wider than ambient microincrements (Chen et al. 2002a). There was a microincrement abrupt accretion after five or six smooth microincrements near the core in otolith sections (Fig. 2a). A prominent check followed 20 to 24 microincrements after this microincrement.

Through the specimen growth record of otolith was examined, checks were grouped in several recurrent patterns, and their periodicity was a week (Fig. 2d), a fortnight (Fig. 2d) and a month (Fig. 2c), respectively (verified by validated microincrement counts). The range of microincrement numbers varied in these patterns among 4–7, 14–17 and 21–25, respectively. The monthly recurrent pattern was not seen every month, and the weekly and fortnightly recurrent patterns occurred at apparently random locations in the growth record. However, these checks often occurred in the first 3 years of the life-history.

Discussion

Validation of daily increments

The monthly changes in the number of microincrements on the margin of the otolith were from 21 to 25, it was approximately equal to monthly periodicity, 28 days. So it was presumed that the microincrements in *O. stewartii* otoliths would deposit at a rhythm of

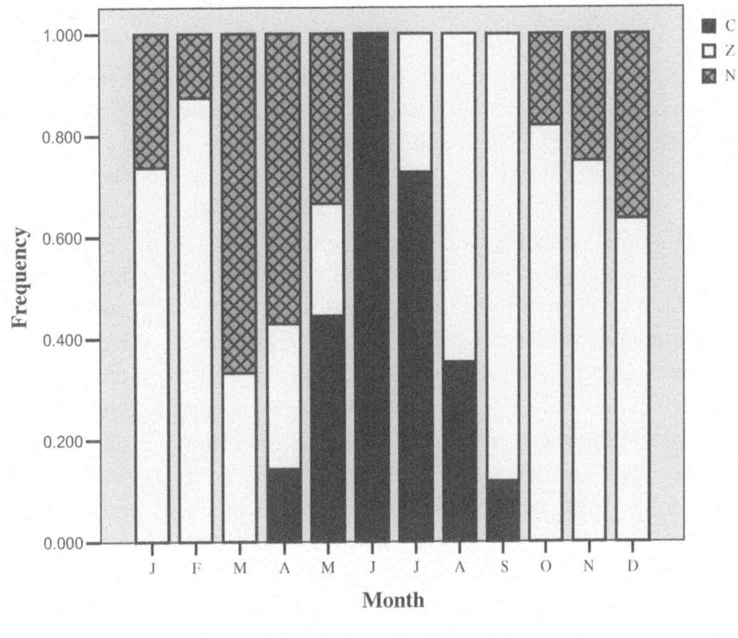

Fig. 3 Percentage of sectioned otoliths with a 'C' (if MIR≤40%), 'Z' (if 40%<MIR<75%), and 'N' (if MIR≥75%) type margin (under transmitted light)

approximately one increment per day. For wild population, microincrement did not deposit at accurately rhythm of one increment per day, because some environmental variables may disturb the rhythm. Accurate interpretation of microstructural growth patterns required knowledge of many factors that might affect the production of one increment per day (Campana and Neilson 1985). Environmental variables such as photoperiod (Taubert and Coble 1977; Radtke and Dean 1982), water temperature (Marshall and Parker 1982), and amount of food (Methot and Kramer 1979) might all fluctuate cyclically; and all had the potential to influence otolith deposition (Campana and Neilson 1985). For example, daily increments were not formed under either constant light or dark condition; low temperature could result in cessation of the increments formation; fish in the starvation may result in cessation of otolith growth. For wild populations, the history of exposure to such variables would generally be unknown (Campana and Neilson 1985). Microincrement formation of *O. stewartii* distributed in the Lhasa River may be subject to these variables, so deposition of microincrements might cease for a few days. Therefore, it was proposed that the microincrements in *O. stewartii* otolith should deposit at the rhythm of approximately one increment per day, though the monthly increase in microincrements number was less than 28.

Growth characteristics

Daily increment width may reflect fish growth (Campana and Neilson 1985). Several factors may influence the daily increment width, especially temperature of water (Marshall and Parker 1982; Neilson and Geen 1982; Campana 1984a). The change of width of daily increments in *O. stewartii*'s otolith section within annulus was consistent with that of the water temperature. The width of daily increments decreased near the translucent zone when water temperature decreased demonstrating that *O. stewartii* growth became slower at that time. However, daily increments are only present in opaque zones which represented the rapid growth period of the year and the fine daily increments in translucent zones with light microscopy couldn't be seen. This could be attributed to the theoretical resolution limit of a light microscope. Due to the diffractive nature of light, objects separated by less than 0.20 μm can't be resolved (Campana and Neilson 1985). The translucent zones which represented a slow or no growth period in the year may be composed of many fine increments, but these increments were too fine to be seen with a light microscope. Perhaps more accurate methods such as scanning electron microscope may be necessary.

The main growth period of *O. stewartii* was only about 7 or 8 months in 1 year. The mean numbers of increments between two annuli represent the number of days when *O. stewarti* grew rapidly. Based on analysis of the edge type, some specimens' growth became slow in October; the mean microincrements number between the first annuli and the second annuli was 171 and the monthly growth of the increments ranged from 21 to 25, so it is concluded that the mean microincrements number was about 7 or 8 months. Then we may estimate that the primary growth period of *O. stewartii* was from late March to October and fish grew slowly or stopped growing from November to the next March. These growth characteristics were adapted to the environment of Lhasa River. There was remarkable change in temperature from March to April and from October to November in Lhasa River; the range of change exceeded 3°C (Fig. 4; Xiong and Tang 1992). The upstream reach of the Lhasa River with an altitude exceeding 4,000 m was covered with ice from December to the next February, and although reaches upstream of the Lhasa with an altitude lower than 4,000 m did not freeze, its temperature was lower than 2°C (wild survey), and fish growth became slow with the lower temperature. The period of lower temperature lasted several months, therefore, the rapid growth period of *O. stewartii* was short.

The mean numbers of increments from the core to the first annuli were 153; therefore it was suggested that the primary growth period of *O. stewartii* was 6 or 7 months in the first year. Meanwhile fish growth became slow in October, so it was estimated that *O. stewartii* hatched between April and May. *O. stewartii* has later sexual maturity period, and we could collect only a few specimens which had matured gonad so we could not confirm it through checking the development of gonad. As a result, hatching time was estimated only based on the character of the otolith microstructure. Our conjecture will be confirmed in the later work. The mean numbers of increments in the first annuli was less than in the second one, which reflects the fact that the primary

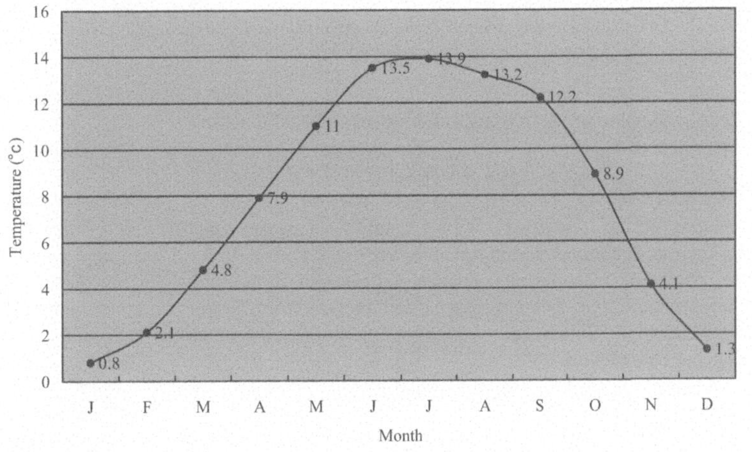

Fig. 4 Mean monthly water temperature of the Lhasa River throughout the year

growth period in the first year was shorter than the next years. Possible reason was that the time of hatching was later than the time when older fish began to grow rapidly.

Checks in otolith sections

Checks, or discontinuities, were characteristic of most otolith growth sequences, and they may record periods of perturbation or stress to the fish (Campana and Neilson 1985). Fish grew slowly and steadily before hatching, and then they may have an abrupt accretion after hatching. The hatch period of *O. stewartii* should be long because they settled in lower water temperature. Conditions in which the subfamily schizothoracine settled were similar to *O. stewartii*: such as *Gymnocypris namensis* and *Gymnocypris przewalskii*, had a long hatch period. *G. namensis* and *G. przewalskii* required about 10 days (Chen et al. 1999) and about 12 days from fertilization to hatching (Hu et al. 1975), respectively. Since they belong to the same subfamily, *O. stewartii* and *G. namensis* might have similar embryonic development processes. *G. namensis* required 5 or 6 days from appearance of otoliths to hatching (Chen et al. 1999). Therefore, the increment which had an abrupt accretion after five or six smooth increments near the core was presumably associated with hatching and we refer to it as the 'hatch check (HC)'. *O. stewartii* settles in the plateau, and its eggs are bigger than the other cyprinid species which settle in the plain. For example, the egg diameter of *Cyprinus carpio* was about 1.7 mm, *Carassius auratus* was about 1.5 mm, *Megalobrama terminalis* was about 1.2 mm, *Megalobrama amblycephala* was about 1.3 mm, *Xenocypris davidi* was about 1.3 mm (Cao 1992) while diameter of adult *O. stewartii* was exceeding 2.2 mm in December. The bigger egg might provide more yolk for the larval fish to survive in the hostile environment. So it was presumed that the check next to the HC may be related to the stress during the transition from endogenous to exogenous nutrition sources, and a few days delay was observed between hatching and nutrition sources transition may be of the results the yolk providing nutrition for delaying start of nutrition sources transition. Therefore, the check was assumed to be 'yolk absorption check'.

The monthly recurrent pattern of otolith growth was reported in some marine fishes and explained by a tidal influence (Pannella 1971; Campana 1984a). This pattern was also reported in members of the subfamily schizothoracinae such as *G. selincuoensis* (Chen et al. 2002a); however, they didn't provide a further explanation. It was proposed the possible explanation was that the light of the full moon changed the photoperiod. Tibet lies on a plateau with thin air and clouds, so the light of the full moon is strong and the water in Lhasa River is clear (wild survey), and the strong light might change the photoperiod and disturb the endogenous circadian rhythm of fish, resulting in the formation of the check.

The microincrement formation is linked to endogenous circadian rhythm. The endogenous circadian rhythm of fish, especially for fish of the early age might be disturbed by certain environment variables (Campana and Neilson 1985). The checks often occurring in the early stage of the life-history may be due to immature rhythm of younger *O. stewartii.*

The rhythm might be disturbed by certain environment variables, which result in formation of checks.

Acknowledgement The project was financially supported by the project of the Knowledge Innovation Project of the Chinese Academy of Sciences (grant no. KSCX2-SW-125; KSCX1-SW-13-04) and the National Natural Science Foundation of China (grant no. 30471345). We are thankful to many people for their contributions to this research. We thank Dekui He for his constructive comments on the manuscript, and thank Songguang Xie for technical assistance. We also thank Dekui He, Feng Chen, Fuhua Hao, and Chengzhi Ding for assistance in sample collection.

References

Campana SE (1984a) Lunar cycles of otolith growth in the juvenile starry flounder, *Platichthys stellatus*. Mar Biol 80:239–246

Campana SE (1984b) Microstructural growth patterns in the otoliths of larval and juvenile starry flounder, *Platichthys stellatus*. Can J Zool 62:1507–1512

Campana SE, Neilson JD (1982) Daily growth increments in otoliths of starry flounder *Platichthys stellatus* and the influence of some environmental variables in their production. Can J Fish Aquat Sci 39:937–942

Campana SE, Neilson JD (1985) Microstructure of fish otoliths (reviews). Can J Fish Aquat Sci 42:1014–1032

Cao WX (1992) The resources of the Chinese freshwater fishes. In: Liu JK, He BW (eds) Cultivation of the Chinese freshwater fishes (3rd edn). Science, Beijing, pp 42–48 (in Chinese)

Chen YF, Cao WX (2000) Schizothoracinae. In: Fauna Sinica Osteicthtyes Cypriniformes III.. Science, Beijing, pp 367–368 (in Chinese)

Chen YF, He DK, Chen ZM, Cai B, Ba Z (1999) The preembryonic development and sperm longevity of *Gymnocypris namensis*. Tibet. Journal of Lanzhou University 35:56–65 (in Chinese)

Chen YF, He DK, Duan ZH (2002a) Annuli characters of Selincuo schizothoracini (*Gymnocypris selincuoensi*) in Selincuo Lake. Tibet. Acta Zool Sin 48(3):384–392 (in Chinese, with English abstract)

Chen YF, He DK, Chen YY (2002b) Age discrimination of Selincuo schizothoracini (*Gymnocypris selincuoensis*) in Selincuo Lake, Tibet. Acta Zool Sin 48(4):527–533 (in Chinese, with English abstract)

Geffen AJ (1987) Methods of validating daily increment deposition in otoliths of larval fish.. In: Summerfelt RC, Hall GE (eds) Age and Growth of Fish. Iowa State University Press, Ames, USA, pp 223–240

Hass RE, Recksiek CW (1995) Age verification of winter flounder in Narragansett Bay. Trans Am Fish Soc 124:103–111

Hu A, Tang SS, Gong SX (1975) Reproductive biology research of *Gymnocypris przewalskii przewalskii (Kessler)*.. In: The fish fauna of Qinghai Lake Region and Biology of *Gymnocypris przewalskii przewalskii* (Kessler). Science, Beijing, pp 49–62 (in Chinese)

Marshall SL, Parker SS (1982) Pattern identification in the microstructure of sockeye salmon (*Oncorhynehus nerka*) otoliths. Can J Fish Aquat Sci 39:542–547

Meisfjord J, Midtøy F, Folkvord A (2006) Validation of daily increment deposition in otoliths of juvenile *Limnothrissa miodon* (Clupeidae). J Fish Biol 69:1845–1848

Methot RD, Kramer DA (1979) Growth of northern anchovy, *Engraulis mordax*, and northern lampfish, *Stenobrachius leucopsarus*. Rapp P-v Reun Cons Int Explor Mer 78: 424–431

Neilson JD, Geen GH (1982) Otoliths of Chinook salmon (*Oncorhynchus tshawytscha*): daily growth increments and factors influencing their production. Can J Fish Aquat Sci 39:1340–1347

Pannella G (1971) Fish otoliths: daily growth layers and periodical patterns. Science 173:1124–1127

Radtke R, Dean J (1982) Increment formation in otoliths of embryos, larvae and juveniles of the mummichog *Fundulus heteroclitus*. Fish Bull 80:201–215

Song ZB, Fu ZD, Yue BS, Zhao EM (2006) Otolith microstructure of larval *Gymnocypris potanini* Herzenstein from the Minjiang River in China. Environ Biol Fish 75:431–438

Taubert BD, Coble DW (1977) Daily rings in otoliths of three species of *Lepomis* and *Tilapia mossambica*. J Fish Res Broad Can 34:332–340

Team of Chinese Academy of Science of the Scientific Expedition to the Qinghai-Xizang Plateau (1983) Fluvial geomorphology. In: Geomorphology of Xizang (Tibet). Science, Beijing, pp 33–60 (in Chinese)

Victor BC (1982) Daily otolith increments and recruitment in two coral-reef wrasses, *Thalassoma bifasciatum* and *Halichoeres bivittatus*. Mar Biol 71:203–208

Vilizzi L, Walker KF (1999) Age and growth of the common carp, *Cyprinus carpio*, in the River Muray, Australia: validation, consistency of age interpretation, and growth models. Environ Biol Fish 54:77–106

Xiong Y, Tang QC (1998) The hydrology of river in cold area of the Tibetan Plateau. In: Xiong Y, Tang QC (eds) The hydrology of river in China. Science, Beijng, p 118 (in Chinese)

Yue PQ, Chen YY (1998) Pisces. In: Wang S (ed) China red data book of endangered animals. Science, Beijing, pp 158–159

Zhu Q, Xia L, Chang J (2002) Computer identification on otolith microstructure of fish. Acta Hydrobiol Sin 26 (6):600–604 (in Chinese, with English abstract)

Environ Biol Fish (2009) 86:53–61
DOI 10.1007/s10641-008-9420-0

Otolith characteristics and age determination of an endemic *Ptychobarbus dipogon* (Regan, 1905) (Cyprinidae: Schizothoracinae) in the Yarlung Tsangpo River, Tibet

Xiuqi Li · Yifeng Chen · Dekui He · Feng Chen

Received: 4 June 2007 / Accepted: 21 May 2008 / Published online: 23 October 2008

Abstract We describe the microincrements, checks and annuli in the lapilli of the schizothoracine *Ptychobarbus dipogon*, an endemic species of the Tibetan plateau. We collected samples in the Yarlung Tsangpo River and its tributaries on a monthly basis (from April 2004 to August 2006). We describe the shape features of the three pairs of otoliths and document the full trajectory of lapillus development. We found that five to seven checks were clearly visible in the opaque zone of the first annulus. The pattern of 21–23 daily growth increments within each check might be explained as a lunar-induced deposition. We counted between 137 and 154 increments within the first annulus. Annuli appeared as a sequence of gradually declining increment widths, whereas false rings were characterized by abrupt checks. Our oldest estimates were 23^+years for males and 44^+ for females. The time of annulus completion was clearly between March and April each year using monthly marginal increments analysis. We consider the factors responsible for daily increment formation as an endogenous circadian rhythm. Environmental information, such as strong sunlight and cold water temperatures in the Tibetan Plateau, could reinforce the endogenous daily cycle. Our results provided important data addressing the ecology and population dynamics of *P. dipogon*.

Keywords *Ptychobarbus dipogon* · Otolith · Age determination · Yarlung Tsangpo River

X. Li · Y. Chen (✉) · D. He · F. Chen
Laboratory of Biological Invasion and Adaptive Evolution, Institute of Hydrobiology, Chinese Academy of Sciences, Wuhan 430072, China
e-mail: chenyf@ihb.ac.cn

X. Li · F. Chen
Graduate School of Chinese Academy of Sciences, Beijing 100039, China

Introduction

Accurate age determination is imperative for understanding population dynamics and thereby for an optimal resource management in ecological studies of various fish species (Hilborn and Walters 1992). Determination of periodic growth increments in many calcified structures is the most common means of determining age (Casselman 1996; Campana 2001). Otolith growth patterns not only provide a chronological record of early life history, stock structure, growth parameters and migration, but can also record ambient environmental information, such as water temperature, sunlight intensity, and food supply. The annual pattern is typically an alternating sequence of opaque and translucent bands represented differential otolith growth as a result of seasonal changes. Campana and Neilson (1985) suggested that daily microincrements in otoliths of teleost fishes is a widespread phenomenon. Daily increment counts between the nucleus and first annulus had been successfully used to validate annular periodicity (Victor and Brothers 1982).

The subfamily Schizothoracinae has a wide distribution in the Tibetan Plateau and adjacent areas.

Degenerated body scales are one of the prominent characteristics of the subfamily (Cao et al. 1981). Age determination in these species has relied mainly on the irregular anal scales (Tsao and Wu 1962; Zhao et al. 1975). However, it has been difficult to obtain precise and reliable age estimates owing to the slow growth and narrow increments in the anal scales. Zhao et al. (1975) pointed out that it was very difficult to determine ages from the anal scales for *Gymnocypris przewalskii* over 10 years of age. Chen et al. (2002a, b) reported that sectioned lapilli provide more reliable age estimation based on assessing the consistency of age interpretations for *G. selincuoensis*. Thus, the otolith growth mark patterns appeared to be more legible and regular than those of scales for schizothoracine species. In addition, Song et al. (2006) also described otolith microstructure of larval *G. potanini* Herzenstein from the Minjiang River in China.

Ptychobarbus dipogon is an endemic species in the Tibetan Plateau, only distributed in the middle and upper reaches of the Yarlung Tsangpo River and its tributaries at an elevation of about 4000 m (Wu and Wu 1992; Chen and Cao 2000). *P. dipogon* generally inhabits the deeper waters of tributaries and river meanders, and have powerful muscular cylindrical bodies. Although this species was locally abundant, the available information refers mostly to taxonomical species description. Therefore, there is an urgent need for studies of the life-history characteristics and adaptability of *P. dipogon*. It is important to understand its age structure for the reasonable conservation and management of *P. dipogon*.

The objective of our study was firstly to describe the morphological characters of the three types of otoliths of *P. dipogon*; secondly, to validate the daily increments and the presence of checks on lapilli; thirdly, to estimate age and investigate the nature of annuli; and finally, to confirm the time of annuli formation using marginal increment analysis. We also considered the mechanisms underlying the formation of checks and annuli in relation to the plateau environment.

Materials and methods

Collection of samples

We randomly collected a total of 610 *P. dipogon* individuals from the Yarlung Tsangpo River and its tributaries monthly from April 2004 to August 2006 (Fig. 1). The majority of *P. dipogon* (90.4%) caught were from the Lhasa River. To ensure most age-classes, we captured the samples using floating gill nets and cast nets. The body weight (W), total length (TL) and standard length (SL) were measured to the nearest 0.1 g and 0.1 mm using digital scales and vernier calipers on fresh specimens, respectively. Specimens were classified as male or female or undetermined by macroscopic examination of gonads. We removed the otoliths (sagitta, asteriscus and lapillus) from the head (vestibular apparatus) of fish using the 'guillotine' method (Secor et al. 1992). After rinsing with water, the otoliths were air-dried, and then stored in labelled tubes.

Otolith preparation and microstructural measurement

The size distribution of *P. dipogon* whose lapilli were extracted ranged from 70.6 to 593.0 mm TL. A preliminary analysis of three types of otolith under light microscopy showed that only sectioned lapilli exhibited clear increments. Therefore, the microstructures of sagittae and asterisci were not further examined. The lapillus was mounted on glass slides using thermoplastic resin, grinded transversely using wetted 800–1200 grit abrasive paper and polished with film by hand until close to the core. The section was remounted and grinded with the polished surface down. The thickness of the sectioned otolith was about 0.1–0.2 mm when annuli became clear. Sectioned otoliths were observed under the microscope with transmitted light, and photographed (Carl Zeiss Axioplan 2 Microimaging System). Magnification varied from ×50 to ×400 according to the size of section.

The microstructural variables were measured on all interpretable sectioned otoliths. Increment width measurements were performed along a very clear increment trajectory from the primordium to the lateral and posterior edges. All counts and increment measurements were made along the same reading axis using an image analyzing system (Zhu et al. 2002). The reader had no prior knowledge of length, sex or time of capture before the estimation.

Otolith reading and age validation

Alternating concentric opaque and translucent zones around the lapillus nucleus were considered as annuli and counted. Under light microscopy, microstructure provided a means of distinguishing between the translu-

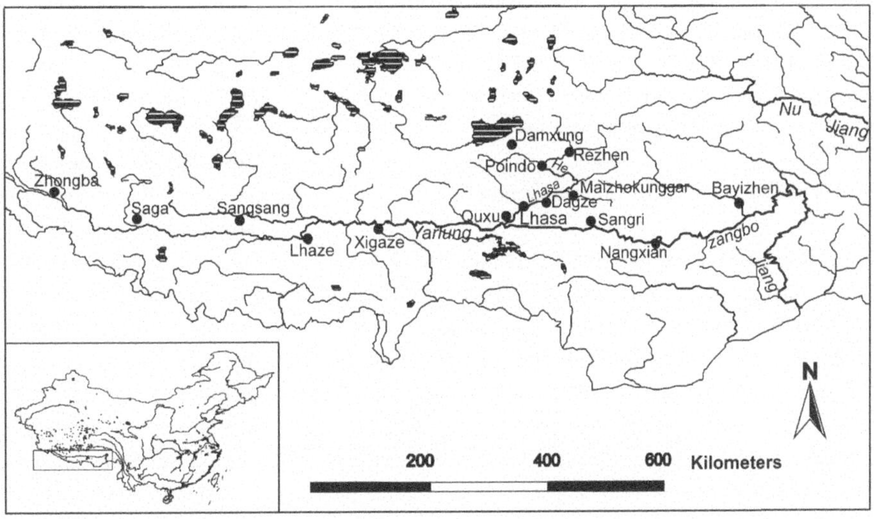

Fig. 1 Sampling locations of *P. dipogon* in the Yarlung Tsangpo River, Tibet

cent material that formed false rings and annuli. Ages were determined twice by the same interpreter after a considerable time (30 days). Age bias plots were used to assess intra-readings bias. Coefficients of variation (CV, %) were calculated as the standard deviation of the corresponding mean estimated ages in each age class between two readings, as it is considered the most valuable coefficient when dealing with comparisons (Campana 2001). The equation (Chang 1982) was expressed as follows:

$$CV_j = 100\% \times \frac{\sqrt{\sum_{i=1}^{R} \frac{(X_{ij}-X_j)^2}{R-1}}}{X_j}$$

Where CV_j is the age precision estimate for the *j*th fish, R is the number of readings, X_{ij} is the *i*th age determination of the *j*th fish, and X_j is the mean age calculated for the *j*th fish. The result is considered good if the CV is lower than 0.05 (Laine et al. 1991). If no precise pairs of readings were found for an otolith, then that otolith was excluded from subsequent analyses.

Marginal increment ratio analysis

The marginal increment (the distal edge of the last translucent zone) ratio analysis (MIR) was used to validate the period of annuli formation. The monthly changes of marginal increment ratio were estimated with the following equation of Hass and Recksiek (1995): $MIR = (O_c - O_n)/(O_n - O_{n-1})$; where O_c (μm) is the otolith radius at the time of capture, O_n (μm) is the radius to the last formed annulus, respectively. O_{n-1} (μm) is the radial distance of the penultimate formed annulus, and n denote the number of rings. Typically, the months of mark formation were usually indicated by a bimodal MI frequency distribution because of the occurrence of '0' MIs and maximum marginal increment. The '0' MI was defined as a very narrow opaque margin in the distal edge of the translucent band of otolith section. The margin increment called '0' MI never exceeded 10 μm. In addition, the date of annulus formation of 1^+fishes was validated by back-calculating the new microincrements according to the date of its capture.

All data were analyzed using the Origin 7.0 package. The images were modified with Photoshop 8.0.

Results

Otolith morphometrics and readability

Each otolith of *P. dipogon* has a relatively stable external character. The lapillus was the largest pair of otoliths of *P. dipogon* (Fig. 2a). The increments in the sectioned lapilli were clearly deposited from the primordium to the proximal margin. The asteriscus was a disk with a serrated edge (Fig. 2b). The wavy characters of annuli rings were observed with an ambiguous core, leading to difficulty in discerning age. The sagitta was shaped like an arrowhead (Fig. 2c). Although the anterior part of the sagitta was readable without the grinding and polishing procedure, the sagitta was not appropriate for growth analysis because it was fragile and often broke during the extracting procedures. Consequently, only the lapillus was appropriate for daily increment analysis and age determination on the basis of structural suitability and visibility.

Microstructural features and daily increments

Lapilli showed a regular pattern of increments along the otolith radius when viewed by light microscopy. We observed and measured otolith microstructure of 13 juveniles with TL ranging from 70.6 to 108.5 mm. For the description of internal features, we followed the terminologies of otolith microstructure defined by Campana and Neilson (1985) and Greely et al. (1999). Three distinct regions were apparent within each sectioned lapillus: the larval zone (LZ), postlarval zone (PLZ) and postmetamorphic zone (PMZ). The larval zone included the core out to the metamorphic check which usually located at the last continuous circular increment. The otolith microstructures were shown in some details of primordium (P), nuclear (N) and hatch check (HC). Two primordia formed one nucleus region in most otoliths. The diameter of the nucleus was approximately 72.56±3.40 μm (mean ± SD). The primordium was spherical in shape, and measured 15.72±1.63 μm in diameter (Fig. 3a). The daily growth microincrements (DGI) and monthly-like growth increments (MGI) under light microscopy were clearly showed in Fig. 4. The translucent L-zone and opaque D-zone, which were respectively equivalent to the accretion zone and the discontinuous zone, composed a daily growth increment. The increment width decreased from 3.21 μm in the innermost to 1.84 μm at the margin of the lapilii (Fig. 4c,d). The checks corresponded to a sequence of narrow translucent zones in a annulus. Monthly-like growth increments (MGI) were deposited at about 50 μm spacing with a succession of wide and narrow microincrements. The daily increments varied from approximately 21–23 cases in MGI. The number of daily increments was close to one synodic month, except for the full moon phase (Fig. 4a,b).

Although the postlarval zone of the lapillus was extremely dark, the checks and daily growth increments encircling the nucleus region were prominently visible. The postmetamorphic zone was the region from the terminal increment of the PLZ to the outermost edge of the lapillus. The PMZ was abruptly demarcated by a transition in elemental composition. Under transmitted light, the PMZ represented a regular pattern of alternating continuous and discontinuous zones. The first annulus was observed at intervals of five to seven distinct discontinuous zones (check C), and had an average semidiameter of 205.11±25.78 μm (Fig. 3b). The number counted from regular microincrements ranged from 137 to 154, with a maximum of 220. The pattern of microincrements was the same in the sectioned lapilli from individuals of different sizes. Checks often clearly occurred on the first three annuli. However, the deposition of daily microincrements and the pattern of checks between successive annulus could not be observed in peripheral regions of the otoliths of adults.

Annuli characteristics and age validation

Macroscopically, the sectioned otolith showed a clear pattern of opaque-and translucent-zones under transmitted light (Fig. 5). Annuli appeared as a sequence of concentric zones of gradually declining increment widths. The annuli faded gradually out from the nucleus to the outmost edge (Fig. 5b). This phenomenon commonly occurred in individuals prior to 10 years of age. The characteristics of lapilli presented a more stable, regular pattern with elapsed years. During the early growth phase, the opaque zone contained wider increments relative to the translucent zone, and was composed of five to seven successive growth interrup-

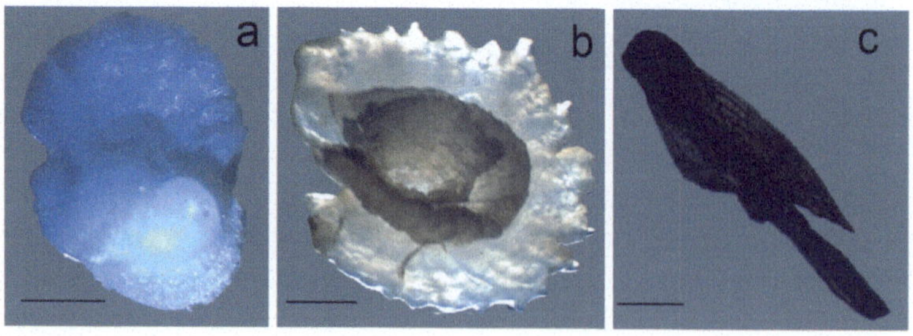

Fig. 2 Morphology of three types of otoliths of *P. dipogon*. **a** lapillus, **b** asteriscus, **c** sagitta (scale bar=1 mn)

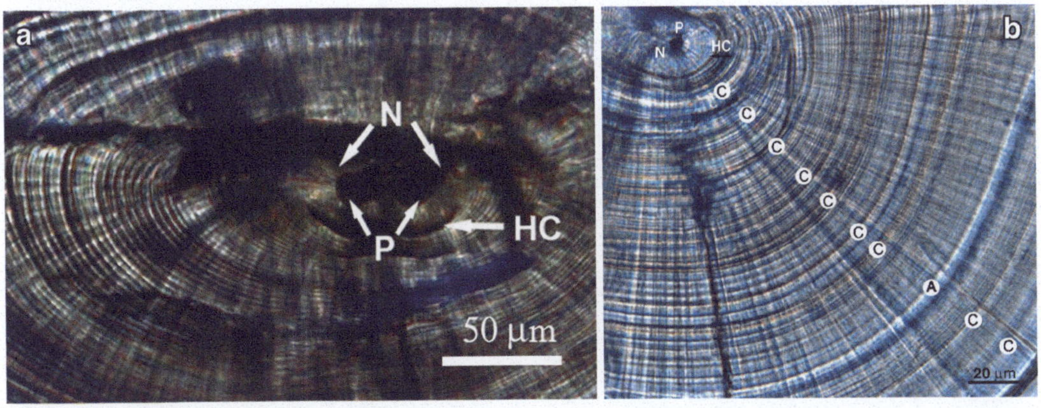

Fig. 3 Microstructures near the nucleus region of the *P. dipogon* on lapillus viewed under light microscopy, showing **a** primordia (P), nucleus (N), hatch check (HC). **b** monthly-like check (C) and annual ring (A) of lapillus from a 70.6 mm TL *P. dipogon*

tions (checks) (Fig. 5a). However, the pattern of depositions was reversed for the oldest age classes. In a word, the translucent zone became much broader than the opaque one (Fig. 5b). The annulus characters in the ventrolateral area of the otolith appeared to be more reliable because of the legibility and regularity of its growth marks (Fig. 5c). Otoliths grew at a measurable rate throughout the life of the fish. However, the increment width abruptly declined, probably at about 19 years old.

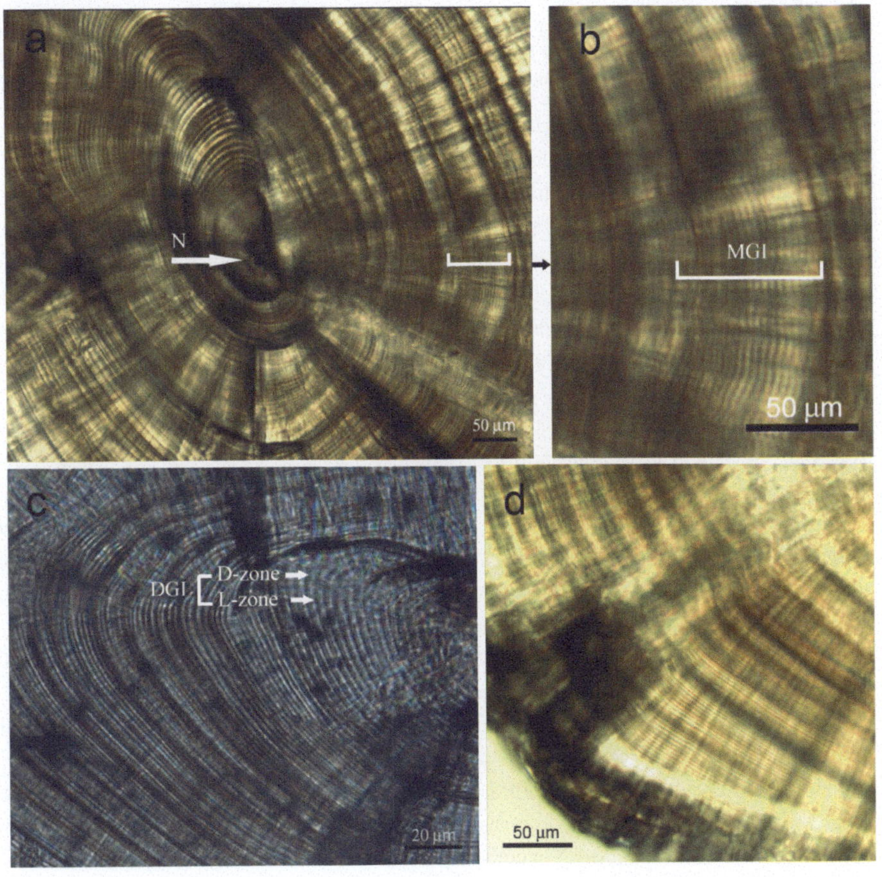

Fig. 4 Daily growth increments and monthly-like growth increment in the lapillus of *P. dipogon* exhibited under transmitted light, showing **a** area surrounding the core. **b** Monthly-like growth increments (MGI). **c** Daily growth increments (DGI) of core region. **d** Daily growth increments (MGI) of peripheral area. *Arrow head* indicates the translucent zones (L-zone) and opaque zone (D-zone)

Fig. 5 The typical magnification appearance of a *P. dipogon* transverse cross-section lapillus under transmitted light. Annual growth increments are denoted as annuli with *triangles*, *black dots* and *arrows* from first to fifth in (**a**), 1st–13th in (**b**) and 5th, 15th, 25th and 35th annulus in (**d**), respectively

Ages were determined in 420 lapilli, only 28 (6.67%) were discarded due to natural deformations and unidentifiable annulus deposition. Ages ranged from 1+ to 23+ years for males and 1+ to 44+ for females. The deviations observed between successive readings were ±3 years for 5≤ ages ≤8 years and >19 years. The low CV (3.14%) of age estimates reflected good concordance in the readings.

Marginal increment ratio and annuli formation

For otolith sections with 1+–8 annuli, their marginal increments followed the same pattern. A fish with four annual marks and a 0 MI was observed in April (Fig. 6c), a new mode of very small increments appeared during June, indicating individuals had formed a new mark during March or April (Fig. 6a,b). Monthly MI frequency plots showed marked deviations through March or April (Fig. 7), showing that 0 MI's and maximum margin growth co-occurred. In general, the progression of the mode was consistent over long periods of time (from May to November) when otolith margin growth resumed. The MI remained stable in winter, indicating very slow winter otolith growth. These consistent patterns of MI growth indicated that the time of mark formation was restricted to the same short time period each year, March and April.

Discussion

The morphometrics and daily rhythm of otolith microincrement of *P. dipogon* are the first successful validation in schizothoracine fishes. Under the microscope, the microincrements of *P. dipogon* lapillar otoliths were very clear. However, the reproducibility of annulus determination relied on the section orientation during otolith preparation and the exclusion of false annuli. The reading errors mainly involved the first annulus because the center of the section was relatively opaque. In addition, marginal annulus estimates became subjective due to the very narrow and limited visibility in the outermost edge of some older ones.

Fig. 6 The edge condition for a lapillus of *P. dipogon* under the light microscope. The *arrowhead* indicates the annual ring formation: **a** Otolith from first year fish with opaque edge (age=1+, 70.6 mm TL and 2.3 g, June 2006); **b** Otolith with a wide opaque edge (age=2+, 162.6 mm TL and 36.1 g, June 2006); **c** Otolith with a translucent edge (age=4, 231.2 mm TL and 98.0 g, April 2004)

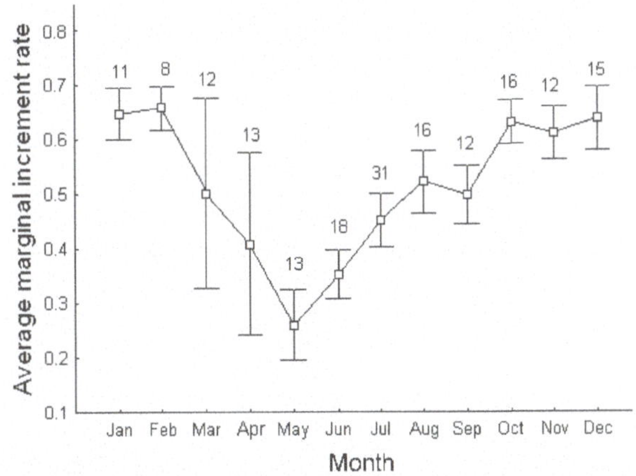

Fig. 7 Monthly changes in mean marginal increments rates of otoliths from pooled sexes and size classes of *P. dipogon*. The *vertical bars* indicated standard errors (n=177)

The microstructure of sectioned otoliths of *P. dipogon* revealed the bipartite nature of otolith increments, each of which consisted of a translucent zone (calcium-rich component) and an opaque zone (organic-rich component) under transmitted light. The sectioned lapilli demonstrated a distinct pattern of five to seven checks in an annulus, and the daily increments of *P. dipogon* contained approximately 21–23 cases within each check. The daily growth increments of *P. dipogon* are regulated by an endogenous circadian rhythm. The otolith checks of *P. dipogon* reflect the influence of lunar cycles on the formation of increments. However, the number of monthly checks never approached 12, nor did the number of daily growth increments within these regions approach 365. Chen et al. (2002b) described this phenomenon in *G. selincuoensis*. The average size per increment would have gradually decreased to about 0.1 μm day^{-1} in winter. It seems that optical resolution would be insufficient to keep track of daily increments.

We presume that the narrower microincrements might be due to decreasing growth rates during the cold winter (Feet et al. 2002; Chen et al. 2002a; Guibbolini et al. 2006). Geffen (1982) stated the growth rate limitation hypothesis that translucent zone formation was triggered at a threshold of metabolic stress, and that the combined energetic requirements of reproduction and migration might maintain translucent zone formation. The climate of the region is strongly influenced by the elevation of the Tibetan Plateau. Water temperatures of the Lhasa River averaged 7.5°C, fluctuating from 0.8 to

13.9°C (Xiong and Tang 1998). It is possible that *P. dipogon* grows slowly for about 5 months in winter and their otoliths also underwent a greatly reduced increment width of the translucent zone when the water temperatures drop below 5°C (Casas 1998). In addition, Hayashi et al. (2001) reported otolith microincrements growth in the myctophid (*Myctophum asperum*) increased during the night and early morning and stopped growth during the day, and also exhibited lunar periodicity in the deposition of growth increments. The mean increment widths around the time of a full moon were significantly narrower than those around a new moon in 29 cases. The moonlight in the Tibetan Plateau is as strong as the daytime sunlight in the full moon period. The appearance of checks could also have resulted from the slower growth caused by staying in deeper and colder water in response to the light of the full moon.

The annual mark consisted of a bipartite set of bands, a wide opaque band formed in summer and a narrow translucent band formed in winter. A full trajectory of lapillus growth was seen with gradually declining increment widths. Age determination of older fish confirmed 23+ years for males and 44+ for females. Thus, we believe that *P. dipogon* is a long-lived and slow growing coldwater fish based on the correlation of the otolith and somatic growth.

Despite the differences in sampling locations and time, all samples presented the same pattern of growth in their otoliths. Therefore, this deposition pattern could be explained that *P. dipogon* did not subject to migration but physiological transformation. In addition, *P. dipogon* in the Yarlung Tsangpo River lived substantially longer and grew more slowly than previously reported in this subfamily (Tsao and Wu 1962; Zhao et al. 1975; Chen et al. 2002b). Therefore, as a highly vulnerable fisheries resource, it is essential to take adaptive management to maintain the sustainability of this species and stability of the fish community.

Acknowledgements We would like to express our heartfelt gratitude to Fuhua Hao, Yintao Jia and Chengzhi Ding provided assistance in collecting specimens. This research was supported by the Knowledge Innovation Project of the Chinese Academy of Sciences (grant no. KSCX2-SW-125; KSCX1-SW-13-04) and the National Natural Science Foundation of China (grant no. 30471345). Additional sincerely thanks are extended to Prof. Songguang Xie and Prof. Weisong Feng for their kind help in laboratory work and several anonymous reviewers for their valuable advice on the manuscript.

References

Campana SE (2001) Accuracy, precision and quality control in age determination, including a review of the use and abuse of age validation methods. J Fish Biol 59:197–242, doi:10.1111/j.1095-8649.2001.tb00127.x

Campana SE, Neilson JD (1985) Microstructure of fish otoliths. Can J Fish Aquat Sci 42:1014–1032, doi:10.1139/f85-127

Cao WX, Chen YY, Wu YF, Zhu SQ (1981) Origin and evolution of schizothoracine fishes in relation to the upheaval of the Xizang Plateau. In: Tibetan Expedition Team of the Chinese Academy of Science (ed) Studies on the period, amplitude and type of the uplift of the Qinghai-Xizang Plateau, Science Press, Beijing, pp 118–130 (in Chinese)

Casas MC (1998) Increment formation in otoliths of slow-growing winter flounder (*Pleuronectes americanus*) larvae in cold water. Can J Fish Aquat Sci 55:162–169, doi:10.1139/cjfas-55-1-162

Casselman JM (1996) Age, growth and environmental requirements of pike. In: Craig JF (ed) Pike: Biology and Exploration. Chapman and Hall, London, pp 70–101

Chang WYB (1982) A statistical method for evaluating the reproducibility of age determination. Can J Fish Aquat Sci 39:1208–1210, doi:10.1139/f82-158

Chen YF, Cao WX (2000) Schizothoracinae. In: Yue PQ (ed) Fauna sinica osteicthtyes cypriniformes III. Scince Press, Beijing, pp 273–388, in Chinese

Chen YF, He DK, Duan ZH (2002a) Annuli characters of *Selincuo schizothoracini* (*Gymnocypris selincuoensi*) in Selincuo Lake, Tibet. Acta Zool Sin 48(3):384–392, in Chinese, with English abstract

Chen YF, He DK, Chen YY (2002b) Age discrimination of *Selincuo schizothoracini* (*Gymnocypris selincuoensi*) in Selincuo Lake, Tibetan Plateau. Acta Zool Sin 48(4):527–533, in Chinese, with English abstract

Feet P, Ugland KI, Moksness E (2002) Accuracy of age estimates in spring spawning herring (*Clupea harengus* L.) reared under different prey densities. Fish Res 56:59–67, doi:10.1016/S0165-7836(01)003130-7

Geffen AG (1982) Otolith ring deposition in relation to growth rate in herring (*Clupea harengus*) Snd turbot (*Scophalmus maximus*) larvae. Mar Boil 71:317–326, doi:10.1007/BF00397048

Greely TM, Gartner JV Jr, Torres JJ (1999) Age and growth of *Electrona antarctica* (Pisces: Myctophidae), the dominant mesopelagic fish of the Southern Ocean. Mar Biol (Berl) 133:145–158, doi:10.1007/s002270050453

Guibbolini M, Borelli G, Mayer-Gostan N, Priouzeau F, Pontual HD, Allemand D, Payan P (2006) Characterization and variations of organic parameters in teleost fish endolymph during day–night cycle, starvation and stress conditions. Comp Biochem Physiol Part A 145:99–107

Hass RE, Recksiek CW (1995) Age verification of winter flounder in Narragansett Bay. Trans Am Fish Soc 124:103–111, doi:10.1577/1548-8659(1995)124<0103:AVOWFI>2.3.CO;2

Hayashi A, Kawaguchi K, Watanabe H, Ishida M (2001) Daily growth increment formation and its lunar periodicity in otoliths of the myctophid fish *Myctophum asperum* (Pisces: Myctophidae). Fish Sci 67:811–817, doi:10.1046/j.1444-2906.2001.00327.x

Hilborn R, Walters CJ (1992) Quantitative fisheries stock assessment: Choice, dynamics, and uncertainty. Chapman & Hall, New York, p 570

Laine AO, Momot WT, Ryan PA (1991) Accuracy of using scales and cleithra for aging northern pike from an oligotrophic Ontario lake. N Am J Fish Manage 11:220–225, doi:10.1577/1548-8675(1991)011<0220:AOUSAC>2.3.CO;2

Secor DH, Dean JM, Laban EH (1992) Otolith removal and preparation for microstructural examination. In: Stevenson DK, Campana SE (eds). Otolith Microstructure Examination and Analysis. Can Spec Publ Fish Aquat Sci 117:19–57

Song ZB, Fu ZD, Yue BS, Zhao EM (2006) Otolith microstructure of larval *Gymnocypris potanini* Herzenstein from the Minjiang River in China. Environ Biol Fishes 75:431–438, doi:10.1007/s10641-006-0029-x

Tsao WX, Wu XW (1962) An investigation of the fish biology and fishery problems in Ganze-Apa region of western Szechwan Province. Acta Hydrobiol Sin 2:79–111, in Chinese

Victor BC, Brothers EB (1982) Age and growth of the fallfish *Semotilus corporalis* with daily otolith increments as a method of annulus verification. Can J Zool 60:2543–2550, doi:10.1139/z82-326

Wu YF, Wu CZ (1992) The fishes of the Qinghai-Xizang Plateau. Sichuan Publishing House of Science & Technology, Chengdu (in Chinese)

Xiong Y, Tang QC (1998) The hydrology of river in cold area of The Tibetan Plateau. In: The hydrology of river in China, Science Press, Beijng, pp 118–130 (in Chinese)

Zhao LH, Wang SH, Zhao TQ (1975) The age and growth of *Gymnocypris przewalskii przewalskii* (Kessler). In: Institute of Biology, Qinghai Province (eds) The Fish Fauna of Qinghai Lake Region and Biology of *Gymnocypris przewalskii przewalskii* (Kessler). Science Press, Beijing, 37–45 (in Chinese)

Zhu Q, Xia LQ, Chang JB (2002) Computer identification on otolith microstructure of fish. Acta Hydrobiol Sin 26 (6):600–603, in Chinese, with English abstract

Environ Biol Fish (2009) 86:63–64
DOI 10.1007/s10641-007-9318-2

Threatened fishes of the world: *Trachidermus fasciatus* Heckel, 1837 (Cottidae)

Ling Cao · Weimin Wang · Chengtai Yang · Youji Wang

Received: 7 April 2007 / Accepted: 1 November 2007 / Published online: 17 January 2008

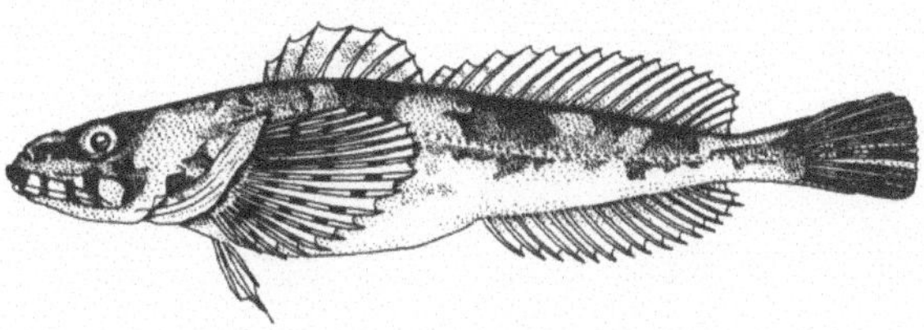

Common name: Roughskin sculpin. *Conservation status*: Not in IUCN Red List; Listed as critically endangered in Category II of the National Key Protected Wildlife List and in China Red Data book of Endangered Animals-Pisces (Yue and Chen 1998). *Identification*: D VIII-IX/18–20, A 16–18, P I/17–18; Pectoral fin fan-shaped; Caudal fin truncate. Depressed body; Head large; Mouth large; Protruding upper jaw and largest spine on pre-operculum; Pseudobranch present; No scale (Right to illustration held by Kim Ik-Soo). *Distribution*: Distributed along southern and western coasts of Korean Peninsula, eastern coast of China, and rivers flowing into Ariake Bay on Kyushu Island of Japan (Norio et al. 2002). *Habitat and ecology*: Benthic, catadromous and carnivorous, feeding on zooplankton, fish and shrimp. Juveniles occur in freshwater lakes and rivers; adults enter estuaries and sea during reproductive season. Euryhaline (0~32‰) indicates flexibility for reproduction and early life (Kinoshita et al. 1999). *Reproduction*: Age at first maturity is 1 year. Spawning sites locate in intertidal zone with oyster reefs. When water temperature reaches 4–5°C, spawn in empty shells of oyster or bivalve during February to March (Takeshita et al. 1999). Eggs adhesive, 1.48~1.58 mm in diameters. Absolute fecundity 5,100~12,800 eggs·fish^{-1} (Wang 1999). *Threats*: Wild populations have seriously declined since 1970s mainly due to the destruction of natural habitats and spawning sites, as well as overfishing and lack of regulation for sustainable fisheries. *Conservation action*: China has established protected areas (Wendeng, Shandong province), and investigated habitat requirements. Japan and South Korea studied spawning nests, distribution and life history. *Conservation recommendations*: Since most populations disappear in traditional habitats (Zhuang et al. 2006), long term protection of the remaining population should be ensured. Industrial water in coastal areas should be treated before discharge and further research needed to access the feasibility of fish passage construction around dams.

L. Cao · W. Wang (✉) · C. Yang · Y. Wang
College of Fisheries, Key Lab of Agricultural Animal Genetics, Breeding and Reproduction of Ministry of Education, Huazhong Agricultural University, Wuhan 430070, China
e-mail: wangwm@mail.hzau.edu.cn

References

Kinoshita I, Azuma K, Fujita S, Takahashi I, Niimi K, Harada S (1999) Early life history of a catadromous sculpin in western Japan. Environ Biol Fish 54:135–149

Norio O, Naohiko T, Seiichi M, Seiro K (2002) Spawning grounds and nests of *Trachidermus fasciatus* (Cottidae) in the Kashima and Shiota estuaries system facing Ariake Bay, Japan. Ichthyol Res 49:198–201

Takeshita N, Onikura N, Matsui S, Kimura S (1999) A note on the reproductive ecology of the catadromous fourspine sculpin *Cottus kazika* (Scorpaeniformes: Cottidae). Ichthyol Res 46:309–313

Wang JQ (1999) Advances in study on the ecology and reproduction biology of *Trachidermus fasciatus* Heckel. Acta Hydrobiologica Sinica 23:729–734 (in Chinese with English abstract)

Yue PQ, Chen YY (1998) China red data book of endangered animals—Pisces. Science Press, Beijing, pp 240–243

Zhuang P, Wang YH, Li SF, Deng SM, Li CS, Ni Y (2006) Fishes of the Yangtze Estuary. Shanghai Science Press, Shanghai, pp 168–172 (in Chinese)

Environ Biol Fish (2009) 86:65–71
DOI 10.1007/s10641-007-9282-x

Detection of hybridization between two loach species (*Paramisgurnus dabryanus* and *Misgurnus anguillicaudatus*) in wild populations

Cuihong You · Xiaomu Yu · Jingou Tong

Received: 22 December 2006 / Accepted: 31 July 2007 / Published online: 21 September 2007

Abstract Artificial interspecific hybrids between large scale loach *P. dabryanus* and tetraploid pond loach *M. anguillicaudatus* (Cobitidae, Cypriniformes) are viable. To detect the occurrence of possible natural hybridization, genetic analyses by using microsatellite markers were performed for natural populations of large scale loach and pond loach, the reciprocal laboratory hybrids, and "supposed hybrids" with ambiguous morphology. The fertility of the artificial hybrids was also tested. At one diagnostic microsatellite (Mac50), one out of 20 "supposed hybrids" was identified to be F_1 hybrid between the two loach species because it had the same genotype as that of the laboratory hybrids. The triploid hybrids between the two species were confirmed to be female-sterile. The results show that rare hybridization has occurred between diploid large scale loach and tetraploid pond loach in nature although it may have little effect in genetic introgression. This study is helpful for fish conservation and encourages further investigation on natural hybridization and introgression of loaches.

Keywords Loach · Natural hybridization · Artificial cross · Fertility · Microsatellite

C. You · X. Yu · J. Tong (✉)
State Key Laboratory of Freshwater Ecology and Biotechnology, Institute of Hydrobiology, the Chinese Academy of Sciences,
Wuhan 430072, People's Republic of China
e-mail: jgtong@ihb.ac.cn

C. You
The Graduate School of the Chinese Academy of Sciences,
Beijing 100039, People's Republic of China

Introduction

Hybridization has been documented as a widespread phenomenon among animals, which provides an opportunity to produce novel genetic variations (Allendorf et al. 2001; Seehausen 2004). Scribner et al. (2001) described 139 artificial or natural hybrids in freshwater fishes, of which 47 intergeneric hybrids were mostly observed in Cyprinidae. In practice, interspecific hybrid fishes have been artificially produced to increase economic profits (Bartley et al. 2000). Interspecific crosses in nature may lead to genetic introgression from one species to another and even to extinction of parental lineages if their hybrids are fertile (Verspoor and Hammar 1991; Epifanio and Philipp 2000). Hybridization between species has become an increasingly important issue in conservation biology.

Commercial farming for loaches has been plentiful in the past two decades, although the majority of the fisheries are harvested from natural populations. Large scale loach, *Paramisgurnus dabryanus*, is endemic to China, and distributed in middle and lower reaches of the Yangtze River, Pearl River and inland waters of

Taiwan. Pond loach, *Misgurnus anguillicaudatus*, can be found in all areas of China except for Tibetan plateau, and is found in rivers, lakes, ponds, swamps and rice fields. These two loach species are found sympatrically in most areas with similar morphological characters and nearly overlapping spawning seasons (March to October). They were assigned to the same genus (*Misgurnus*) of the Cobitidae until large scale loach was nominated as a new genus *Paramisgurnus* in 1981, because a few distinct anatomic traits were recorded (Chen and Zhu 1984).

Pond loach in most areas of China has 100 chromosomes (Li et al. 1983), and artificial hybridization between large scale loach and pond loach can produce viable reciprocal hybrids (Zhao et al. 1999, 2002; Long et al. 2000). The hybrids were speculated to be sterile due to their remarkably different numbers of chromosomes (large scale loach=48 and pond loach=100) but data to support this is not available (Zhao et al. 1999). A suspicion has been raised as to whether or not hybridization between these closely-related loach species can occur in natural populations. There has been some evidence to support this suspicion (Yin et al. 2005).

Molecular evidence from diagnostic markers allows for the identification of potential hybrids in wild populations of teleosts. Microsatellites are particularly suitable because of they are codominant, taxon-specific, and can be used for detecting genome-wide processes such as hybridization and introgression (Gross et al. 2004; Hānfling et al. 2005). In the present study, we analyzed the natural populations of large scale loach, pond loach and phenotype-ambiguous individuals with microsatellite markers for detecting possible interspecific hybridization. The fertility of the artificial hybrids of the two loach species was also tested. The aims of the study were to expand our understanding of potential hybridization and introgression in freshwater fish, and to provide necessary genetic information for fish conservation.

Materials and methods

Sample collection and artificial crosses

Large scale loach and pond loach populations were collected from basins along the Yangtze River from Wuhan, Shashi and Poyang Lake areas, based on their morphological characteristics as described by Yao (1989). Ambiguous individuals with intermediate phenotypic characteristics from both loach species were also collected as probable hybrids (H).

Mature females and males of the two loach species were sampled from local fish markets in Wuhan as candidates for artificial fertilization. A single intraperitoneal injection of human chorionic gonadotropin (HCG) with a dose of 6–8 IU/g body weight (BW) for females and 2 IU/g BW for males, was used for the preparation of artificially reciprocal and control crosses between the two loach species. All experiments were repeated three times, and fertilization (up to the gastrula stage), hatching (up to yolk sac absorption) and survival rates (up to feeding) were counted, respectively.

After hatching, 50 reciprocal hybrids from one replicate were continuously reared. They were initially cultivated in 50 l glass aquaria and then transferred to 1 m diameter fiberglass tanks until maturity, feeding on rotifers and subsequently commercial carp feed (39% crude protein).

Chromosome analyses

For all the reciprocal and control crosses, chromosome spreads were prepared from 20 embryos randomly sampled at the gastrula stage using the method described by Yu et al. (1989) with slight modifications. Forty clear metaphase plates on each slide were chosen to count the chromosome numbers by using an optical microscope under oil immersion at a magnification of 350×.

Genetic analyses with microsatellite DNA

Microsatellite markers from pond loach (Morishima et al. 2001) were used in this study. Five of the 15 microsatellite primers, namely Mac9, Mac15, Mac24, Mac45 and Mac50 were polymorphic and showed a Mendelian segregation in two full-sib families of large scale loach. Fourteen reciprocal hybrids were genotyped as the references of the artificial hybrids at these five loci. The polymorphic information of the five loci (Table 1) was based on 15 test individuals of each species (five individuals from each of the following three populations, Wuhan, Shashi and Poyang Lake, respectively).

The shared alleles were classified as typical for one species or another according to their frequencies, and

Table 1 Cross-species amplification in *Paramisgurnus dabryanus*, (*P. dabr*) using microsatellites from *Misgurnus anguillicaudatus* (*M. angu*)

Locus	DDBJ	Annealing temperature (°C)		Number of alleles		Allele size	
		M. angu	*P. dabr*	M. angu	*P. dabr*	M. angu	*P. dabr*
Mac9	AB060175	54	52	2	5	92–107	92–104
Mac15	AB060177	60	56	6	9	142–168	124–184
Mac24	AB060178	54	54	3	9	98–120	92–116
Mac45	AB060185	60	52	5	5	80–102	80–96
Mac50	AB060182	60	52	5	4	88–106	84–110

then each individual was assigned to a certain genotypic category based on the relative distribution of alleles of each species among the multi-locus genotypes (Hānfling et al. 2005; Ibarra et al. 2005). The five loci had largely overlapping allelic ranges between large scale loach and pond loach. Previous investigation on genetic diversity of these three populations (Wuhan, Shashi and Poyang Lake populations, 109 individuals) showed that the large scale loach had the typical alleles at Mac9, Mac45 and Mac50 but not at Mac24 and Mac15 (unpublished data). Mac9, Mac45 and Mac50 were also useful for classifying individuals when different typical alleles for the two loach species were observed. Therefore these three microsatellite loci were used for the analyses of pond loach and probably hybrids (20 individuals each population), together with our previous analyses of the large scale loach.

Genomic DNA was extracted from fin tissue or fry according to a traditional phenol-chloroform method. PCR was performed in a 12.5 μl reaction mixture, which contained 1×buffer (with 1.5 mM Mg^{2+}), 0.4U *Taq* polymerase, 100 μM dNTP, 2.5 pmol primers, 50–100 ng template DNA, in the following conditions: an initial denaturation for 5 min at 94°C followed by 37 cycles of denaturation at 94°C for 40 s, annealing at 50–60°C for 35 s and extension at 72°C for 35 s, and a final extension at 72°C for 10 min. PCR products were separated by 8% polyacrylamide gels, and fragments were visualized by ethidium bromide staining and UV transillumination (GeneGenius, SYNGENE).

The software package FSTAT (Goudet 1995) was utilized to calculate the number of alleles, distributions of allele and genotype frequencies, and to test the linkage disequilibrium (LD) for the three loci in the wild populations of large scale loach, pond loach and the probable hybrids, respectively. The program Micro-Checker v 2.2.3 (Van-Oosterhout et al. 2004) was used to detect null alleles that could exist in the microsatellite analyses.

Results

Cytology and morphology of artificial hybrids

There were no differences in mean fertilization, hatching and survival rates between the reciprocal and control crosses in the two loach species (data not shown). The chromosomal analyses showed a bulk of 74 pairs of chromosomes from the reciprocal crosses (Fig. 1), indicating that the offspring were hybrids rather than gynogenetic or androgenetic diploids of large scale loach ($2n$=48) or pond loach ($2n$=100). The reciprocal hybrids exhibited normal growth and had no significantly reduced survival rates compared to self-mating individuals within each species.

The progenies of large scale loach ♀×pond loach ♂ (PM) were more morphologically similar to pond loach rather than large scale loach. After being injected with HCG, 150–250 eggs were extruded from each of the two female progenies. The eggs were significantly smaller in size than those from normal females of large scale loach or pond loach. After these eggs were fertilized with sperms of large scale loach or pond loach, aberrant divisions were observed and no embryos survived to 24 h post-fertilization.

Genetic analyses of artificial hybrids

All the hybrids, either from PM (large scale loach ♀× pond loach ♂) or from MP (pond loach ♀×large scale loach ♂) groups, had heterozygous genotypes con-

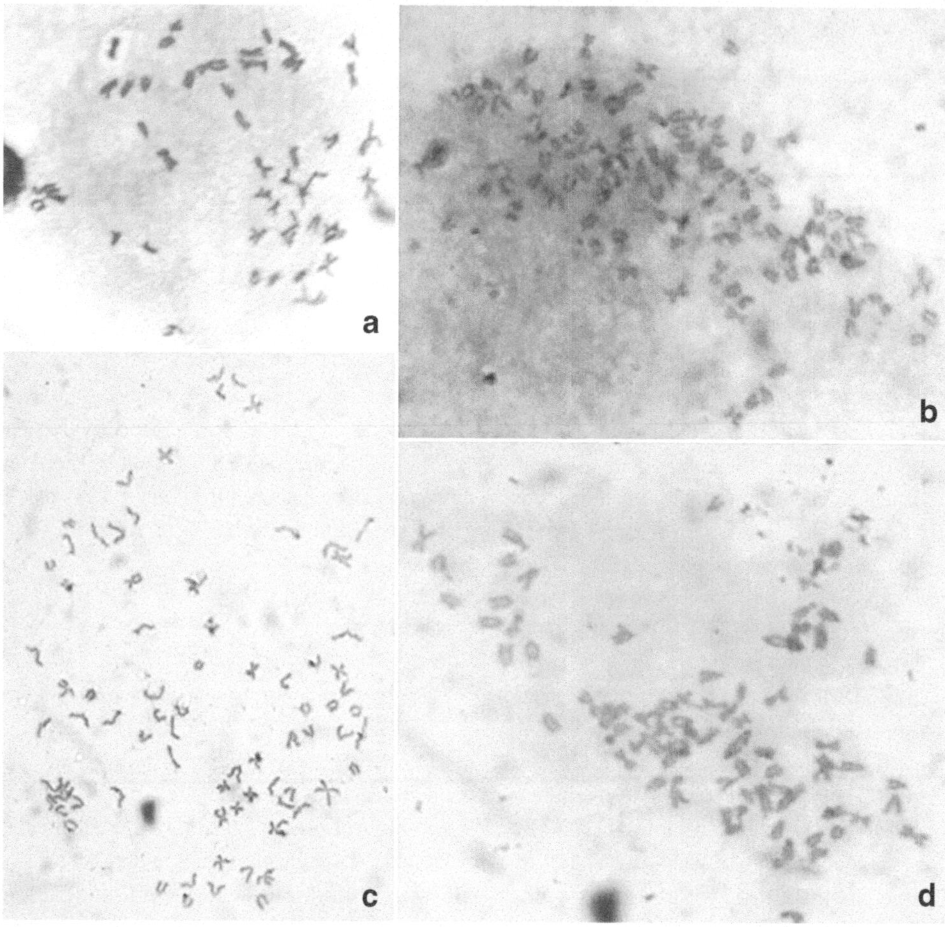

Fig. 1 The chromosomal set of the two loach species and their reciprocal hybrids. **a** *P. dabryanus*, $2n=48$; **b** *M. anguillicaudatus*, $4n=100$; **c** *P. dabryanus* ♀×*M. anguillicaudatus* ♂, $3n=74$; **d** *M. anguillicaudatus* ♀×*P. dabryanus* ♂, $3n=74$

taining alleles from both parents at Mac9, Mac15 and Mac45 (Fig. 2a). However, at the locus Mac24, all the reciprocal hybrids only possessed parental alleles from pond loach while the alleles from large scale loach were absent (Fig. 2b). The paternal pattern of pond loach in the MP group at Mac24 exhibited a constant four-allele type, indicating a duplicated locus (Fig. 2b). At Mac50, the band intensity of all the

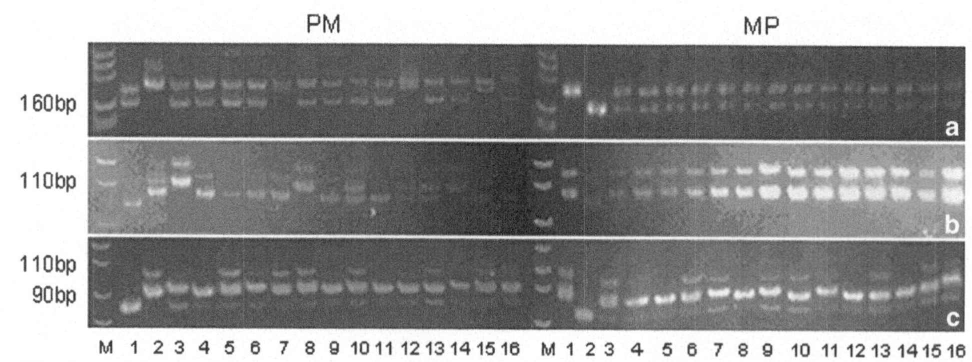

Fig. 2 Amplification profile for the F_1 reciprocal hybrids of *P. dabryanus* and *M. anguillicaudatus* at the microsatellite Mac15 (**a**), Mac24 (**b**), and Mac50 (**c**). *PM*: the laboratory hybrids of *P. dabryanus* ♀×*M. anguillicaudatus* ♂, *MP*: laboratory hybrids of *M. anguillicaudatus* ♀×*P. dabryanus* ♂; *M*: pBR322 DNA/*Msp* I molecular marker; *1*: maternal parent; *2*: paternal parent; *3–16*: F_1 laboratory hybrids

Table 2 Genotypic segregations for three microsatellite loci in 20 individuals of the wild "supposed hybrids"

Supposed hybrids	Genotype		
	Mac9	Mac45	Mac50
H1	92/92	82/82	110/110
H2	92/92	82/96	84/90[a]
H3	92/92	82/82	88/90/106[b]
H4	92/92	82/92	90/92
H5	92/92	82/82	90/90
H6	92/92	82/82	90/96
H7	92/92	82/82	90/90
H8	92/104	82/82	88/90/106[b]
H9	92/92	82/82	90/106
H10	92/92	82/82	90/90
H11	92/92	82/82	90/106
H12	92/92	82/82	90/106
H13	92/92	82/94	90/96/106[b]
H14	92/92	82/82	90/90
H15	92/92	82/82	84/110
H16	92/92	82/82	90/90
H17	92/92	82/82	90/90
H18	92/92	82/82	90/96/106[b]
H19	92/104	82/92	90/90
H20	92/92	82/92	90/90

[a] F_1 hybrid genotype

[b] Individuals exhibiting three bands

reciprocal hybrids was stronger in pond loach than in large scale loach, and the three-band pattern of the hybrids maybe an indication of triploids (Fig. 2c).

Analyses of the wild populations using microsatellite DNA

Three microsatellite loci, Mac9, Mac15 and Mac45, were all in linkage equilibrium (P>0.05) in natural populations of two reference species and probable hybrids (H). No null alleles were detected in the probable hybrids (H) at the three loci.

At Mac9 and Mac45, the two species held the same typical alleles e.g. Mac9-92 with 0.972 in large scale loach and with 0.925 in pond loach, and Mac45-82 with 0.894 in large scale loach and with 0.500 in pond loach, and provided little information for classifying 20 probable hybrids (H), though the other exclusive alleles of pond loach e.g. alleles Mac9-104, Mac45-92 and Mac45-94 were observed in five individuals (H_4, H_8, H_{13}, H_{19} and H_{20}). At Mac50, the two loach species exhibited polymorphism with different alleles e.g. alleles 84 and 110 for large scale loach and alleles 88, 90, 92, 96 and 106 for pond loach, suggesting their potential use as diagnostic markers. Therefore, two of these individuals, H_1 and H_{15}, were assigned to large scale loach, and the others were assigned to pond loach (Table 2). However, a weakly amplified band with the same size as the diagnosed allele (Mac50-84) of the large scale loach attached to the bright diagnosed allele (Mac50-90) of the pond loach in individual H_2. This individual exhibited the same genotype as the laboratory reciprocal F_1 hybrids, and was identified as a natural F_1 hybrid (Fig. 3). Additionally, four individuals, H_3, H_8, H_{13} and H_{18}, showed different three-band patterns at Mac50 (Table 2, Fig. 3).

Discussion

Generally, morphological examination was commonly used to identify large scale loach and pond loach hybrids, which was shown to be correct in most cases (Scribner et al. 2001). However, it could be misleading in cases where morphological measurements were used as the sole source of inference, particularly for hybrid individuals beyond the F_1 generation (Nuff and Smith 1979). The development of molecular markers helped to resolve this problem. In this study, from a total of 20 individuals of probable hybrids with ambiguous morphological traits, only one individual

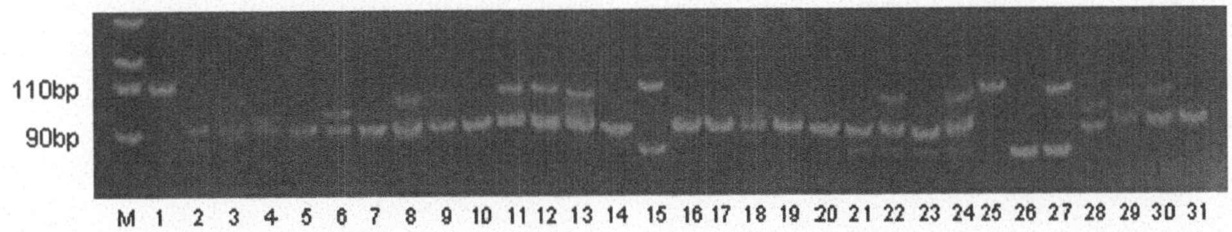

Fig. 3 Amplification profile for wild "supposed hybrids" (H) at microsatellite Mac50. M: pBR322 DNA/*Msp* I molecular marker; *1–20*: individuals of the wild "supposed hybrids" (H_1–H_{20}), *21–24*: four typical genotypes of the hybrids from the PM group or MP group, 25–27: typical genotypes of *P. dabryanus*, *28–31*: typical genotypes of *M. anguillicaudatus*, the *arrow* indicated a "supposed hybrid" loach (H_2) which had the same genotype as the distinct genotype of some laboratory F_1 reciprocal hybrids

(H_2) was clearly identified as F_1 hybrid, and the others were assigned to be either large scale loach or pond loach by using a single diagnostic microsatellite locus. These results suggested that it was important to combine both morphologic and genetic data when identifying pure species and hybrids.

Thorgaard and Allendorf (1988) indicated that fish might be less susceptible to severe developmental incompatibilities than interspecific hybrids in other vertebrates under comparable levels of genetic divergence. In the present study, one out of 20 suspected individuals of the wild populations was identified as F_1 hybrid, suggesting natural hybridization between the two loach species occurred with a low frequency, probably due to spatial, temporal or behavioral isolating mechanisms (Scribner et al. 2001). Rare hybridization between pond loach and large scale loach was also observed in one of the wild individuals in Wuhan area, with 74 chromosomes (Yin et al. 2005). In the present study, the microsatellite genotype of natural F_1 hybrid showed a weak band from large scale loach allele and a strong band from pond loach allele, a pattern resembling that of the artificial hybrids. The similar phenomenon of amplification asymmetry was also observed in the hybridization between silver carp, *Hypophthalmichthys molitrix*, and bighead carp, *Aristichthys nobilis* (Mia et al. 2005), although it was noted that they should be recorded with caution to avoid incorrect scoring of unknown stutter bands. More diagnostic microsatellite loci or other molecular markers of the two species are required to identify hybridization events in natural populations in the future.

In fish, great variations in the fertility of F_1 hybrids have been reported (Nikoljukin 1971). The artificial F_1 hybrids of large scale loach ♀×pond loach ♂ appeared to be female-sterile. However, we could not deduce the conclusion that all F_1 hybrids were sterile as the male hybrids were not tested. Atlantic salmon, *Salmo salar*, and brown trout, *Salmo trutta*, also differed in terms of chromosome numbers ($2n$=58 and 80, respectively), and 0.1 to 13% hybridization rates between two species in wild populations were observed (Verspoor 1988; Jansson et al. 1991; Garcia-Vazquez et al. 2001). The females of these F_1 hybrids were backcrossed with both parental species in artificial conditions, suggesting that they were actually fertile to various degrees (Garcia-Vazquez et al. 2004). A similar case was observed in another fish species e.g. sturgeons (Nikoljukin 1971). In this study, gene introgression may have occurred due to transmission of whole exotic chromosomes or chromosome sections (Verspoor and Hammar 1991). However, in interspecific, intergeneric and even in more remote crossings, reciprocal and backcrossed hybrids may vary significantly in viability or fertility because of the cytogenetic compatibility (Nikoljukin 1971; Gui et al. 1993; Garcia-Vazquez et al. 2004).

A historical event of genome duplication from diploidy ($2n$=50) to tetraploidy (100 chromosomes) has been proposed for pond loach in most areas of China including Wuhan (Li et al. 1983). This was confirmed by the amplification pattern of four bands of pond loach at Mac24 in the MP group. The controlled laboratory hybrids in the present study were cytogenetically deduced to be triploids ($3n$=74), which was consistent with their three-banded patterns at Mac50. Four individuals from the probable hybrids, with other three-band patterns at Mac50, might be an indicator of triploid hybrids of pond loach (Table 2, Fig. 3). Natural triploids of pond loach were found in Japan (Oshima et al. 2005). The triploid males were generally sterile and showed little potential to produce haploid sperm, while the triploid females were fertile and laid a large triploid egg and a normal-size haploid egg (Matsubara et al. 1995; Oshima et al. 2005). Viable reciprocal diploid and triploid hybrids between mud loach, *Misgurnus mizolepis* ($2n$=48) and pond loach, *M. anguillicaudatus* ($2n$=50) were also recorded in South Korea (Kim et al. 1995), in which diploid hybrids ($2n$=49) were fertile, and triploid hybrids ($2n$=74), with two sets of pond loach and one set of mud loach chromosomes, were sterile in both sexes (Nam et al. 2004). The triploid hybrids of the two loach species were female-sterile. However, introgressions would be possible if the hybrids between large scale loach and diploid pond loach were fertile, because the F_1 hybrids might backcross to their parental generation as shown in other loaches (Kim et al. 1995; Nam et al. 2004). Diploid pond loaches were also reported in some areas of China (Yu et al. 1989).

In summary, the results of the present study demonstrated that rare hybridization occurred between large scale loach and tetraploid pond loach, although there may be little danger in genetic introgression due to their sterile offspring. For the purposes of fish ecology and conservation, further studies on the fertility of F_1 hybrids, occurrence of natural hybridization and possible introgressions between large scale loach and diploid pond loach in China are necessary by using controlled matings and additional molecular markers.

Acknowledgements The authors would like to thank X. Liao and C. Wang for technical assistance. This study was supported by NSFC (30370225) and NSF of Hubei Province (2004ABC005).

References

Allendorf FW, Leary RF, Spruell P et al (2001) The problems with hybrids: setting conservation guidelines. Trends Ecol Evol 16:613–622

Bartley DM, Rana K, Immink AJ (2000) The use of inter-specific hybrids in aquaculture and fisheries. Rev Fish Biol Fish 10(3):325–337

Chen J, Zhu S (1984) The subfamily classification of the Cobitidae fishes and their phylogenetic relationships. Acta Zootaxonomica Sinica 9:201–207

Epifanio J, Philipp D (2000) Simulating the extinction of parental lineages from introgressive hybridization: the effects of fitness, initial proportions of parental taxa, and mate choice. Rev Fish Biol Fish 10:339–354

Garcia-Vazquez E, Moran P, Martinez JL et al (2001) Alternative mating strategies in Atlantic salmon and brown trout. J Heredity 92:146–149

Garcia-Vazquez E, Perez J, Ayllon F et al (2004) Asymmetry of post-F1 interspecific reproductive barriers among brown trout (*Salmo trutta*) and Atlantic salmon (*Salmo salar*). Aquaculture 234:77–84

Goudet J (1995) FSTAT (version 1.2): a computer program to calculate F-statistics. J Heredity 86:485–486

Gross R, Gum B, Reiter R et al (2004) Genetic introgression between Arctic charr (*Salvelinus alpinus*) and brook trout (*Salvelinus fontinalis*) in Bavarian hatchery stocks inferred from nuclear and mitochondrial DNA markers. Aquac Int 12:19–32

Gui J, Liang S, Zhu L et al (1993) Cytogenetic analysis of developmental difference in hybrid embryos between reciprocal crosses in distant hybridization of fishes. Zoological Research 14(2):171–177

Hänfling B, Bolton P, Harley M et al (2005) A molecular approach to detect hybridisation between crucian carp (*Carassius carassius*) and non-indigenous carp species (*Carassius* spp. and *Cyprinus carpio*). Freshw Biol 50:403–417

Ibarra AM, Hernández-Ibarra NK, Cruz P et al (2005) Genetic certification of presumed hybrids of blue × red abalone (*Haliotis fulgens* Philippi and *H. rufescens* Swainson). Aquac Res 36:1356–1368

Jansson H, Holmgren I, Wedin K et al (1991) High frequency of natural hybrids between Atlantic salmon, *Salmo salar* L., and brown trout, *S. trutta* L., in a Swedish river. J Fish Biol 39(Suppl. A):343–348

Kim DS, Nam YK, Park I-S (1995) Survival and karyological analysis of reciprocal diploid and triploid hybrids between mud loach (*Misgurnus mizolepis*) and pond loach (*Misgurnus anguillicaudatus*). Aquaculture 135:257–265

Li K, Li Y, Zhou D (1983) Comparative studies on the karyotypes of two species of loach (*Misgurnus anguillicaudatus* and *Paramisgurnus dabryanus*. Zoological Research 4:75–80

Long L, Zhao ZS, Tang B et al (2000) Analysis of genetic variation in hybrids between *Misgurnus anguillicandatus* cross *Paramisgurnus dabryanus* reciprocally by using RAPD. Acta Hydrobiol Sin 24:659–662

Matsubara K, Arai K, Suzuki R (1995) Survival potential and chromosomes of progeny of triploid and pentaploid females in the loach, *Misgurnus anguillicaudatus*. Aquaculture 131:37–48

Mia MY, Taggart JB, Gilmour AE et al. (2005) Detection of hybridization between Chinese carp species (*Hypophthalmichthys molitrix* and *Aristichthys nobilis*) in hatchery broodstock in Bangladesh, using DNA microsatellite loci. Aquaculture 247:267–273

Morishima K, Nakayama I, Arai K (2001) Microsatellite-centromere mapping in the loach. *Misgurnus anguillicaudatus*. Genetica 111:59–69

Nam YK, Park I-S, Kim DS (2004) Triploid hybridization of fast-growing transgenic mud loach *Misgurnus mizolepis* male to pond loach *Misgurnus anguillicaudatus* female: the first performance study on growth and reproduction of transgenic polyploid hybrid fish. Aquaculture 231:559–572

Nikoljukin NI (1971) Fundamentals of hybridization in fish culture. In: Lectures of the seminar/ study tour in the USSR on genetic selection and hybridization of cultivated fishes. Project reports (not in a Series) – TA 2926:365

Nuff NA, Smith GR (1979) Multivariate analysis of hybrid fishes. Syst Zool 28:176–196

Oshima K, Morishima K, Yamaha E et al (2005) Reproductive capacity of triploid loaches obtained from Hokkaido Island, Japan. Ichthyol Res 52:1–8

Scribner KT, Page KS, Bartron ML (2001) Hybridization in freshwater fishes: a review of case studies and cytonuclear methods of biological inference. Rev Fish Biol Fish 10:293–323

Seehausen O (2004) Hybridization and adaptive radiation. Trends Ecol Evol 19:198–207

Thorgaard GH, Allendorf FW (1988) Developmental genetics in fishes. In: Malacinski GM (ed) Developmental genetics of animals and plants. MacMillan, New York, pp 369–391

Van-Oosterhout C, Hutchinson W, Wills D et al (2004) MICRO-CHECKER: software for identifying and correcting genotyping errors in microsatellite data. Mol Ecol Notes 4 (3):535–538

Verspoor E (1988) Widespread hybridization between Atlantic salmon, *Salmo salar*, and introduced brown trout, *S. trutta*, in eastern Newfoundland. J Fish Biol 32:327–334

Verspoor E, Hammar J (1991) Introgressive hybridization in fishes: the biochemical evidence. J Fish Biol 39:309–334

Yao W (1989) The applied technique of breeding loach. Biol J 6:19–22

Yin J, Zhao ZS, Chen X et al (2005) Karyotype comparison of diploid and tetraploid loach *Misgurnus anguillicandatus*. Acta Hydrobiol Sin 29(4):469–472

Yu X, Zhou D, Li Y et al (1989) Chromosomes of Chinese fresh-water fishes. Science Press, Beijing

Zhao ZS, Gao G, Sun W et al (1999) Studies on insemination cytology in hybridization between *Paramisgurnus dabryanus* and *Misgurnus anguillicandatus*. Journal of Huazhong Agriculture University 18:75–77

Zhao ZS, Gao G, Wu Q (2002) The chromosome set variation in hybridization of *Misgurnus anguillicandatus* with *Paramisgurnus dabryanus* and in artificial androgenetic *P. dabryanus*. Journal of Dalian Fisheries University 17:15–19

Environ Biol Fish (2009) 86:73–74
DOI 10.1007/s10641-007-9323-5

Threatened fishes of the world: *Onychostoma alticorpus* (Oshima, 1920) (Cyprinidae)

I-Shiung Chen

Received: 30 May 2007 / Accepted: 19 October 2007 / Published online: 28 January 2008

Abstract The conservation status, species identification, distribution, specific habitat and ecology of a threatened, endemic cyprinid fish, *Onychostoma alticorpus* (Oshima 1919) is reported herein. It is merely found in five river basins of Taiwan. The protection of natural habitat would be very essential.

Keywords *Onychostoma* · Critically endangered · Cyprinidae · Freshwater fish · Conservation · Taiwan

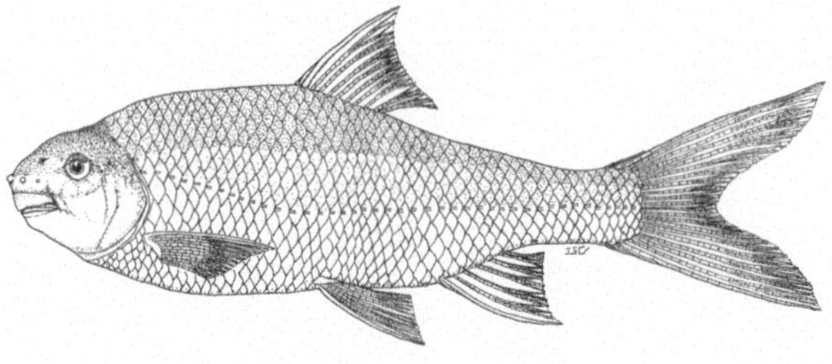

Common names: Deep-body shovel-jaw carp; Kao-Shen-Ku-Yu (Mandarin); Shar-Ven (Taiwanese). *Conservation status*: Critically endangered (Ministry of Interior), endangered (IUCN). *Identification*: D 3+8; V 1+8; A 3+5; LL 43–44; PreD 13–15. Body oval, snout rounded, mouth subterminal, two tiny barbels, upper lip thick, prominent, lower lip with a horizontal shovel margin. Dorsal unbranched rays without serration. Body grayish silver dorsally; silver white ventrally. Dorsal, anal and caudal fins pinkish. Size 15–30 cm (SL), maximum 50 cm (Oshima 1920; Shen 1993; Chen and Fang 1999). *Distribution*: Endemic in middle reaches of the Kao-Ping, Hua-Lien, Shiu-Ku-Luan, Pei-Nan, and Tai-Mar-Li Rivers (Tzeng 1986; Chen and Chang 2005). *Abundance*: Adults/100 m^2, 35–50, in the densest populations. *Habitat and ecology*: Prefers weak alkaline water (pH 8.0–8.8) with high DO(>70%), depth(0.5–7 m), moderate temp (19–24°C), clear water(0–50 NTU) and moderate flow(20–100 $cm{\cdot}s^{-1}$), around 30–800 m altitude. Adult ascending migration during floods (May to August); descending movement is in September to February. Feeds diurnally on periphyton mainly, Bacillariophycae (Fang et al. 1995; Han et al. 2000). *Reproduction*: Larvae (1–3 cm) appears after rainy season, October to February at low altitudes, 100–350 m (Chen and Fang 1994; Fang et al. 1996). *Threats*: Habitat destruction from road construction, damage of natural vegetation and heavy silt; occasional illegal fishing by poison and electro-fishing; dam construction blocked seasonal fish migration; and water pollution (Chen 2001). *Conservation action*: Natural protective areas established in three

I-S. Chen (✉)
Institute of Marine Biology and Department of Life Science,
National Taiwan Ocean University,
Keelung 202, Taiwan, Republic of China
e-mail: isc@mail.ntou.edu.tw

basins have significantly improved for fish populations. *Conservation recommendation*: Prohibit damage of natural vegetation and river habitat, prohibit illegal fishing and avoid new dam construction. *Remarks*: The recovery of fish populations suggests the ranking of 'endangered' instead of 'critically endangered' in recent years. However, more field work on population dynamics may be necessary for further re-evaluation. *Varicorhinus alticorpus* is a synonym for *Onychostoma alticorpus* (Chen and Chang 2005).

References

Chen I-S (2001) Research history, geographical distribution, conservation strategy and future perspective of the freshwater fishes in Taiwan. Taiwan Ref 52:45–60 (in Chinese)

Chen I-S, Chang YC (2005) A photographic guide to the inland-water fishes of Taiwan, Vol. I. Cypriniformes. The Sueichan, Keelung

Chen I-S, Fang LS (1994) The current status and risk of endemic freshwater fishes of Kao-Ping River, Taiwan. Quart Mag Nature 45:42–45 (in Chinese)

Chen I-S, Fang LS (1999) The freshwater and estuarine fishes of Taiwan. National Museum of Marine Biology/Aquarium, Pingtung (in Chinese)

Fang LS, Han CC, Chen I-S (1995) Deep-body Ku Fish, a very rare and endemic native of mountain stream in Taiwan. The Preparatory Office of National Museum of Marine Biology/Aquarium, Kaoshiung (in Chinese)

Fang LS, Su LY, Chen I-S, Han CC, Chen YH (1996) The morphology, distribution, and biology of endemic cyprinid, *Varicorhinus alticorpus* from Taiwan. Chinese Biosci 39:78–87 (in Chinese with English abstract)

Han CC, Tew KS, Chen I-S, Su LY, Fang LS (2000) Environmental biology of an endemic cyprinid, *Varicorhinus alticorpus*, in subtropical mountain stream of Taiwan. Environ Biol Fish 59:153–161

Oshima M (1920) Notes on freshwater fishes of Formosa, with descriptions of new genera and species. Proc Acad Nat Sci Philad 72:120–135

Shen SC (1993) Fishes of Taiwan. National Taiwan University Press, Taipei (in Chinese)

Tzeng CS (1986) Distribution of the freshwater fishes of Taiwan. J Taiwan Mus 39:127–146

Environ Biol Fish (2009) 86:75–88
DOI 10.1007/s10641-008-9342-x

Age composition, growth, and reproductive biology of yellow catfish (*Peltobagrus fulvidraco*, Bagridae) in Ce Lake of Hubei Province, Central China

Ling Cao · Biyu Song · Jinmiao Zha · Chengtai Yang · Xinfu Gong · Jianbin Li · Weimin Wang

Received: 28 March 2007 / Accepted: 12 February 2008 / Published online: 20 May 2008

Abstract This study presents fundamental information on the reproductive biology, size and age composition, and growth of yellow catfish, *Peltobagrus fulvidraco*. Sampling was carried out monthly from Ce Lake in Hubei province from January 2004 and December 2004. Overall female/male (*F*/*M*) sex ratio was 1.08, not significantly different from expected 1.0 ratio. But sex ratio (*F*/*M*) of the spawning group was 1.14:1 with skewness towards females ($p<0.05$). The smallest sexually mature male and female massed 20.51 and 16.93 g, and measured 100.1 and 94.3 mm in total length, respectively. Reproductive period for this species is from April to August when GSI is considerably higher than the rest months ($p<0.05$). The highest GSI values were measured from May to July for both sexes and this period was presumably the peak spawn. First maturation occurred at age 1^+ for both sexes. Fish of age 1^+ and 2^+ comprised the greatest portion of spawners. The absolute fecundity (AF) of *P. fulvidraco* ranged from 657 to 11,696 eggs per fish, with mean AF value of 3,570±122 (S.D.) eggs per fish, varying considerably at given length, weight and age. The fecundity relative to total length (*F*/*L*) fluctuated from 58 to 680 eggs/mm, with a mean value of 246±34 (SD) eggs/mm, and fecundity relative to total weight (*F*/*W*) varied from 29–133 eggs/g, with a mean value of 67±7 (SD) eggs/g. *P. fulvidraco* is a single spawner, with low fecundity compared with other Bagrid fish.

Keywords *Peltobagrus fulvidraco* · Age · Growth · Reproductive biology · Fecundity

L. Cao · C. Yang · W. Wang (✉)
College of Fisheries, Key Lab of Agricultural Animal Genetics, Breeding and Reproduction of Ministry of Education, Huazhong Agricultural University, Wuhan 430070, China
e-mail: wangwm@mail.hzau.edu.cn

L. Cao
School of Natural Resources and Environment, University of Michigan, Ann Arbor, MI 48109, USA

B. Song
School of Resource and Environmental Science, Wuhan University, Wuhan, Hubei 430072, China

J. Zha
State Key Laboratory of Environmental Aquatic Chemistry, Research Center for Eco-Environmental Sciences, Chinese Academy of Sciences, P.O. Box 2871, Beijing 100085, China

X. Gong · J. Li
Fuer Fish Nursery Limited Company of Huangshi, Huangshi City, Hubei 435000, China

Introduction

Yellow catfish, *Peltobagrus fulvidraco*, belongs to Bagridae family. With both high economic and nutritional value, feeding upon cladocerans and zooplankton, *P. fulvidraco* is one of the promising freshwater species for both semi-intensive and intensive culture in East and South Asia (Liu 1997). It is widely distributed in lakes along the midstream and downstream of Yangtze River Basin in central China, with most abundance in Hubei Province (Du 1963). Based on recent assessment of this species, the natural stock is probably at a reduced reproductive capacity and at risk of being harvested unsustainably (Liu 1997; Luo et al. 2005). Knowledge of the reproductive biology and the estimation of fecundity and egg production are fundamental in the study of the population dynamics of fish species (Hunter et al. 1992; Murua et al. 2003), as they are basal to quantify reproductive capacity at both individual and population levels of a fish species. Manipulation of fish's reproductive system under artificial conditions also requires an understanding of natural spawning patterns and other influential factors based on information of the species' reproductive biological characters such as size-at-sexual maturity, the duration and periodicity of spawning.

However, the information available on the reproductive biology of *P. fulvidraco* is still incomplete. Detailed descriptions of fecundity, egg production, reproductive season, and the relationships between somatic and gonadal growth have not been well documented. Previous studies on the reproductive biology of *P. fulvidraco* have indicated that this fish of each geographical population spawns at different seasons with different levels of spawning activity; being maximum during peak spawning from middle May to Middle July in Liangzi Lake (Du 1963), from June to July in Honghu Lake (Ma 1991), from April to June in Poyang Lake (Liu 1997) and at lower levels during the rest months of the spawning period. Moreover, there were inconsistent reports on the spawning pattern of this species. Some indicated that *P. fulvidraco* was batch spawner with asynchronous oocyte development and, therefore, the population egg production should be estimated combining batch fecundity, spawning fraction and the number of active females in the adult population (Hu et al. 1993; Wang 1999). According to Hunter et al. (1992), multiple spawning fish are called indeterminate serial or batch spawners, if the annual fecundity is not fixed at the beginning of spawning and the standing stock of present oocytes in the ovary is continuously replaced by new developing oocytes during one spawning cycle. On the other hand, fish that have a fixed fecundity prior to the onset of spawning are called single spawners or determinate serial spawners. Yang (1994) indicated that this species belonged to determinate serial spawner. Similar results were reported by Liu and Zhang (2003). Therefore, determining the spawning patterns is of great importance for estimating the potential annual fecundity of this fish and its ability to recover from human fishery impact. Spawning strategy, fecundity and egg production of this fish species varied annually and/or spatially among stocks in relation to condition and environmental factors, which is of considerable importance because it influences stock reproductive potential, and thus has a direct implication on a population's ability to sustain harvesting.

The aim of this research was to investigate the spawning activity and spawning pattern of yellow catfish in Ce Lake from January to December 2004, thus providing comprehensive information for its age, growth, reproductive biology study and future sustainable use. This was achieved by investigating and assessing important reproductive parameters such as sexual maturity, sex ratio, gonadosomatic index values, spawning seasonality, monthly specific variation of oocyte size-frequency distribution and fecundity of this fish species. In addition, condition factor (*K*) is explored as a potential factor governing variation in fecundity (Murua et al. 2006).

Materials and methods

Study site and sampling procedure

Ce Lake, with an area of 9.4 km^2, located in the east of Hubei Province in central China (Fig. 1). The width of Ce Lake ranges from constrictions of less than 2 m to large pools several hundred meters across. Mean water depth of this lake is about 2.6 m. Mean annual DO (dissolved oxygen) concentration in this freshwater lake is above 4 $mg^{-1}l^{-1}$ throughout of the year. The pH varies from 6.5~7.5. The lake was surrounded by rice fields and fish ponds. The exposed parts of the lakeshore of Ce Lake contain a plant community with mostly small herbaceous species. The fine sediment is nutrient-rich and consists mainly of clay. Vegetation cover is generally low and decreasing towards the lake.

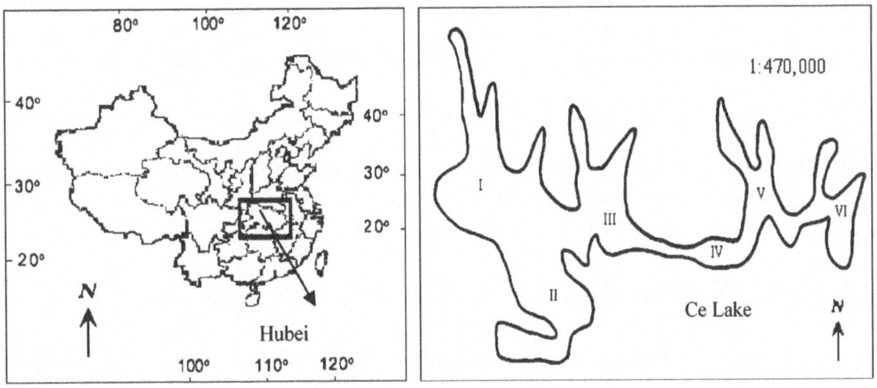

Fig. 1 Map of the study location: the Ce Lake in Hubei Province, central China. Six roman numerals indicate six sampling sites locating at upstream, mid-stream and downstream, respectively in Ce Lake

Monthly sampling was carried out from January to December 2004. From April to August, sampling frequency was twice a month because this was the time when *P. fulvidraco* actively spawned. A total of 768 samples were obtained in Ce Lake. Fish were collected using seine nets, consisting of two 40 m long wings with a 3 cm mesh and a 100 m long packet with a 2 cm mesh, each of which had a length of 10 m. Specimens collected were anesthetized using tricaine methanesulfonate (MS 222), and fixed in 10% buffered formalin for later analysis.

Total length, caudal peduncle length (the oblique distance from the insertion of the anal fin to the midpoint of the end of the hypural plate) and caudal peduncle depth (the least depth of this structure from the mid-line of the ventral surface) of each sample were measured to the nearest 0.1 mm with dial calipers. Total weight of fish was weighed to the nearest 0.01 g. Sex of each fish was differentiated by sexual characteristics (genitals of males) or by examining its gonads under a dissecting microscope. Size at maturity was determined for each sex by assessing samples collected during the reproductive season.

Age and growth

The sagittal otoliths of each fish were removed, cleaned and stored dry in gelatin capsules. They were later placed in methyl salicylate in a black dish and examined under a dissecting microscope at ×40 magnification using reflected light (Morgan et al. 2000). The distance between the outer edge of the translucent zone and the periphery of each otolith was measured. Each distance was always measured along the same axis and recorded to the nearest 0.025 mm. For convenience, the relative marginal increment was subsequently referred to as marginal increment. The age of each fish at the time of capture was estimated using otolith information. Growth curves were then fitted to the length at body weight of each fish using a non-linear technique.

Gonadosomatic index

Gonads were separated from fish body and weighed to the nearest 0.0001 g in the laboratory (Johnston and Knight 1999) and subsequently preserved in 10% formalin within 24 h of capture for subsequent histological analyses following a standard histological protocol (Pearce 1965). Seasonal change in gonad mass for both sexes was determined using the gonadosomatic index (GSI). GSI was calculated as GSI (%)=100×gonad weight (total body weight-gonad weight)$^{-1}$ for each fish and all values were averaged monthly (Anderson and Gutreuter 1983). Spawning periods were determined from monthly evolution of GSI.

Oocyte size-frequency distributions

Ovarian condition of females was categorized using the classification of Heins and Baker (1993): (1) latent (stage I), (2) early maturing (stage II), (3) late maturing (stage III), (4) maturing (stage IV), (5) ripening (stage V) and (6) spent (stage VI). Ovarian stages at IV and V were considered reproductive, while stages I, II, III and VI were considered non-reproductive (Heins and Baker 1993). Reproductive condition of males was determined by visual examinations of testes. Mature males have large and opaque testes. Males that have small and transparent testes were considered latent. The numbers of oocytes in ovaries at spawning period (from April to August) were estimated by the gravimetric hydrated-oocyte method (Hunter et al. 1985), as the average of three replicates. Subsamples of 1 g each, were extracted from anterior, middle and posterior parts of the right ovary, and weighed to the nearest 0.0001 g. The number of hydrated oocytes was counted in each subsample under a binocular microscope (Murua and Motos 2006). The diameter of oocytes was measured as the average of major and minor axes. The size-frequency distribution of the oocytes in each ovary was recorded and histograms were constructed from pooled data for estimating spawning pattern and frequency.

Fecundity

The condition factor (CF) of the sampled fish (n= 218) were analyzed. Monthly samples of fish captured within the same study area and the same period were analyzed to relate to seasonal variations in fecundity. The values of condition factor was calculated as CF= 100 (total body weight-gonad weight)/l^3.

Mature ovaries were identified from their external appearance and the extent to which they filled the body cavity. Fifty-nine ovaries at stage IV and V were chosen from fish collected from April to August of 2005, when GSI values were highest, to estimate absolute fecundity and relative fecundity. The oocyte packing density in the ovary was investigated to ensure that the subsamples were representative of the entire ovary (Murua et al. 2003). Natural logarithm transfer of oocyte density was used to comply with the assumption of homoscedasticity. The numbers of oocytes per gram of ovarian tissue were examined using ovaries of 59 females in six sub-samples taken at different locations (anterior, mid and posterior) of each lobe (left and right). All ovaries used to estimate were histologically screened to check for the occurrence of postovulatory follicles. Ovaries with recent postovulatory follicles were eliminated for fecundity calculations, since their presence indicated part of the batch had ovulated and the fecundity thus would be underestimated.

The number of oocytes in these chosen ovaries was estimated by the gravimetric hydrated-oocyte method (Hunter et al. 1985) as mentioned above. The fecundity of each fish was then estimated as the weighted mean density of the three subsamples multiplied up to determine the total ovary mass. If the coefficient of variation was higher than 5%, then additional samples were counted (Murua et al. 2003). The absolute fecundity of yellow catfish was recorded as the number of mature eggs in the ovaries of ripe samples. Relative fecundity (number of eggs per unit weight) was determined as the number of mature oocytes in relation to total weight or total length (Hotosa et al. 2000).

$$\text{Absolute fecundity} = \text{egg numbers per 1 gram ovary} \times \text{ovary weight (g)}$$

$$\text{Relative fecundity}(F/W) = \text{absolute fecundity}/\text{total weight (g)}$$

Statistical analysis

Sex ratio was determined for each population by using pooled fish from monthly collections. Deviations from the expected 1:1 sex ratio were determined using a chi-square test. The relationship between fecundity and body length, ovary-free body weight and age were analyzed with linear or non-linear regression using Pearson's correlation analysis (Bagenal 1978). All values were $\log_{10}$ transformed prior to analysis and tested for normality and homogeneity of slopes to satisfy the assumptions of statistical analysis. T test was performed to test if there were significant differences between the ratio of caudal peduncle length (CPL) and total length (L) of females and that of males. Data were analyzed statistically by using SPSS (version 11.5) statistical software package (SPSS Inc., Chicago, USA). Differences were considered significant at an alpha level of 0.05.

Results

Sex ratio

According to dissecting microscope examination and secondary sexual characteristics, a total of 768 fish was sexed, of which 398 (51.8%) and 370 (48.2%) were females and males, respectively. Overall female/male (F/M) sex ratio was 1.08, not significantly different from expected 1:1 ratio (χ^2=0.098, df=1, p=0.325; Table 1).

But sex ratio (F/M) of the spawning group was 1.14:1 with skewness towards females ($p<0.05$). The highest sex ratio of 1.74 (F/M) was observed in May. From April to June, deviations from 1:1 ratio were found in *P. fulvidraco* with skewness towards females ($p<0.05$). However, from July to August, deviations from 1:1 ratio were found with skewness towards males ($p<0.05$). There were no significant differences from expected 1:1 ratio for the rest months ($p>0.05$).

Size and age distribution

There were significant differences of the length of mature *P. fulvidraco* between sexes. The sizes of females ranged from 92.7 to 175.7 mm, while that of males ranged from 91.5 to 216.5 mm in total length. Gonads were macroscopically visible for individuals >100 mm (total length) for males and >90 mm (total length) for females. The distribution of sexes by size class (Table 2) showed a significant predominance of females in specimens smaller than 150 mm in TL, while a significant predominance of males larger than 160 mm ($p<0.05$).

Based upon otolith checking, it was found that age 0^+ fish were primarily of stage I and had a very low GSI and gonad weight of almost zero. Age 1^+ and higher fishes were represented in all other gonad stages (II-VI), with values of GSI and gonad weight indicative of reproductive potential. In all gonad stages, especially IV and V, the predominance of ages 1^+ and 2^+ was very evident. Spawning stock of *P. fulvidraco* ($N=332$) is comprised mostly of age 1^+ and age 2^+ fish during reproductive season (Table 3). Age 1^+ fish were less than 120 mm TL (both sexes combined accounts for 32.9% of total), age 2^+ fish ranged from 120 to 170 mm (49.7%), age 3^+ or higher fish exceeded 130 mm (15.0%). First maturation occurred at age 1^+ for both sexes. The smallest sexually mature male and female massed 20.51 and 16.93 g, respectively, and measured 100.1 and 94.3 mm in total length, respectively.

Table 1 Sex ratio of females to males (F/M) of *P. fulvidraco* in Ce Lake in 2004

Sampling months	No. (females)	No. (males)	Sex ratio (F/M)	Chi-square test
JN	34	33	1.03	$p>0.05$
FE	30	31	0.97	$p>0.05$
MA	31	29	1.07	$p>0.05$
AP	47	35	1.34	$p<0.05$
MA	40	23	1.74	$p<0.05$
JU	48	32	1.50	$p<0.05$
JL	22	33	0.67	$p<0.05$
AU	27	38	0.71	$p<0.05$
SE	37	40	0.93	$p>0.05$
OC	23	21	1.10	$p>0.05$
NV	34	31	1.10	$p>0.05$
DE	25	24	1.04	$p>0.05$
Total	398	370	1.08	$p>0.05$

Table 2 The total length distribution of both sexes by size class of *P. fulvidraco* ($n=768$) collected in this study

Total length (mm)	Number of specimens		
	Female	Male	Total
90–100	72	10	82
100–110	97	34	127
110–120	47	23	70
120–130	54	15	69
130–140	36	17	53
140–150	55	22	77
150–160	30	11	41
160–170	9	45	54
170–180	2	89	91
180–190	0	68	68
190–200	0	29	29
200–210	0	6	6
210–220	0	1	1
Sex ratio	398	370	768

Sex characters

Through T test, there were significant differences between the ratio of caudal peduncle length (CPL) and total length (L) of females ($n=50$) and that of males ($n=50$; $t=-15.6019$, $p<0.0001$). In males, the CPL/L of males during the whole year ranged from 0.1310 to 0.1662 which were all higher than 0.13; while those of females ranged from 0.1120 to 0.1299, all less than 0.13. No difference was detected between the ratios of caudal peduncle depth and body depth of both sexes ($p>0.05$).

Following Keys formula, the relationship between total weight (W, g) and total length (L, mm) is typically expressed as $W=aL^b$ (a, b indicate constants; Maack and George 1999). Based on data pooled from the whole year, the length–weight exponential relationship of *P. fulvidraco* was shown in Fig. 2.

Table 3 Age composition of spawning stock of *P. fulvidraco* (N=332) in Ce Lake during reproductive season

Age	Total length (mm)		Total weight (g)		Oocyte diameter (mm)		Percentage (%)
	Mean±95%	Range	Mean±95%	Range	Mean±95%	Range	
1	106.3±8.4	94.3–119.8	26.87±9.03	16.93–36.28	1.25±0.16	1.019–1.449	32.9
2	146.6±4.1	124.8–168.5	51.67±10.59	36.90–74.40	1.27±0.24	1.105–1.575	49.7
3	151.2±9.2	132.1–171.5	55.44±9.65	38.13–70.46	1.31±0.09	1.159–1.397	10.2
4	169.0±7.5	151.9–203.9	76.58±21.27	61.46–119.50	1.45±0.22	1.282–1.684	4.8
Not aged	107.7±6.5	102.1–114.9	27.24±8.36	21.33–36.85	1.12±0.18	0.999–1.325	2.4

Spawning season and patterns

As an indicator of reproductive condition, GSI were calculated for both sexes using all available fish and for each sex separately (Fig. 3). Water temperature was also shown in Fig. 3 as a potential factor governing variation in GSI. For females, the mean GSI increased from April, reached a peak in May (17.29%). It thereafter decreased slightly in June. In July and August, it increased again to form the second peak (12.77%) which was obviously smaller than the first peak. In September, it lowered down rapidly to approach the lowest values and remained at a low level between October and March with a maximum

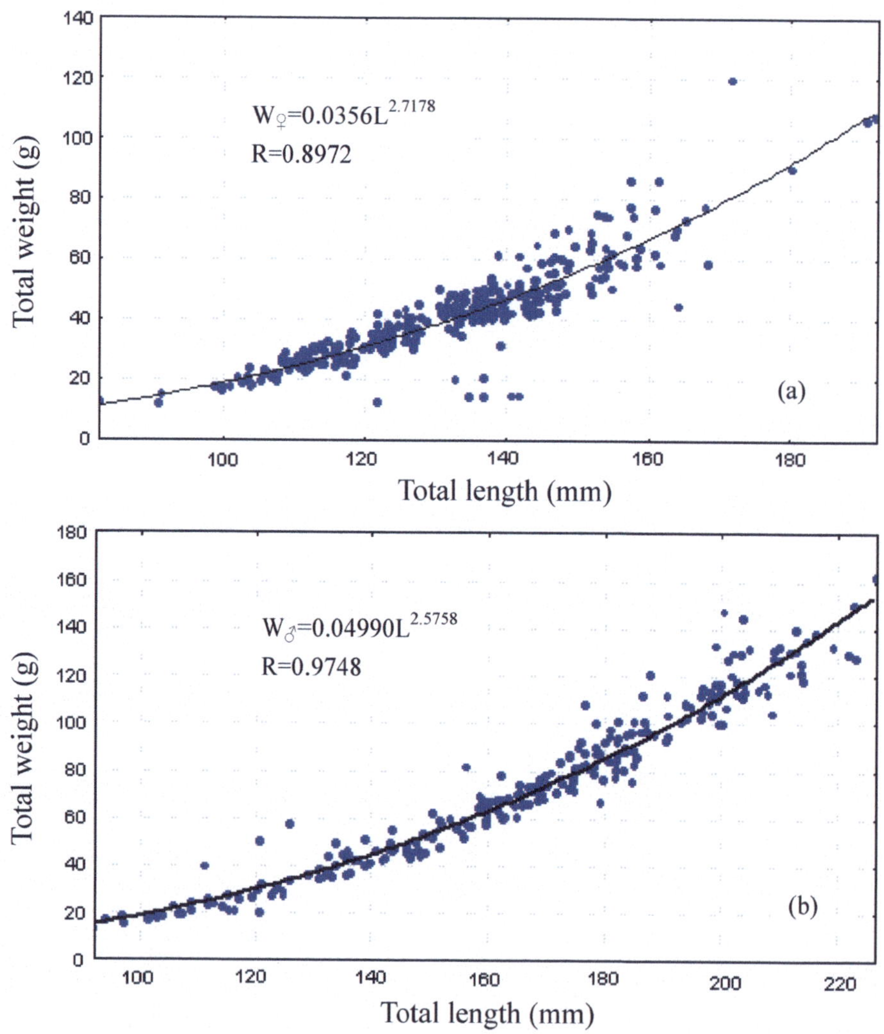

Fig. 2 The length-weight relationship of *P. Fulvidraco*: **a** female: n=398; **b** male: n=370

Fig. 3 Annual changes in gonadosomatic index (mean values±SD) for male and female *P. fulvidraco* in Ce Lake: **a** females: n=398; **b** males: n=370; **c** monthly changes of mean water temperature in Ce Lake

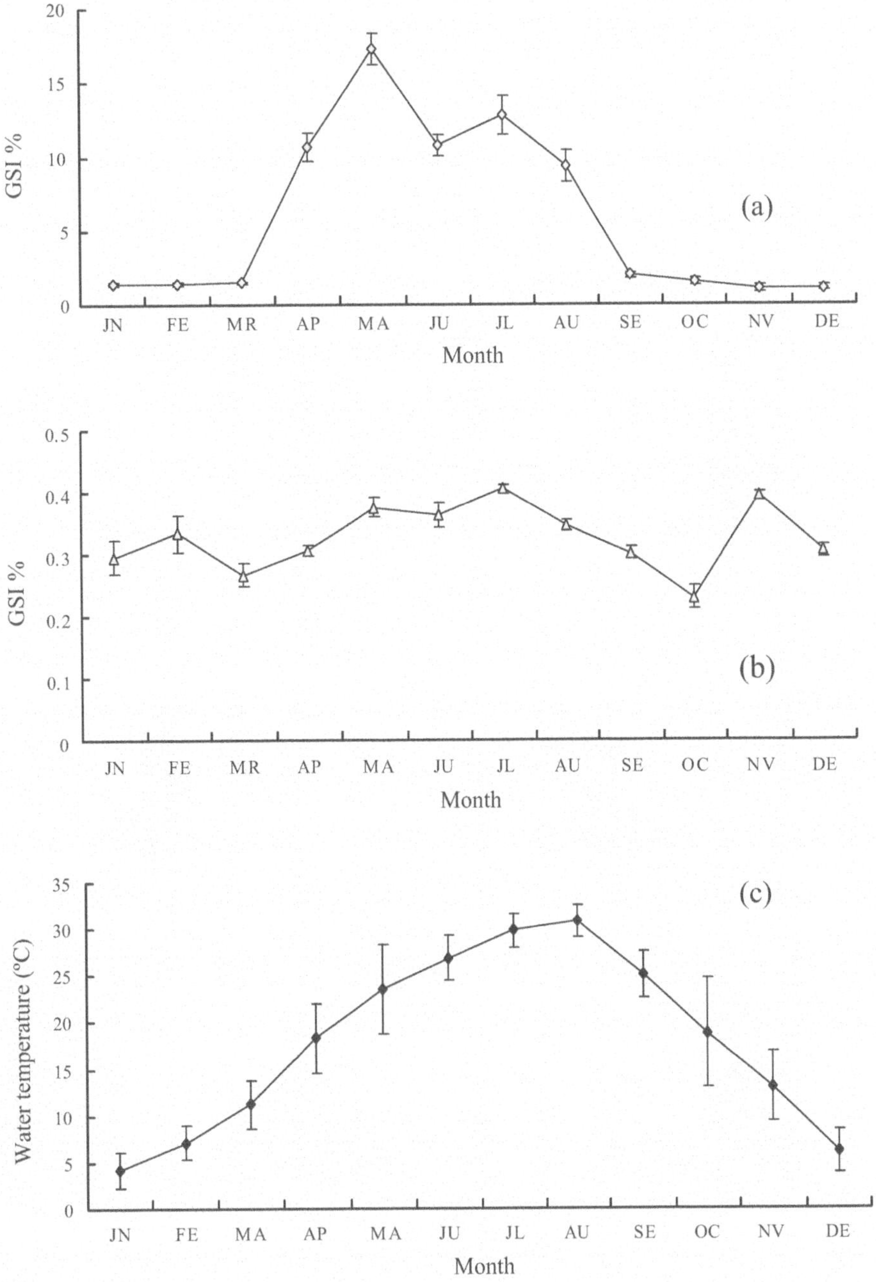

value of 2%. The GSI curve of females fluctuated more obviously than that of males. For males, there were two peaks of GSI in the spawning season. One was from May to August, which was concurrent with that of females. The other was shaped in November. The minimum individual GSI was in October (0.23%). Besides, the GSI of males had another two peaks in the non-spawning season, from February to March and from November to December, respectively.

Thus, the reproductive period for this species in this particular environment was from April to August when GSI was considerably higher than the rest months. During spawning season, GSI values were 0.31–0.41% for males and 9.38–17.29% for females, respectively. The highest female GSI values were measured from May to July and this period was presumably the peak spawn.

Ovaries from 156 females were prepared for histological analyses. There were two distinct phases during oocyte growth. The primary growth phase (PGP) included stage I, II and III. Secondary growth phase (SGP) included stage IV, V and VI. Oocyte size-frequency distribution was studied in the mature fish (Fig. 4a–l). It showed the developmental sequence of maturation. In stage II and stage III, oocyte diameter ranged from 0.5–0.6 and 0.7–0.9 mm, respectively. Yolked oocytes ranged from 1.1–1.85 and 1.4–1.9 mm for maturity stages IV and V, respectively. The frequency of larger oocytes ranging from 1.4–1.9 mm for the V stage was higher than that of stage IV. The oocyte diameter for stage VI dropped to the range of 0.3–1.3 mm, which largely belonged to the small oocytes in the range of 0.3–0.7 mm. The

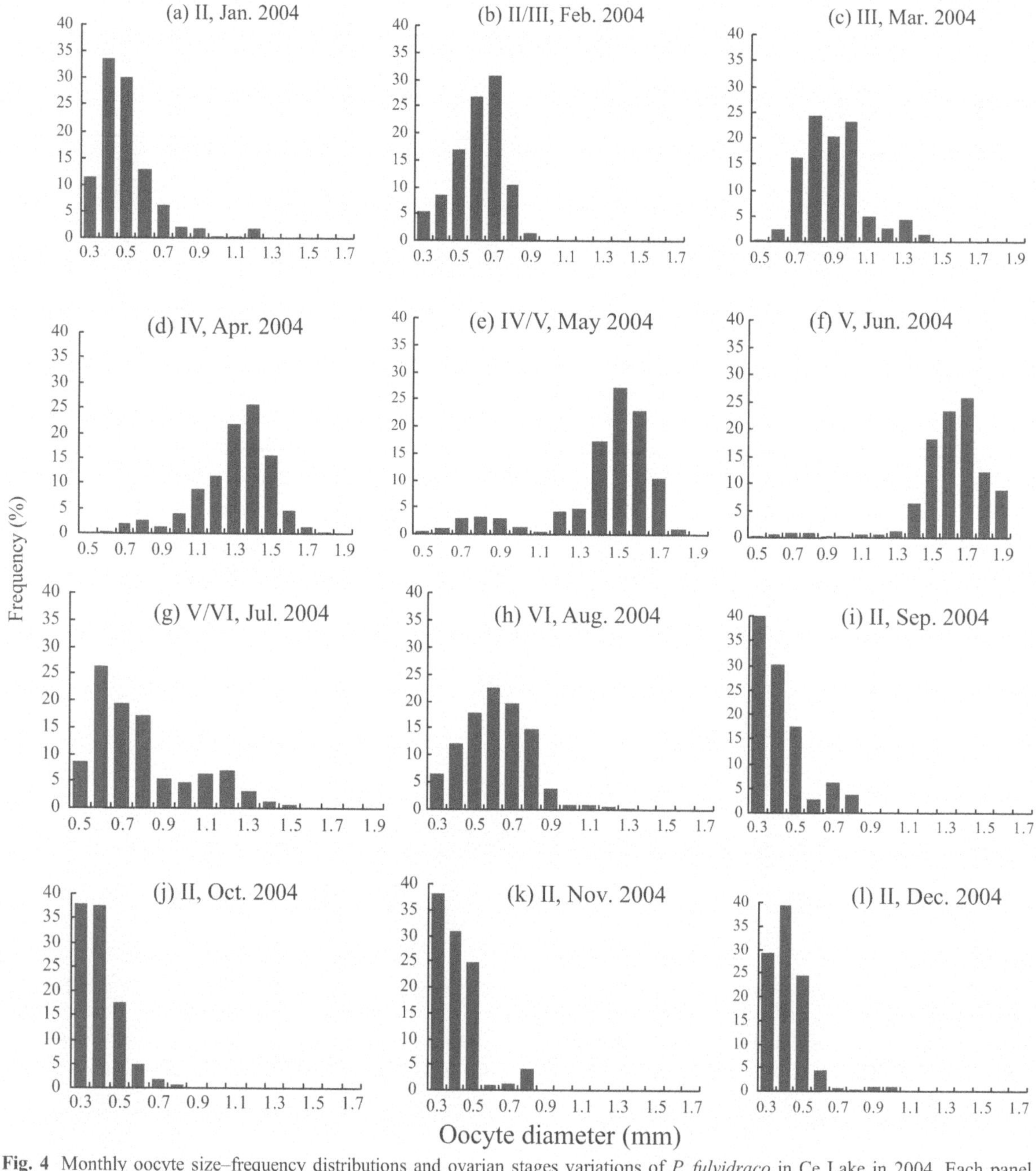

Fig. 4 Monthly oocyte size–frequency distributions and ovarian stages variations of *P. fulvidraco* in Ce Lake in 2004. Each panel corresponds to an individual fish

period from April to August was considered as spawning period. After spawning, individuals returned to stage II for overwintering and never developed into stage III till next March. This distribution pattern of the developing oocytes, indicated that *P. fulvidraco* is a determinant spawner or single spawner, spawning only once during its reproductive period. Based on this, the estimated absolute fecundity is the annual fecundity. None of total weight, total length and age of this species was significantly related to oocyte diameter ($p>0.05$). After logarithmic transformation of all variables, there were no significant correlations among them, neither.

Fecundity

Annual changes of condition factor (CF) of *P. fulvidraco* were shown in Fig. 5. It peaked in February and October for both sexes. The CF of females ranged from 11.65–15.58 with a mean value of 13.29, in which April-August were obviously lower than those of the rest ($p<0.05$). The CF of males ranged from 11.79–15.35 with a mean value of 13.82 and reached nadir in July. Except in January and October, there was no significant difference of CF for males in the rest months, while CF of females fluctuated fiercely throughout the year.

All the oocytes were counted but only the more advanced ones (size>1.1 mm) were used for fecundity estimates. The fecundity was estimated from yolked oocytes in 100 ovaries macroscopically staged as IV and V from April to August. The absolute fecundity (AF) of *P. fulvidraco* ranged from 657 (total length, 94.3 mm) to 11, 696 (total length, 175.7 mm) eggs per fish, with mean AF value of 3,570±122 (S.D.) eggs per fish, varying considerably at given length, weight and age. The fecundity relative to total weight (F/W) varied from 29–133 eggs/g, with a mean value of 67±7 (SD) eggs/g.

All the correlation coefficients calculated between fecundity and each of these independent variables, were judged to be either moderately or high and statistically significant ($p<0.05$) indicative of a very good model for a fecundity study, considering additionally that fecundity was not completely depended on the size of the fish, but is affected by various other factors such as the aquatic environment and more importantly the condition of the fish (Hotosa et al. 2000). The relationships of AF and relative fecundity (F/W) with total weight (W) were presented in Fig. 6. There was a positive linear correlation between AF and relative fecundity (F/W) with total weight ($p<0.05$). The correlation can be best described as: $AF=108.0624W-1611.2737$, $R=0.8765$, $r^2=0.7683$, $n=74$; $F/W=3.082101W^{1.11342065}$, $R=0.8110$, $r^2=0.6578$, n=74, respectively. The relationships between fecundity and age (t) were presented in Fig. 7. There was positive linear correlation existed between absolute fecundity (F) and relative fecundity (F/W) with age ($p<0.05$), though relative fecundity (F/W) dropped when fish reached age 4. Their correlation can be best described as: $AF=2{,}282.2890\times t-769.0522$, $R=0.8510$, $r^2=0.7242$, $n=74$; $F/W=128.6138\times t+6.987002$, $R=0.8025$, $r^2=0.6440$, $n=74$, respectively.

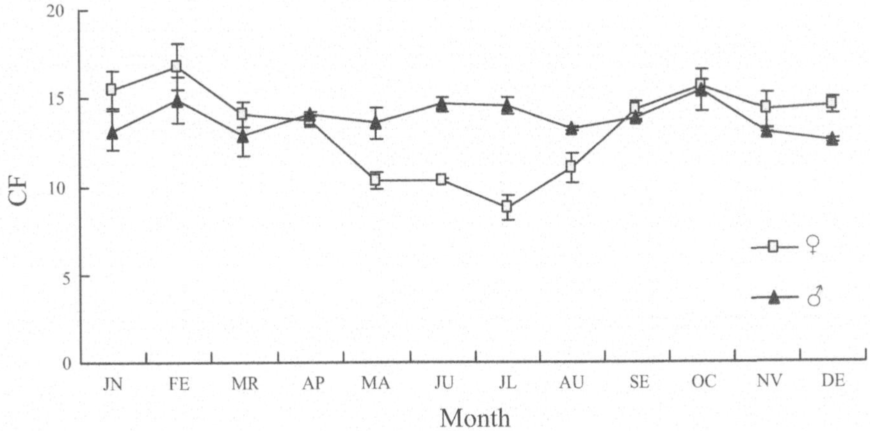

Fig. 5 Annual changes of condition factor (*CF*) for both sexes of *P. fulvidraco* in Ce Lake

Fig. 6 Relationships between fecundity and total weight of *P. fulvidraco* (*N*=74) in Ce Lake: **a** absolute fecundity (*AF*) and total weight (*W*); **b** relative fecundity (*F*/*W*) and total weight (*W*)

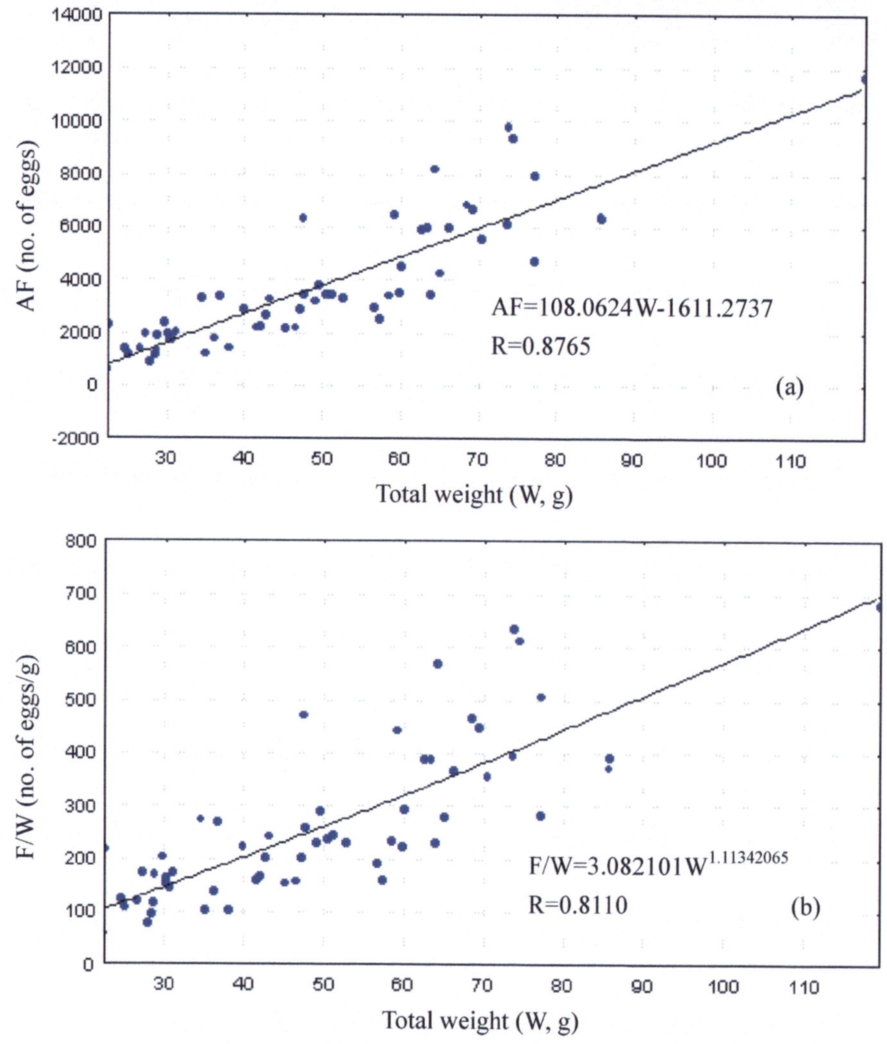

Discussion

Comparison on the reproductive biology of *P. fulvidraco* among different parts of their range during reproductive season was presented in Table 4. Differences do exist for the Ce Lake *P. fulvidraco* population relative to populations in the North. Variations in spawning period, size and age of spawners and differences in absolute fecundity provide biological evidence for specific spawning characteristics in the Ce Lake.

Sex differentiation is one of the most important indexes in fish biology study and also the preliminary step for artificial breeding. Both morphological differentiation and ultrasonography methods can be used to determine fish sex (Mattson 1991). Three types of sexual characteristics differentiation could be summarized in fishes: (1) body size of females > that of males; (2) body size of females < that of males; (3) similar size of both sexes with partial differences on morphology (Lin and Lei 2004). Type I is most common in favor of high production of females. The rest two types make males easier to chase females and give better protection to their offspring, thus enhancing their fertilization rates, hatching rates and survival rates. In the present study, based on our observation, *P. fulvidraco* belonged to type II, since males were always much bigger than females at the same age. Larger size of males than females, higher abdominal capability of females than males and significant differences on contrasting total weight and net body weight indicated that female *P. fulvidraco* utilized more energy for reproduction, while males distributed more energy for growth performance. Our observa-

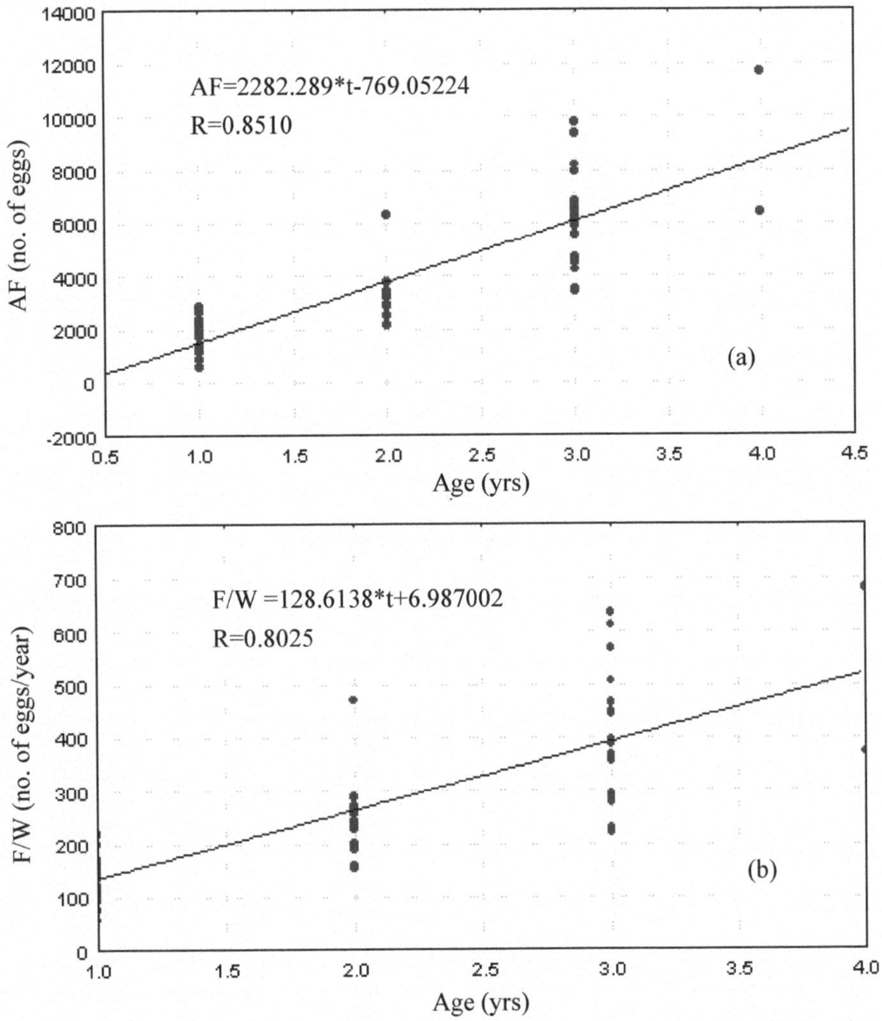

Fig. 7 Relationships between fecundity and age of *P. fulvidraco* (N=74) in Ce Lake: **a** absolute fecundity (*AF*) and age; **b** relative fecundity (*F*/*W*) and age

tions on the morphology of external sex organs such as genital in the male *P. fulvidraco* could be also used as an efficient way to differentiate sex, which were in good consistence with the conclusions from the earlier studies (Wang et al. 2002; Lin and Lei 2004). Moreover, the CPL/*L* ratios were all higher than 0.13 during the whole year in males, while the reverse were found in females. Compared with other bagrid fish (Wang et al. 2002; Xiao et al. 2002), this would be a useful standard for sex differentiation of *P. fulvidraco*.

Sex ratio of a certain population is one of the most important parameters for population ecological study. Though overall sex ratio was close to expected 1.0 ratio (p>0.05), there has significant difference between the sex ratios in the spawning season and non-spawning season. The sex ratio for the spawning group was 1.14:1 with skewness towards females (p< 0.05). During early spawning season (April–June), the number of female of *P. fulvidraco* was higher than that of male, while at the end of spawning season (July–August), males were more. In contrast, during non-spawning season, sex ratios were not bias significantly from 1:1. Sex ratio difference was also found in *P. fulvidraco* at different size. Males were predominant in larger size group (>160 mm), which was consistent with previous study (Lei and Lin 2004). This could be considered as one of the reproductive strategies of *P. fulvidraco*. During post-spawning period, since females already consumed lots of energy for spawning, more male fish would better safeguard the fertilized eggs and offspring (Wang et al. 2002). However, different sex ratios of spawning group were reported before in other lakes. In Honghu

Table 4 Comparison of spawning population demographics of *P. fulvidraco* in Ce Lake (present study) and throughout their geographic range in China, listed from south to north

Location	Sampling no.	Fish size[a] (mm)	Fish weight[b] (g)	Fish age[c] (year)	Spawning season	Absolute fecundity	Sex ratio (*F*/*M*)
Ce Lake (present study)	768	94.3	16.93	1^+, 2^+	Apr.~Aug.	657~11,696	1.14:1
Honghu Lake (Ma 1991)	221	78.0	11.90	1^+, 2^+	May~Aug.	794~4,873	0.87:1
Liangzi Lake (Du 1963)	376	–	–	–	May~Jul.	1,086~4,496	–
Zhangdu Lake (Luo et al. 2005)	370	90.0	10.00	2^+, 3^+	Apr.~Jun.	–	–
Dongting Lake (Xiao et al. 2003)	155	102.9	30.67	1^+, 2^+	Apr.~Aug.	1,345~7,208	–
Poyang Lake (Liu 1997)	587	80.0	13.90	0^+, 1^+, 2^+	Apr.~Jun.	1,134~12,412	1.23:1
Oujiang River (Lei and Lin 2004)	212	95.4	–	–	Apr.~Jun.	–	1.79:1
Taihu Lake (Zou and Tian 1998)	165	112.0	28.50	2^+, 3^+	May~Jul.	–	–
Tanghu Lake (Qiu et al. 2000)	3,241	126.0	28.00	1^+, 2^+, 3^+, 4^+	Jun.~Aug.	1,405~14,960	1.81:1

[a] Fish size at first sexual maturity

[b] Fish weight at first sexual maturity

[c] Fish age refers to the age of the majority of spawning fish in that site.

Lake, the sex ratio (*F*/*M*) was 0.87:1 with skewness towards males during reproductive season (Ma 1991). On the reverse, it was 1.23:1 with skewness towards females for Poyang Lake (Liu 1997). For Oujiang River and Tanghu Lake, it was 1.79:1 and 1.81:1, respectively, with extremely bias toward females (Qiu et al. 2000; Lei and Lin 2004). In general, the sex ratio of reproductive-aged fishes is known to vary with the age and size at which fish begin to participate in breeding, resistance to physical environmental stress in each sex, water temperature and/or pH during the sex determination period, population density, and other factors (Liu 1997).

The most common practice for the determination of a species spawning season is the establishment of its GSI, oocyte diameter frequency, as well as microscopic and histological examination of the gonads (Wang et al. 2002). The fact that, during January, February, September to December 2004, the mean GSI of female fish were less than 2% and the ovary of no fish exceeded stage III, demonstrates that *P. fulvidraco* undergoes little gonadal development in the early spring, autumn and winter. However, the marked increase that occurs in mean GSIs, oocyte diameters and the stage in ovarian development from April to August shows that the gonads mature rapidly between late spring and summer (April~August). The data on the monthly trends exhibited by the oocyte size-frequency and gonadal stages provide strong evidence that *P. fulvidraco* spawns between April and August, with spawning activity peaking from May to July, which was consistent with previous study in Donting Lake (Xiao et al. 2003). For Liangzi Lake, reproduction was found to occur from May to late July (Du 1963). The reproductive season of *P. fulvidraco* from Honghu Lake, occurred from May to August with a peak in June and July (Ma 1991).

To verify the spawning pattern of *P. fulvidraco* is of great importance for fecundity estimation. However, there was no agreement in former studies. Wang et al. (2002) reported it as a batch spawner and spawned twice during one spawning season per year. Liu and Zhang (2003) insisted that it is a single spawner. In the present study, we agree with the second viewpoint. Though there were two peaks of GSI for females and four for males, the ovaries returned to stage II after spawning and never developed till the new spawning cycle. Thus there was little chance for *P. fulvidraco* to return stage IV after first spawning and spawn the second time.

Analysis of age distribution of *P. fulvidraco* during reproductive season indicated that age 1^+ and age 2^+ fish dominate reproduction in Ce Lake, which was in accordance with the results in Honghu Lake and Dongting Lake (Ma 1991; Xiao et al. 2003). This may be related to over-exploitation which leads to the degradation of natural resources. However, in Zhangdu Lake and Taihu Lake, the majority of spawning fish are age 2^+ and age 3^+ (Zou and Tian 1998; Luo et al. 2005). Surprisingly, for Poyang Lake, fish at age 0^+ is one of the major populations among spawning fish (Liu 1997). The age structure in

Tanghu Lake was random and no majority existed (Qiu et al. 2000). First maturation size and weight of *P. fulvidraco* in Ce Lake was lower compared to Dongting Lake, Taihu Lake and Tanghu Lake, which further proved that the natural stock of this species was under pressure and rational management for protection should be proposed.

The fecundity estimation showed that *P. fulvidraco* had a reproductive strategy common to many other bagrid catfish (Qiu et al. 2000; Wang et al. 2002). It had a low absolute fecundity ranged from 657 to 11,696. According to previous studies, absolute fecundity of *P. fulvidraco* in Honghu Lake, Poyang Lake, Liangzi Lake and Tanghu Lake ranged 794~4,873, 1,132~12,412, 1,086~4,496, and 1,405~14,960, respectively (Du 1963; Ma 1991; Liu 1997; Qiu et al. 2000). Among five geographical groups, absolute fecundity of *P. fulvidraco* in Ce Lake was close to that of Poyang Lake and Tanghu Lake, and higher than the rest two groups. In response to environmental differences, the ability of changing its reproductive biology is usually thought of as an adaptive strategy to differing environments. These changes may have effects beyond the individuals, and may alter ecological processes in the environment by changing many direct and indirect ecological interactions. Local adaptation is a process that also may produce differences in reproduction. Life history parameters, reproductive traits in particular, are known to vary across environmental gradients in some fishes (Mazzoni and Iglesias-Rios 2002). The proposed reasons for these findings include changes in stream hydrology, sediment input and temperature. Relative fecundity (29–133 eggs·g^{-1}) of *P. fulvidraco* in Ce Lake was close to other bagridea catfish (Yang 1994; Wang et al. 2002). Though absolute fecundity is low, with high quality eggs, high fertilization rates and careful protection for offspring, *P. fulvidraco* gain high survival rates of their juveniles (Wang et al. 2002). This ecological strategy proved that under the pressure of over fishing, *P. fulvidraco* could still survive as the predominant species, which was consistent with the study in Honghu Lake (Ma 1991).

Management of many exploited fish populations is based on spawner-recruitment models. Traditional recruitment models assume that the reproductive potential of a population is proportional to its spawning stock biomass (Murua and Motos 2006). This implies that the survival rates of offspring are independent of parental age, body size or condition (Cardinale and Arrhenius 2000), and that total relative fecundity and thus annual egg production per unit mass of adult stock are invariable over time. However, there is increasing evidence indicating that direct proportionality between spawning stock biomass and reproductive potential may not exist. This could even be more evident when the production of eggs per unit stock spawning biomass may vary substantially among years depending upon environmental conditions (e.g. temperature and food availability) during the spawning season (Murua and Motos 2006). Based on present study, four protection recommendations for *P. fulvidraco* are proposed accordingly. First, larger mesh size (>3 cm) of seine nets should be used to replace the old smaller ones during common harvesting. Second, harvesting time should avoid the peak reproductive season (May to July) of *P. fulvidraco* and harvest fish size larger than 15 cm with a minimum age of 2^+. Third, artificial breeding, larvae releasing to natural water bodies and artificial culture of *P. fulvidraco* should be conducted to reinforce the population and lessen the pressure of natural stocks (Xiao et al. 2002). Finally, sewage and industrial wastewater discharging to these lakes should be controlled and monitored and better waste treatment methods should be investigated (Xiao et al. 2003).

Acknowledgement We would like to thank the four master students from Huazhong Agricultural University for help in collection samples and providing data. Thanks are further extended to Zhiguo Zhang for help in the processing of samples. Special thanks go to the workers of Ce Lake aquafarm, without whose generosity and cooperation, the entire work would not have been possible. This study is funded by "National Key Technology R&D Program in Eleventh-Five-Year Plan" titled as "High quality and high efficiency freshwater aquaculture technology research program". The accession number is 2006BAD03B07. And "The integrated demonstration of safe fishery production". The accession number is 2006BAK02A22.

References

Anderson RO, Gutreuter SJ (1983) Length, weight, and associated structural indices. In: Nielsen LA, Johnson DL (eds) Fisheries techniques. American Fisheries Society, Bethesda, pp 283–300

Bagenal TB (1978) Aspects of fish fecundity. In: Gerking SD (ed) Ecology of fresh water fish production. Blackwell, London, pp 75–101

Cardinale M, Arrhenius F (2000) The relationship between stock and recruitment: are the assumptions valid? Mar Ecol Prog Ser 196:305–309

Du JR (1963) Studies on the reproduction and feeding habit of Bagrid Catfish in Liangzi Lake. J Zool Res 2:74–77, (in Chinese)

Heins DC, Baker JA (1993) Clutch production in the darter *Etheostoma lynceum* and its implications for life-history study. J Fish Biol 42:819–829

Hotosa GN, Avramidoua D, Ondrias I (2000) Reproduction biology of *Liza aurata* (Risso, 1810), (Pisces Mugilidae) in the lagoon of Klisova (Messolonghi, W. Greece). Fish Res 47:57–67

Hu CY, Hong YJ, Lin GH (1993) Gonad study of *Pelteobagrus fulvidraco* in spawning season. Jiangxi Science 11:159–163, (in Chinese)

Hunter JR, Lo NCH, Leong R, Macewicz B (1985) Batch fecundity in multiple spawning fishes. In an egg production method for estimating spawning biomass of pelagic fish: application to the northern anchovy, *Engraulis mordax* (Lasker R ed.). NOAA Tech Rep NMFS 36:67–78

Hunter JR, Macewicz BJ, Lo NCH, Kimbrell CA (1992) Fecundity, spawning, and maturity of female Dover sole, *Microstomus pacificus*, with an evaluation of assumptions and precision. Fish Bull 90:101–128

Johnston CE, Knight CL (1999) Life-history traits of the bluenose shiner (*Pteronotropis welaka*) (Cypriniformes: Cyprinidae). Copeia 1999:200–205

Lei HZ, Lin ZH (2004) The sexual ratio of *Pseudobagrus fulvidraco* population in Oujiang River during reproductive time. Journal of Wenzhou Normal College 2:77–79, (in Chinese with English abstract)

Lin ZH, Lei HZ (2004) Sexual dimorphism and female reproductive characteristics of *Pseudobagrus fulvidraco*. Chinese Journal of Zoology 39:13–17, (in Chinese with English abstract)

Liu SP (1997) A study on the biology of *Pseudobagrus fulvidraco* in Poyang Lake. Chinese Journal of Zoology 32:10–16, (in Chinese with English abstract)

Liu WB, Zhang XJ (2003) Study on the development and annual change in the ovary of *Pseudobagrus fulvidraco*. Journal of Natural Science in Hunan Normal University 26:73–78, (in Chinese with English abstract)

Luo HB, Li LY, Yao WZ, Jiang YM (2005) Biological characters and resources management of *Pseudobagrus fulvidraco* in Zhangdu Lake. Freshwater Fisheries 5:25–27, (in Chinese with English abstract)

Ma J (1991) Research on biology of *Pelteobagrus fulvidraco* in Honghu Lake. Bio-productivity exploitation and environment optimizing investigation of Honghu Lake. Ocean Press, Beijing, pp 150–160

Maack G, George MR (1999) Contributions to the reproductive biology of *encrasicholina punctifer* Fowler, 1938 (engraulidae) from West Sumatra, Indonesia. Fish Res 44: 113–120

Mattson NS (1991) A new method to determination sex and gonad size in live fishes by using ultrasonography. J Fish Biol 39:673–677

Mazzoni R, Iglesias-Rios R (2002) Environmentally Related Life History Variations in *Geophagus brasiliensis*. J Fish Biol 61:1606–1618

Morgan DL, Gill HS, Potter IC (2000) Age composition, growth and reproductive biology of the salamanderfish *Lepidogalaxias salamandroides*: a re-examination. Environ Biol Fisches 57:191–204

Murua H, Motos L (2006) Reproductive strategy and spawning activity of the European hake *Merluccius merluccius* (L.) in the Bay of Biscay. J Fish Biol 69:1288–1303

Murua H, Kraus G, Saborido-Rey F, Witthames PR, Thorsen A, Junquera S (2003) Procedures to estimate fecundity of marine fish species in relation to their reproductive strategy. J Northwest Atl Fish Sci 33:33–54

Murua H, Lucio P, Santurtun M, Motos L (2006) Seasonal variation in egg production and batch fecundity of European hake *Merluccius merluccius* (L.) in the Bay of Biscay. J Fish Biol 69:1304–1316

Pearce A (1965) A manual for histological technicians, 2nd edn. Little, Brown and Co., Boston, MA

Qiu CG, Liu JZ, Liu BY, Xu SQ, Zhang GQ, Luo XN (2000) Biology and resource management of *Pelteobagrus fulvidraco* in Tanghu Lake. Fish Sci 2:28–30, (in Chinese)

Wang WM (1999) Artificial breeding trial of *Pseudobagrus fulvidraco*. Fish Sci 18:9–12, (in Chinese)

Wang WM, Yan AS, Zha JM (2002) Comparison of two artificial breeding methods of *Pseudobagrus fulvidraco*. Freshwater Fisheries 32:7–8, (in Chinese with English abstract)

Xiao XL, Ou YM, Zhang M, Tao QH, Wei HM, Lv L (2002) Study on some biological characters of *Pelteobagrus fulvidraco* in Poyang Lake. Acta Agriculturae Jiangxi 14:18–22, (in Chinese with English abstract)

Xiao TY, Zhang HY, Wang XQ, Xiao KY, Dai KY (2003) Biological characters of *Pseudobagrus fulvidraco* in Dongting Lake. Chinese Journal of Zoology 38:83–88, (in Chinese with English abstract)

Yang JY (1994) Reproductive biology of *Pelteobagrus fulvidraco* in Jialing River. Journal of Southwest Normal University (Science) 19:639–645, (in Chinese with English abstract)

Zou SX, Tian BP (1998) Study on biology and fishery of *Pelteobagrus fulvidraco* in Taihu Lake. Journal of Sichuan Institute of Animal Husbandry and Veterinary Medicine 12:36–41, (in Chinese with English abstract)

Environ Biol Fish (2009) 86:89–96
DOI 10.1007/s10641-008-9363-5

The reproductive traits of brown trout (*Salmo trutta fario* L.) from the Yadong River, Tibet

FuHua Hao · YiFeng Chen

Received: 10 August 2007 / Accepted: 5 May 2008 / Published online: 21 June 2008

Abstract Some key aspects of the reproductive strategy of the brown trout (*Salmo trutta fario* L.) in the Yadong River, Tibet, including spawning season, age at sexual maturity, fecundity and egg size, have been studied. The majority of the samples were less than 215 mm and age ranged from 1 to 4 in both sexes, indicating that the majority of the fish were younger and the pressure by overfishing was high. The spawning periodicity was determined to be between the end of October and January, mainly in November and December. The ratio of male to female brown trout population (1.29:1 with $P>0.05$) suggested no sex significant differences, although males were significantly more abundant than females in October ($P<0.0001$) on monthly basis. Age and size of males and females at maturity was different and males matured earlier than females. Fecundity was markedly correlated with their body weight ($P<0.001$, $r=0.9255$), standard length ($P<0.01$, $r=0.8879$), and gonad weight ($P<0.001$, $r=0.9366$). The mean size of mature eggs in the spawning season was: 4.0±0.45 mm and tended to increase along with the female spawners size ($P<0.001$, $r=0.9641$). Further researches about the brown trout population in the Yadong River should be conducted on issues such as artificial reproduction, culture, conservation, management, and restocking.

Keywords *Salmo trutta fario* L. · Reproduction · Tibet

F. Hao · Y. Chen (✉)
Institute of Hydrobiology, Chinese Academy of Sciences,
No. 7 Donghu South Road, Wuchang,
Wuhan 430072, People's Republic of China
e-mail: chenyf@ihb.ac.cn

F. Hao
Graduate School of Chinese Academy of Sciences,
100039 Beijing, People's Republic of China

Introduction

The brown trout (*Salmo trutta*), family Salmonidae, is indigenous to Europe, North Africa, and western Asia (MacCrimmon et al. 1970), but it is now found worldwide (Klemetsen et al. 2003). The reproductive traits of brown trout have been described in geographically different populations previously (Hegge et al. 1991; Crisp 1994; Sorensen et al. 1995; Garcia-Vazquez et al. 2001; Nicola and Almodovar 2002; Pender and Kwak 2002; Alp et al. 2003; Estay et al. 2004; Olsen and Vollestad 2005; Rubin et al. 2005; Arslan and Aras 2007).

The brown trout (*Salmo trutta fario* L.) was first reported from the Yadong River, Tibet, in 1962 (Zhang and Wang 1962) although it was probably introduced into the river more than a hundred years ago. The population of this brown trout has decreased because of water pollution, overexploitation and damming for waterpower. Studies of such trout from the Yadong River were limited to their morphology (Zhang and Wang 1962) and our recent study of their embryonic development (Hao et al. 2006). Given the local economic importance of this fish (the most important economic source of local fishery and travel industry), the reproductive biology of the brown trout from the Yadong River, Tibet, has not hitherto been studied thoroughly.

The major objective of this study is to reveal the reproductive traits of the wild brown trout population in the Yadong River, Tibet, by investigating spawning season, age and size at sexual maturity, fecundity and egg size. We believe that such studies are deemed vitally important for the management of the local fishery and environmental protection.

Methods and materials

Sampling locations for brown trout were the mid-section of the Yadong River (Fig. 1), that originates from the south slope of Himalayas and flows into Brahmaputra River, India. The river length is over 60 km. The width and depth varied from 5 to 16 m and from 20 cm to 2.5 m, respectively. It runs throughout a sharp and narrow valley. There is bushy vegetation in both sides. The river bed is covered by boulders and cobbles. Water temperature fluctuated between 0°C (in winter) and 15°C (in summer). The water flow was between 1 and 12 m^3 s^{-1} during the study period. The sampling region has altitudes from around 2,700 to 3,700 m above sea level. No fishes were captured above altitude 3,700 m.

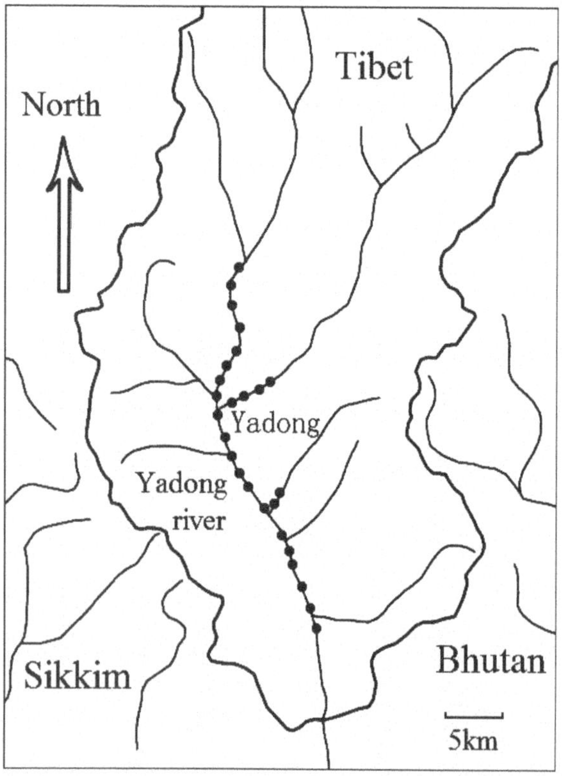

Fig. 1 Schematic map of the Yadong River. *Black dots* indicate the sampling section

We gathered data to investigate size and age at maturity for both male and female, spawning periodicity, sex ratios and fecundity. Brown trout samples were collected by using casting net (1 cm mesh size) for about 20 km along the riversides of the Yadong River from August 1999 to January 2003. Four hundred sixty-two individuals were collected. They were measured for total length (TL) and standard length (SL), with an accuracy of 0.5 mm (putting fish into a box and measuring the length with steel tape), body weight (Wb) with an accuracy of 0.5 g (weighing fish with an electronic balance). Three hundred two individuals were sacrificed and sex was determined visually or by microscopic examination of the gonads. After the gonads were weighted (Wg) with an accuracy of 0.1 g (weighing sample with an electronic balance), they were preserved in 95% ethanol and Bouin's solution, respectively. The other 160 individuals were released after measuring the TL, SL and Wb. The developmental status of the gonads was investigated in six different stages, in which stage I was not observed in the current study; stage II and III were defined as immature, consisting of very young individuals; stage IV and V were considered as mature and stage VI was considered as post-mature (Yin 1995). In the laboratory, total egg number and average egg size (egg diameter, D) were determined for each female from a random sample of 20 eggs to nearest 0.1 mm with a digital micrometer for three times. Scales were used for estimating age (Rifflart et al. 2006). All specimens were preserved in 10% formalin and kept in the Museum of Freshwater Fishes, Institute of Hydrobiology, Chinese Academy of Sciences (IHB, CAS).

Spawning season was estimated from the gonad-somatic index (GSI) and gonad development status.

Fecundity also was quantified as the GSI. The GSI was calculated from the equation: GSI=(Wg/Wb)× 100, where Wg is the gonad weight (gram) and Wb is the body weight (gram). Body weight, standard length, gonad weight and fecundity relationships were determined from the equations according to Alp et al. (2003):

$$\ln F = p + q \times \ln \mathrm{Wb}, F = \mathrm{pl} \times \mathrm{Wb}^q;$$
$$\ln F = p + q \times \ln \mathrm{SL}, F = \mathrm{pl} \times \mathrm{SL}^q;$$
$$\ln F = p + q \times \ln \mathrm{Wg}, F = \mathrm{pl} \times \mathrm{Wb}^q;$$

Where F is the number of eggs (fecundity), SL, Wb, and Wg are the standard length (millimeter), body weight (gram) and gonad weight (gram), respectively, and p and q are constant parameters in linear regression analysis and pl=e^p.

The data analysis was conducted using Statistic 6.0 and Excel. The number of fish was given for each test unless the whole data set was used. The chi-squared test was also performed to compare the sex ratio. One-way ANOVA and Duncan's multiple comparison test were used to determine the difference of GSI. An independent t test was used to compare fecundities of different populations.

Results

Of the 462 individuals examined, 170 were males, 132 were females and 160 were not determined before they were released into the river. The standard length frequency distributions of these individuals were summarized in Fig. 2. Males ranged from 103.0 to 364.0 mm, with a mean standard length of 188.91±52.259 mm and females ranged from 93.0 to 362.0 mm, with a mean standard length of 192.60± 68.581 mm. The majority of the samples (accounting for 68.40%) were comprised of individuals between 120.0–215.0 mm in length group (Fig. 2). Age ranged between 1 and 4 years old in males and females. The group of age 2 was dominant in both sexes (71.2% and 54.5% in male and in female, respectively).

Seasonal distribution of the maturity stages is shown in Table 1. Immature individuals (stage II and III) were observed in most months. Females with maturing ovaries (stage IV) appeared from June to December whereas males with maturing testis (stage IV) were seen from June to October. Fewer female fish were observed with mature ovaries (stage V) in October–December compared with male with mature testis (stage V) observed at the same seasons. The

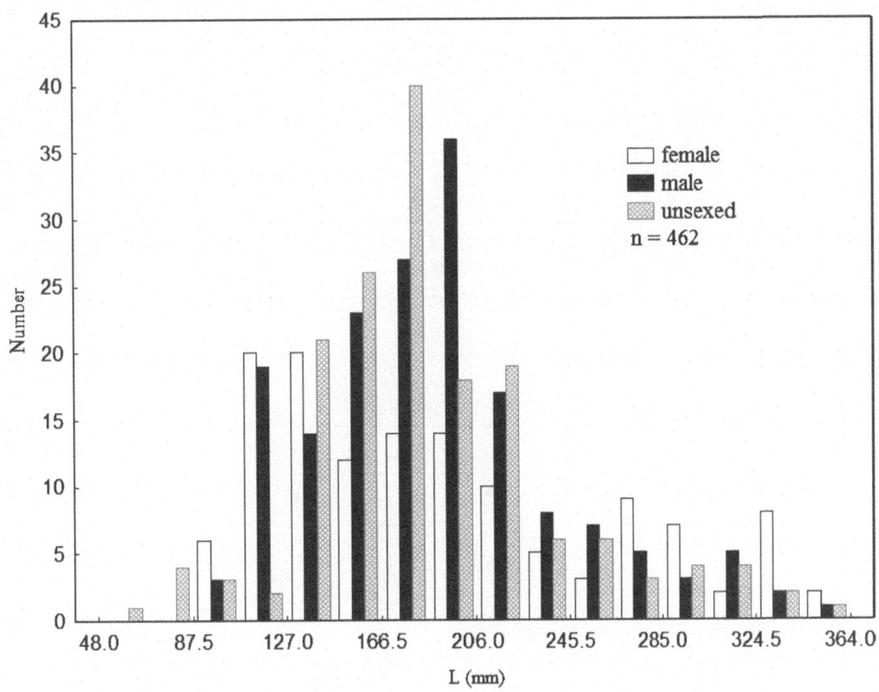

Fig. 2 Standard length frequency histogram of brown trout from the Yadong River between 1999 and 2003 (n=462)

Table 1 The monthly frequency of maturity stages in female (n=132) and male (n=170) of brown trout from the Yadong River

Months	Maturity stages of the females						Maturity stages of the males					
	II	III	IV	V	VI	Mat	II	III	IV	V	VI	Mat
Apr	5	3					13	2				
May	7	7					3	6				
Jun	8	5	1			1	2	5	2			2
Jul	9	5	6			6	11	11	4			4
Aug	8	19	5			5	5	23	1			1
Sep	3	5	2			2	3	9	5			5
Oct		4	2	1		3			1	30		31
Nov		1	2	11	1	14				18		18
Dec			2	8	3	13				10	6	16
Total	40	48	20	20	4	44	37	56	13	58	6	77

Mat Number of mature individuals

monthly distribution of the maturity stages confirmed that brown trout began to mature during the autumn (only one sample in October) and spawned mainly between November and December, although some individuals were observed to spawn in January (data not shown).

For the 302 specimens examined for sex determination, 33.3% females and 45.3% males were sexually mature. For females ages at first maturity were 3 to 4 years old, whereas for males ages at first maturity were about 2 to 3 years old. The youngest female mature brown trout was 3 years old, 242 mm standard length, in the Yadong River, whereas the youngest male mature fish was 2 years old, 168 mm standard length (Table 2).

In the current study, the sex ratio (males/females) of the Yadong brown trout population was 1.29:1 (chi-squared test, P>0.05) and this is not significantly different from 1:1 ratio. The monthly changes of sex ratios were illustrated in Fig. 3. The sex ratios ranged from 1.2:1 to 4.43:1 in spawning seasons. There were no significant differences in sex ratios on monthly basis, with an exception in October (4.43:1, chi-squared test, P<0.0001).

GSI values between October and December were significantly higher (P<0.05) than other months (Fig. 4). The GSI in female brown trout ranged from 0.89 to 17.66. The maximum GSI value in females was observed in December, whereas the minimum value was in April. These results showed that sexual development accelerated in autumn (September, October and November) and reached the maximum in December.

Fecundity varied from 312 to 1,655 with a mean of 946.35±363.07 based on 23 females. The size of eggs from 23 spawners ranged from 3.3 to 4.5 mm with a mean 4.0±0.45 mm. The size of mature eggs trended to increase along with the size of the female spawners (Fig. 5). Some significant correlations were found between fecundity and body weight (P<0.001),

Table 2 The age and size of both sexes brown trout at maturity

Age	Female						Male					
	N	Imm	Mat	% Mat	Min	Max	*N*	Imm	Mat	% Mat	Min	Max
II	72	72					121	89	32	26.4	168	227
III	33	4	29	88	242	313	37		37	100	174	315
IV	15		15	100	245	362	8		8	100	270	364

N Number of the fish, *Imm* number of immature individuals, *Mat* number of mature individuals, *% Mat* the percentage of mature individuals, *Min* minimum value of standard length at maturity (millimeter), *Max* maximum value of standard length at maturity (millimeter)

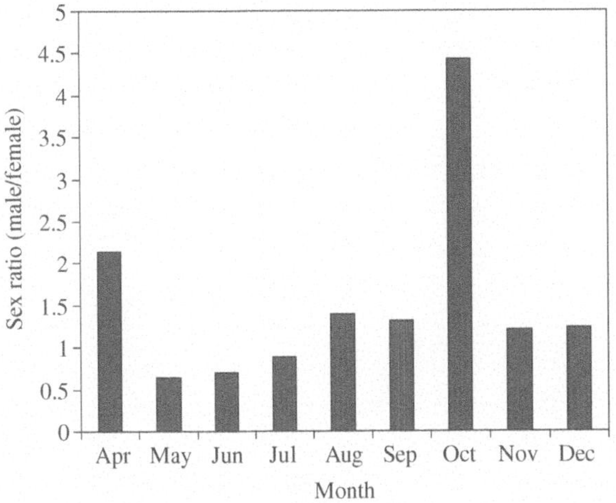

Fig. 3 Monthly sex ratios of brown trout from the Yadong River (n=302)

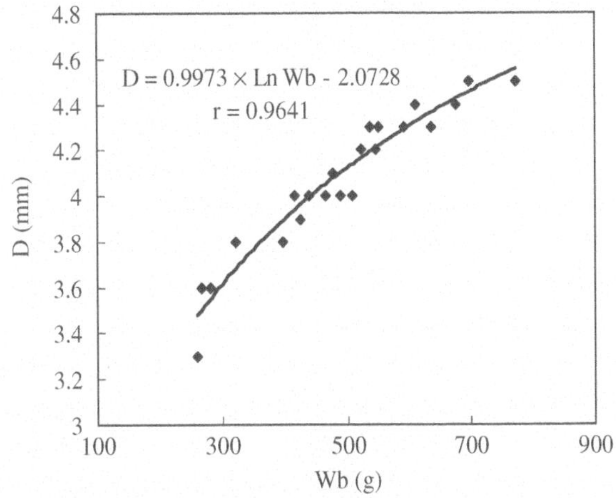

Fig. 5 Relationships between egg size and body weight of brown trout spawners from the Yadong River

standard length (P<0.01) and gonad weight (P< 0.001) (Fig. 6). These linear relationships may be expressed by the following regression equations:

$$\ln F = 1.3345 \times \ln \mathrm{Wb} - 1.4596,$$
$$F = 0.2323 \times \mathrm{Wb}^{1.3345} (r = 0.9255, n = 23)$$
$$\ln F = 3.7972 \times \ln L - 14.954,$$
$$F = 3E - 07 \times L^{3.7972} (r = 0.8879, n = 23)$$
$$\ln F = 1.8485 \times \ln \mathrm{Wg} + 1.1006,$$
$$F = 6.3502 \times \mathrm{Wg}^{1.1006} (r = 0.9366, n = 23)$$

Discussion

In the Yadong River, the standard length of the majority of the samples were less than 215 mm and age ranged from 1 to 4 years old in both sexes, indicating that the majority of the fish were younger and the pressure of overfishing was high. A similar situation has been cited for overfished brown trout populations elsewhere (Almodovar and Nicola 1999; Almodovar et al. 2002; Arslan and Aras 2007).

In the present study, age and size at maturity was different between the males and females in the Yadong River and males matured earlier than females, since females have higher energetic to mature (Euzenat et al. 1999). Similarly, there was considerable variation in age and size at maturity in brown trout among different populations in previous studies (Lobon-Cervia et al. 1986; Hegge et al. 1991; Olsen and Vollestad 2005). According to Olsen and Vollestad (2005), age and size at maturity varied between the sexes and males usually matured earlier than females, which was also agreeable with the early studies (Johnson 1989). There was evidence showing that within an area that a correlation between size and age was a good predictor of maturation, but for a given age and size a salmon parr was more likely to become sexually mature if it came from a high-altitude site (Baum et al. 2004). Some studies showed males may attain maturity from less than 100 mm in length at age 1 year old (Dellefors and Faremo 1988; L'Abée-Lund et al. 1990). Alp et al. (2003) reported that age of sexual maturity of brown trout living in Fırnız Stream was between 2 and 3 years old, the smallest mature male and female fishes were 174 and 178 mm, respectively. Recently, Arslan and Aras (2007) described that male and female brown trout reached

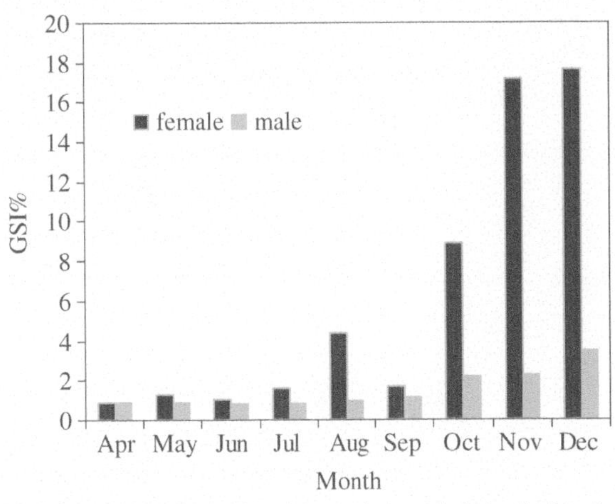

Fig. 4 Seasonal fluctuations in gonad–somatic index in female and male brown trout from the Yadong River

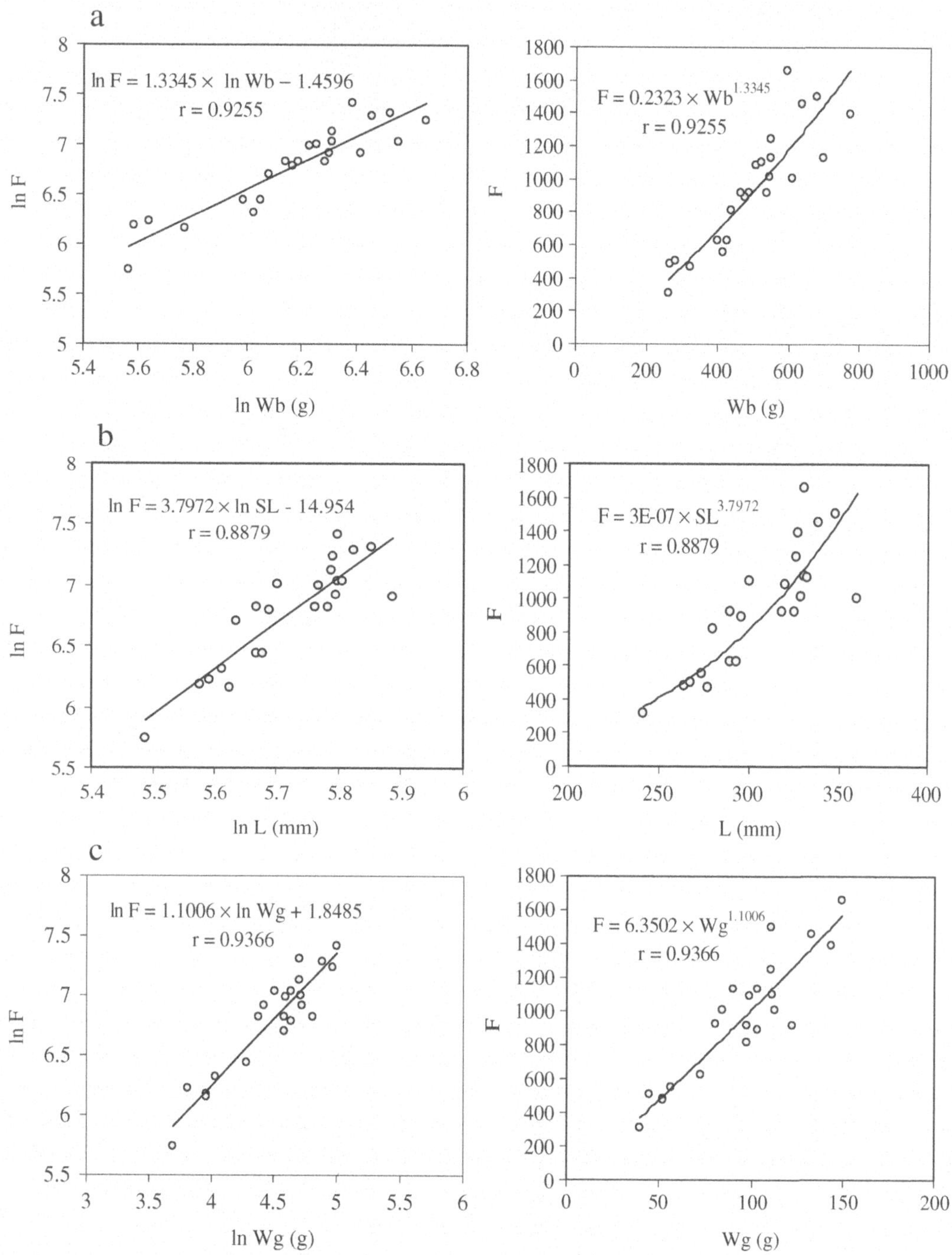

Fig. 6 Relationships between fecundity and body weight (**a**), standard length (**b**) and gonad weight (**c**) of female brown trout from the Yadong River

maturity at the age of 1.99 and 3.19, and when they were 141 and 172 mm in the Anuri Stream while those in the Cenker Stream attained at age of 1.99 and 3.22 when they were 140 and 173 mm length, respectively.

The sex ratio males to females of Yadong River brown trout population was 1.29:1 ($P>0.05$), suggesting no significant difference. In general, the sex ratio is expected to be 1:1 in closed populations(Nikolsky 1963), though some factors (such as food availability, spawning season, spawning ground and so on) could influence sex ratio (Alp et al. 2005). Arslan and Aras (2007) also reported that the sex ratios for brown trout were 0.94:1 and 0.97:1 in the Anuri and the Cenker streams, respectively, indicating a significant difference in the numbers of males and females. However, males were significantly more numerous than females in October ($P<0.0001$) on monthly basis in our study. For the case, the limited number of samples was insufficient to draw a reliable conclusion.

Our data displayed that the spawning periodicity was between the end of October and January (autumn–winter period), mainly concentrated in November and December, being consistent with the general pattern described. As has been documented in the literature on this subject, the spawning time in brown trout occurs during the autumn–winter period (Hobbs 1937; Needham et al. 1945; Horton 1961; Thomas 1964; Hopkins 1970). Some more recent reports also indicate similar spawning periodicity (Pender and Kwak 2002; Alp et al. 2003; Estay et al. 2004; Rubin et al. 2005; Arslan and Aras 2007).

This study observed positive correlations between female length, body weight, gonad weight with fecundity in Yadong brown trout populations. Similarly, high correlation was also found between female size and salmonid fecundity in relation to fish length in an earlier study (Taube 1976). Nicola and Almodovar (2002) also described a significant relationship between female length and fecundity in brown trout populations in seven study streams in Spain. According to Olsen and Vollestad (2003), fecundity increased with fish length and there was no consistent difference in fecundity, based on comparing all eight populations. Estay et al. (2004) also reported that analyses of female body weight during spawning and total fecundity over three seasons revealed a positive linear correlation, that is heavier females produced more eggs. However, the fecundity for Yadong brown trout was different (higher or lower) from other reports (Nicola and Almodovar 2002; Alp et al. 2003; Olsen and Vollestad 2003; Estay et al. 2004; Arslan and Aras 2007), possibly caused by different factors like altitude, water temperature, age, and food availability.

In addition, this study showed the size of mature eggs tended to increase along with the female spawners size. Variation in egg mass and fecundity among the populations studied may result largely from selectively different environmental factors, of which water temperature was considered the most probable selective factor responsible for this variation (Jonsson and Jonsson 1999). However, egg size was strongly influenced by environmental factors, and individual fish produces eggs of different size at different reproductive conditions (Kamler 1992; Chambers and Waiwood 1996). Variation in egg size among salmonid populations may represent local adaptations (Fleming and Grossm 1990). There were some evidences that such variation was heritable (Gall 1975; Gjerde 1986), but egg size was also a plastic trait and it can be influenced by maternal feeding conditions (Jonsson and Jonsson 1999) and maternal size (L'Abée-Lund and Hindar 1990; Heath et al. 1999).

In conclusion, the majority of fish from the Yodong River were younger and age classes were simple, suggesting that this population is threatened, so some protective measures should be taken to prevent capturing the fish in spawning period (from October to January). Further studies should be conducted on artificial reproduction, culture, conservation, management, and restocking to improve our understanding of how to protect, restore and enhance brown trout population in that river.

Acknowledgements This study was supported financially by the National Natural Science Foundation of China (Grant No. 30123004), the Knowledge Innovation Program of the Chinese Academy of Sciences (Grant No. KSCX2-SW-125) and Tibet Autonomous Region Program. Huiru Tang and Fuqiang Xu of Wuhan Institute of Physics and Mathematics are acknowledged for his help in critical reading of the manuscript. We also acknowledge helpful comments and suggestions from referees and editors which enable the manuscript to be improved substantially.

References

Almodovar A, Nicola GG (1999) Effects of a small hydropower station upon brown trout *Salmo trutta* L. in the River Hoz Seca (Tagus basin, Spain) one year after regulation. Regul Rivers Res Mgmt 15:477–484

Almodovar A, Nicola GG, Suarez J (2002) Effects of fishery management on populations of brown trout *Salmo trutta*, in Central Spain. In: Collares-Pereira M, Coelho M, Cowx I (eds) Conservation of freshwater fishes: options for the future. Fishing News Books, Oxford, pp 337–345

Alp A, Kara C, Buyukcapar HM (2003) Reproductive biology of brown trout, *Salmo trutta macrostigma* Dumeril 1858, in a tributary of the Ceyhan River which flows into the eastern Mediterranean Sea. J Appl Ichthyol 19:346–351

Alp A, Kara C, Buyukcapar HM (2005) Age, growth and diet composition of the resident brown trout, *Salmo trutta macrostigma* Dumeril 1858, in Flrnlz Stream of the River Ceyhan, Turkey. Turk J Vet Anim Sci 29:285–295

Arslan M, Aras NM (2007) Structure and reproductive characteristics of two brown trout (*Salmo trutta*) populations in the Coruh river basin, north-eastern Anatolia, Turkey. Turk J Zool 31:185–192

Baum D, Laughton R, Armstrong JD, Metcalfe NB (2004) Altitudinal variation in the relationship between growth and maturation rate in salmon parr. J Anim Ecol 73:253–260

Chambers RC, Waiwood KG (1996) Maternal and seasonal differences in egg sizes and spawning characteristics of captive Atlantic cod, *Gadus morhua*. Can J Fish Aqua Sci 53:1986–2003

Crisp DT (1994) Reproductive investment of female brown trout, *Salmo trutta* L., in a stream and reservoir system in northern England. J Fish Biol 44:343–349

Dellefors C, Faremo U (1988) Early sexual maturation in males of wild sea trout, *Salmo trutt*a L., inhibits smoltification. J Fish Biol 33:741–749

Estay FJ, Noriega R, Ureta JP, Martin W, Colihueque N (2004) Reproductive performance of cultured brown trout (*Salmo trutta* L.) in Chile. Aqua Res 35:447–452

Euzenat G, Fournel F, Richard A (1999) Sea trout (*Salmo trutta* L.) in Normandy and Picardy. In: Baglinière JL, Maisse G (eds) Biology and ecology of the brown trout and sea trout. Springer-Praxis Series in Aquaculture and Fisheries, Chichester, UK, pp 175–205

Fleming IA, Gross MR (1990) Latitudinal clines: a trade-off between egg number and size in Pacific salmon. Ecology 71:1–11

Gall GAE (1975) Genetics of reproduction in domesticated rainbow trout. J Anim Sci 40:19–28

Garcia-Vazquez E, Moran P, Martinez JL, Perez J, De Gaudemar B, Beall E (2001) Alternative mating strategies in Atlantic salmon and brown trout. J Heredity 92:146–149

Gjerde B (1986) Growth and reproduction in fish and shellfish. Aquaculture 57:37–55

Hao FH, Chen YF, Cai B (2006) Embryonic development of *Salmo trutta fario* from Yadong River, Tibet. J Fish China 30(3):289–296

Heath DD, Fox CW, Heath JW (1999) Maternal effect on offspring size: variation through early development of chinook salmon. Evolution 53:1605–1611

Hegge O, Dervo BK, Skurdal J (1991) Age and size at sexual maturity of heavily exploited Arctic char and brown trout in Lake Atnsjo, southeastern Norway. Trans Am Fish Soc 120(2):141–149

Hobbs DF (1937) Natural reproduction of quinat salmon, brown and rainbow trout in certain New Zealand waters. Fish Bull New Zealand 6:1–104

Hopkins CL (1970) Some aspects of the bionomic fish in the brown trout nursery stream. Fish Res Bull New Zealand 4:1–38

Horton PA (1961) The bionomics of brown trout in a Dartmoor stream. J Anim Ecol 30:311–338

Johnson L (1989) The anadromous Arctic charr, *Salvelinus alpinus* of Nauyuk Lake, N.W.T., Canada. Physiol Ecol Japan 1(sp 1):201–228

Jonsson N, Jonsson B (1999) Trade-off between egg mass and egg number in brown trout. J Fish Biol 55:767–783

Kamler E (1992) Early life history of fish. An energetics approach. Chapman and Hall, London

Klemetsen A, Amundsen PA, Dempson JB, Jonsson B, Jonsson N, O'Connell MF, Mortensen E (2003) Atlantic salmon *Salmo salar* L., brown trout *Salmo trutta* L. and Arctic charr *Salvelinus alpinus* (L.): a review of aspects of their life histories. Ecol Freshw Fish 12:1–59

L'Abée-Lund JH, Hindar K (1990) Interpopulation variation in reproductive traits of anadromous female brown trout, *Salmo trutta* L. J Fish Biol 37:755–763

L'Abée-Lund JH, Jensen AJ, Johnsen BO (1990) Interpopulation variation in male parr maturation of anadromous brown trout (*Salmo trutta*) in Norway. Can J Zool 68:1983–1987

Lobon-Cervia J, Montanes C, De Sostoa A (1986) Reproductive ecology and growth of a population of brown trout (*Salmo trutta* L.) in an aquifer-fed stream of Old Castile (Spain). Hydrobiologia 135(1–2):81–94

MacCrimmon HR, Marshall TL, Gotos BL (1970) World distribution of brown trout, *Salmo trutta*: further observations. J Fish Res Board Can 27:811–818

Needham PR, Moffert JW, Slater AW (1945) Fluctuations in wild brown trout populations in Convict Creek, California. J Wildl Manage 9:9–15

Nicola GG, Almodovar A (2002) Reproductive traits of stream-dwelling brown trout *Salmo trutta* in contrasting neighbouring rivers of central Spain. Freshwat Biol 47: 1353–1365

Nikolsky GW (1963) The ecology of fishes. Academic, London

Olsen EM, Vollestad LA (2003) Microgeographical variation in brown trout reproductive traits: possible effects of biotic interactions. Oikos 100:483–492

Olsen EM, Vollestad LA (2005) Small-scale spatial variation in age and size at maturity of stream-dwelling brown trout, *Salmo trutta*. Ecol Freshw Fish 142:202–208

Pender DR, Kwak TJ (2002) Factors influencing brown trout reproductive success in Ozark tailwater rivers. Trans Am Fish Soc 131:698–717

Rifflart R, Marchand F, Rivot E, Bagliniere JL (2006) Scale reading validation for estimating age from tagged fish recapture in brown trout (*Salmo trutta*) population. Fish Res 78:380–384

Rubin JF, Glimsater C, Jarvi T (2005) Spawning characteristics of the anadromous brown trout in a small Swedish stream. J Fish Biol 66:107–121

Sorensen PW, Cardwell JR, Essington T, Weigel DE (1995) Reproductive interactions between sympatric brook and brown trout in a small Minnesota stream. Can J Fish Aqua Sci 52:1958–1965

Taube CM (1976) Sexual maturation and fecundity in brown trout of the Platte river, Michigan. Trans Am Fish Soc 4: 529–533

Thomas JD (1964) Study on the growth of trout, *Salmo trutta* from four contrasting habitats. Proc Zool Soc London 142:459–510

Yin MC (1995) Fish ecology. China Agriculture Press, Beijing, pp 105–131

Zhang CL, Wang WB (1962) A preliminary report on the fishes from Tibet. Acta Zoologica Sinica 14:529–536

Environ Biol Fish (2009) 86:97–105
DOI 10.1007/s10641-008-9371-5

Age structure, growth and mortality estimates of an endemic *Ptychobarbus dipogon* (Regan, 1905) (Cyprinidae: Schizothoracinae) in the Lhasa River, Tibet

Xiuqi Li · Yifeng Chen

Received: 4 June 2007 / Accepted: 26 May 2008 / Published online: 9 July 2008

Abstract *Ptychobarbus dipogon* is an endemic fish in the Yarlung Tsangpo River, but its biology is poorly known. We sampled 582 specimens (total length, TL, between 70.6 and 593.0 mm) from April 2004 to August 2006 in the Lhasa River, Tibet. We estimated ages based on the counts of alternating opaque and translucent zones (annuli) in thin transverse sections of lapilli otoliths. Ages ranged from 1^+ to 23^+ years for males and 1^+ to 44^+ for females. The observed 44^+ years was the oldest reported for schizothoracine fishes. Females attained a larger size than males. The TL–weight relationship was $W{=}7.12\times10^{-6}TL^{3.006}$ for combined sexes. The growth parameters fitted von Bertalanffy growth functions were $L_\infty = 598.66$ mm, $k{=}0.0898$ year^{-1}, $t_0{=}-0.7261$ year and $W_\infty = 1585.38$ g for females and $L_\infty = 494.23$mm, $k{=}0.1197$ year^{-1}, $t_0{=}-0.7296$ year and $W_\infty = 904.88$g for males. The longevities of 32.7 year for females and 24.3 year for males were similar to the observed ages. Using an empirical model we estimated the instantaneous rate of total mortality (Z) at 0.28 per year in the lower reaches. Z in the upper and middle stocks was close to the M because of unexploited or lightly exploited stock. Protracted longevity, slow growth, low natural mortality and large body size were typical characteristics of *P. dipogon*. The current declining trend of *P. dipogon* could be prevented by altering fishing regulations.

Keywords *Ptychobarbus dipogon* · Age · Growth · Mortality · Tibet

X. Li · Y. Chen (✉)
Laboratory of Biological Invasion and Adaptive Evolution, Institute of Hydrobiology, Chinese Academy of Sciences, Wuhan 430072, China
e-mail: chenyf@ihb.ac.cn

X. Li
Graduate School of Chinese Academy of Sciences, Beijing 100039, China

Introduction

The subfamily Schizothoracinae (family Cyprinidae) is the predominant group of endemic fish living in the high-elevation rivers and lakes in the Qinghai-Tibet Plateau (Cao et al. 1981). However, biological information of schizothoracine fishes is seriously lacking in comparing with their wide distribution. Existing publications mainly represent the genus *Schizothorax* (Tsao and Wu 1962) and *Gymnocypris* (Hu et al. 1975; Zhao et al. 1975; Ren and Sun 1982; Chen et al. 2001, 2002b; Yang et al. 2002; Chen et al. 2004). The schizothoracine species are characterized by low growth rate, low fecundity and late sexual maturity as adaptation to their rigorous environment (Wu and Wu 1992; Chen and Cao 2000). These life-history characteristics make them particularly sensitive to the intense exploitation. The conservation of natural population and exploitation of sustainable resources of schizothoracine fishes have been become increasingly matters of concerns.

Ptychobarbus dipogon mainly resides in the Yarlung Tsangpo River and its tributaries at altitudes above 3,000 m. Although there are an abundant fisheries resource because of the religion convention of the local population and remote location, the catches have recently shown a slightly depletion in length-based assessment in the lower reaches of the Lhasa River. The threats are mainly from the indiscriminate fishing as a result of the increasing public demand and the construction of hydropower projects. However, attempts to develop an effective population management strategy have been obstructed by lack of basic biological information.

Accurate age determination and estimates of growth and mortality parameters are fundamental requirements for understanding population dynamics and maintaining sustainable yields in fisheries biology (Campana and Thorrold 2001). Life history variations of *P. dipogon* in the upper-middle reaches and lower reaches are indispensable for constructing age-structured population dynamic. The objective of this study was to examine whether the age structure and life-history traits of *P. dipogon* were influenced by fishery activities in the Lhasa River. In addition, life history characters of *P. dipogon* were compared with those of other schizothoracine fishes in the Qinghai-Tibet Plateau. Finally, we propose some management guidelines that could improve the conservation of *P. dipogon*.

Study area

The Lhasa River (29°20′–30°15′N, 90°05′–93°20′E) with its length of 551 km, is located on the left of the Yarlung Tsangpo River in Tibet Autonomous Region, at an altitude of 3,580–5,200 m, with a mean river bed gradient of 2.9‰. At Zhigung the river course is divided into the upper-middle reaches and lower reaches. Based on the relative abundance and population segmentation due to the construction of a power plant, the division of *P. dipogon* populations was distinct at Zhigung, and respectively representing the unexploited and exploited stocks (Fig. 1).

The climate of the region is strongly influenced by high elevation. The mean temperature of the water in Lhasa is 7.5°C (range 0.8°C to 13.9°C). In the headwaters of the Lhasa River above 4,000 m, the river is covered by a thick layer of ice from November to April. In the moderate reaches, the ice freezing occurs sporadically. The river system is supported by rainfall and snow melt. Nearly 80% of the annual rainfall occurs between mid-June and mid-September (Xiong and Tang 1998). Among all the hydrological factors, flood impulse is probably the strongest factor that regulates adequate water cover, food supply and breeding grounds, especially in its upper stretches and piedmont.

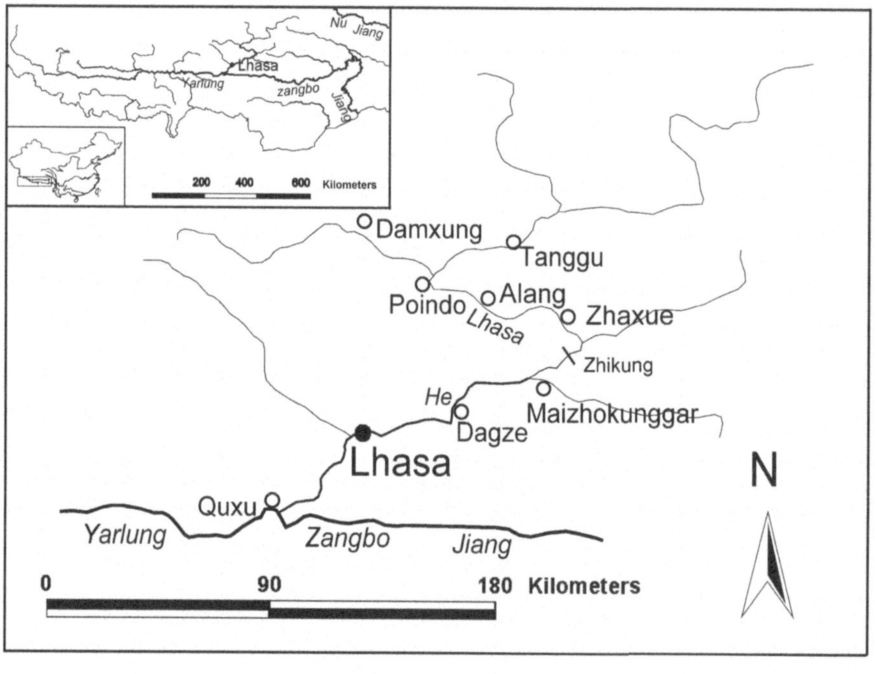

Fig. 1 Sampling location and division of *P. dipogon* stocks in the Lhasa River. Zones were distinguished according to the differences in fish abundance and stocks segmentation

Materials and methods

Sample collection

We collected opportunistically *Ptychobarbus dipogon* from various sites from a variety of fisheries dependent and independent sources from April 2004 to August 2006. Most (90.4%) *P. dipogon* catches were in the Lhasa River with floating gillnets and cast nets (Table 1). We recorded the full range of biological information from all specimens. In addition, we extracted the otoliths from all specimens, and then prepared and measured them (Li et al. 2008).

Age estimates

Based on position of the first increment and the assumption that each opaque and translucent zonal

Table 1 Samples collected for growth estimation of *P. dipogon* in the Lhasa River

Locality	Altitude (m)	Date	Total length (mm)	Total weight (g)	Number of samples (male: female: undetermined sex)
Tanggu	4,200	27/11/2004	355–550	275–1,121	7:13:3
Damxung	4,230	17/06/2004	275–564	157–1,269	6:18:0
		1/11/2004	391–460	486–725	4:4:0
		23/06/2006	178–390	40–436	3:4:2
Poindo	4,100	22/06/2004	175–288	40–192	0:0:5
		13/11/2004	99–199	6.8–55.8	1:6:9
		4/01/2005	203–461	48–699	11:15:0
		18/03/2005	322–430	280–680	2:2:0
		29/05/2006	135–490	14.6–881	5:5:1
Alang	3,980	30/11/2004	256–350	123–286	8:12:0
Zhaxue	3,950	17/07/2004	377–450	406–668	8:0:0
Maizhokunggar	3,850	19/06/2004	250–455	120–724	3:8:1
		10/072004	232–357	84–324	6:5:1
		5/01/2005	218–445	70–630	1:4:0
		15/05/006	333–420	283–543	1:2:0
Pingcuo	3,830	23/09/2004	250–365	230–344	2:6:2
		11/12/2004	220–393	200–368	1:7:1
		5/01/2005	218–445	200–442	1:4:0
Dagzê	3,800	29/03/2004	252–453	101–727	20:16:2
		25/04/2004	277–593	139–1,234	9:15:1
		13/06/2004	268–376	155–422	7:3:0
		5/07/2004	234–326	93–246	2:3:0
		22/09/2004	230–348	87–303	3:19:4
		12/10/2004	240–353	99–333	2:7:1
		14/06/2006	70–314	2.3–231	0:3:9
Zhangdo	3,860	8/07/2004	195–478	54–785	12:8:0
		21/05/2006	390–567	500–1,422	0:7:0
		10/06/2006	126–510	12–1,176	6:10:9
		8/07/2006	187–530	43–1,267	6:10:4
		4/08/2006	235–470	89–756	9:11:3
Lhasa	3,650	24/04/2004	200–289	66–172	2:10:6
		18/08/2004	249–575	105–1,351	10:12:3
		19/11/2004	227–359	68–350	5:6:2
		2/12/2004	278–352	148–294	0:3:0
		18/02/2005	347–450	350–645	18:2:0
		23/03/2005	156–543	25–1,326	12:7:1
		15/04/2005	250–310	113–228	1:4:0
		17/05/2005	210–320	66–233	6:14:3
Qüxü	3,580	20/05/2004	283–388	190–471	5:5:1
		10/12/2004	355–428	352–536	2:1:0
		22/05/2006	163–392	31.8–503	3:6:2
		21/06/2006	147–238	24.1–99.3	1:2:5

pair represents 1 year, otolith reading was done along the inner and lateral axis in the thin transverse sections of lapilli. The reader had no prior knowledge of length, sex or time of capture before age and growth estimation. All ages were determined twice by the same interpreter after a considerable time (30 days). The consistency of age interpretation was assessed by calculating the coefficient of variation (CV) between the repeated readings (Campana 2001).

Growth modeling

The total length–weight relationships were calculated by the power relationship of: $W=aTL^b$ (1). The standard von Bertalanffy growth function (VBGF 1938) was used to express the observed total length-at-age $t : L_t = L_\infty\{1 - \exp[-k(t - t_0)]\}$ (2). Where L_∞ is the asymptotic length-at-age, which represents the average length-at-age of individuals would attain if they grew indefinitely. The k is a curvature parameter determining the rate at which the fish reach the asymptotic length, t_0 is a position parameter defining the initial condition on the time axis when fish length is zero.

The growth performance index was calculated by the equation of Munro and Pauly (1983): $Ø = \log_{10} k + 2\log_{10} L_\infty$ (3). The value of phi-prime (Ø) was used to compare growth parameters obtained in the present work with those reported by others. The utmost age ($A_{\max}$) was defined by Fei and He (1983) according to the following formula: $A_{\max} = 3/k + t_0$ (4).

Mortality estimation

Beverton and Holt (1956) showed that the total mortality rate (Z) could be estimated from length frequency information. The Beverton and Holt's equation was described as: $Z = k\,(L_\infty - \acute{L})\,/\,(\acute{L} - L_c)$ (5) where Lc is the smallest length at first capture and $\acute{L}$ is the mean length of fish calculated from L_c upward. Hoenig's method (1983) was based on the empirical relationship between mortality and longevity. The relationship between the instantaneous rate of total mortality (Z) and maximum age was expressed as:$\ln(Z) = 1.46 - 1.01\ln(T_{\max})$ (6). Longevity ($T_{\max}$) was estimated from the oldest observed age during this study. M was calculated by Pauly (1980) empirical equations. The equation is described as:

$$\log_{10} M = -0.0066 - 0.279\log_{10} L_\infty + 0.6543\log_{10} k + 0.4634\log_{10} T \quad (7)$$

where L_∞ and k are parameters derived from the VBGF; T is the average annual temperature at the surface (in °C), which was 7.5°C in the Lhasa River. Estimates of fishing mortality (F) were calculated by subtracting natural mortality from total mortality.

Data analysis

We used the Origin 7.0 and Statistica 6.0 packages for data analysis. We used Kolmogorov–Smirnov tests (α= 0.05) to determine whether significant differences existed between males and females. The residual sums of squares analysis (ARSS) was used to determine whether any significant difference existed in the length–weight relationships and von Bertalanffy equations for male and female (Chen et al. 1992).

Results

Length–frequency distributions

Of the 582 *P. dipogon* sampled, 212 were males, 308 were females, and 90 were immature or of undetermined sex. Length–frequency distributions were not significant difference between sexes (Kolmogorov–Smirnov; D=0.1128, P=0.076). The largest male registered in this study was 490 mm TL, the largest female was 593 mm TL, whereas the smallest specimen was 70 mm TL. Female attained the larger size than males. There was a significant difference in Length–frequency distributions between the two stocks from the upper-middle reaches and lower reaches (D=0.1756, P=0.001) (Fig. 2).

Length–weight relationship

The total length–weight relationships were depicted separately for males and females (Fig. 3). No significant differences were found between the length–weight for males and females (F=1.4973, P>0.05). Therefore, the regression equation was developed from pooled data as W=7.12×$10^{-6}TL^{3.006}$, r=0.9869

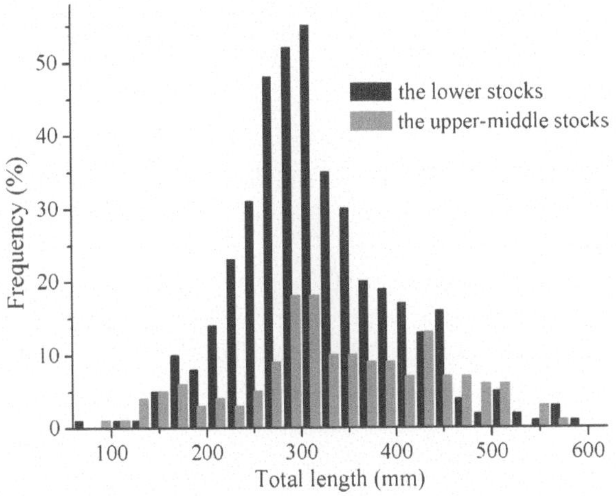

Fig. 2 Length–frequency histogram of *P. dipogon* between the two stocks (n=582)

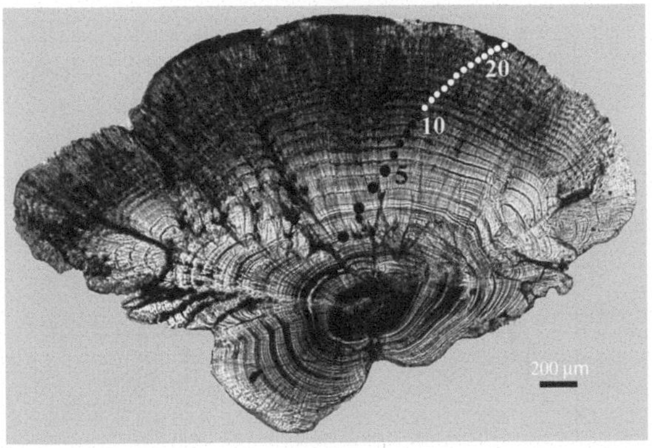

Fig. 4 The typical overlook of *P. dipogon* cross transverse section of lapilli under transmitted light

(n=582). The hypothesis of isometric growth for this species was accepted, as the obtained allometric index value (b) was not significantly different from 3 (t-test, Pauly 1984; t=0.78<$t_{0.05}$=1.96).

Age structure

The total number of translucent zones was recorded and considered to be the number of annuli by alternating broad opaque and narrow translucent zones outwards the nuclei of lapillus (Fig. 4). The ages were successfully determined for 203 females between 158 and 493 mm TL, 141 males between 157 and 490 mm TL, and 49 undetermined specimens between 70 and 550 mm TL. The reliability of the age estimates had low CV (3.14%), reflecting good concordance in the readings. There was a slight difference in the age frequency composition of the pooled samples based on sex (Kolmogorov–Smirnov; D=0.0915, P=0.469). The youngest fish (70.6 mm TL) was 1^+ years old. The oldest female *P. dipogon* was 532.0 mm TL and was estimated to be 44^+ years, and the oldest male was 460.0 mm TL that was estimated as 23^+ years old. The estimated age of 44^+ years was the oldest for schizothoracine fishes, suggesting that *P. dipogon* was an exceptionally long-lived fish. There were significant difference of age structure for pooled sexes between the upper-middle stocks and lower stocks of the Lhasa Rivers (D= 0.3993, P=0.000) (Fig. 5). A much higher proportion

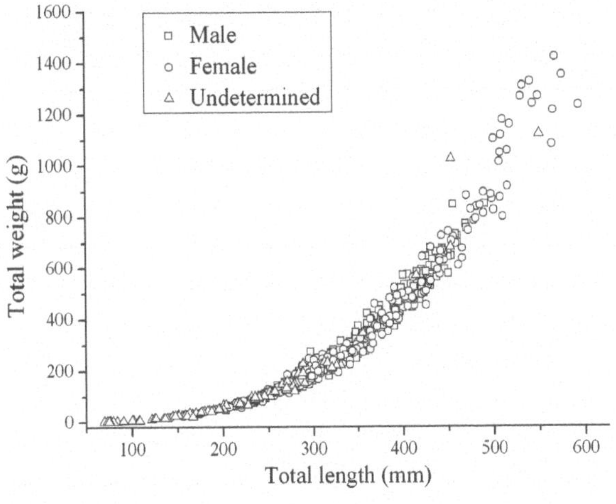

Fig. 3 The total length–weight relationships of *P. dipogon* for males and females

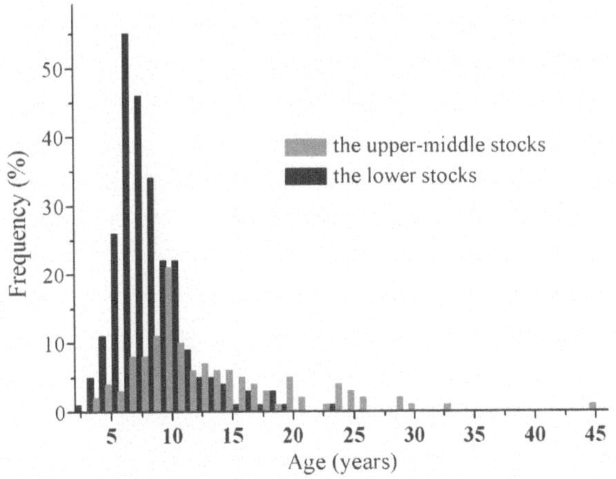

Fig. 5 Comparison of age structure for pooled sexes between the upper-middle stocks and lower stocks of the Lhasa Rivers (n=393)

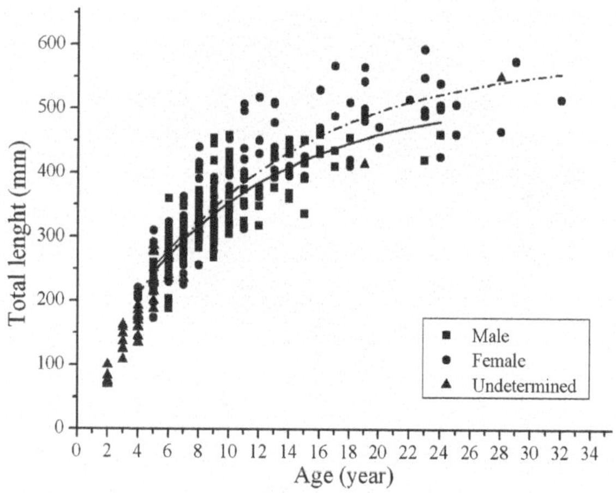

Fig. 6 The von Bertalanffy growth curve of *P. dipogon* with the observed TL at age derived from otolith reading

of older *P. dipogon* were sampled from the catches of the upper-middle headwaters undisturbed (300–500 mm TL; 32.6% over 10 years).

Growth characteristics

Growth models fitted the observed TLs-at-age data were calculated by the least square method for each sex. The von Bertalanffy equations were attained as:

$$L_t = 494.23\{1 - \exp[-0.1197(t + 0.7296)]\} \quad r^2 = 0.8856; (♂)$$

$$L_t = 598.66\{1 - \exp[-0.0898(t + 0.7261)]\} \quad r^2 = 0.8115. (♀)$$

The growth curve described a trend of relatively slow growth based on the observed TLs-at-age data between sexes, respectively (Fig. 6). Females attained a greater theoretical asymptotic TL in comparison to males (F=3.5694, P<0.05). Growth parameters indicated that the growth rate of females was lower than males. In addition, the growth patterns of different stocks were fitted the VBGF were shown when the individuals of above 23 years were excluded because of the poor specimens (Table 3). A comparison derived from the two group parameters clearly demonstrated that *P. dipogon* caught in the lower reaches exhibit relative rapid growth and longer asymptotic total length than those in the upstream zones (F=3.1326, P<0.05).

The asymptotic weights based on the von Bertalanffy growth equations were estimated as W_t=904.88 for males and W_t=1,585.38 for females. As the above depiction in the growth curve showed a rapid increase in weight before 11.6 years for females and 8.5 years for males, and then annual growth rate increments declined gradually. The A_{max} of *P. dipogon* was 32.7 year for females and 24.3 year for males, respectively. The results were similar to observed maximum age. The growth performances of *P. dipogon* were 4.5076 for females and 4.4659 for males.

Mortality estimation

Using Hoenig's longevity-mortality relation method, estimates of total mortality (Z) based on A_{max} were 0.13 for females and 0.17 for males per year except for abnormal 44^+ years. The estimate of M based on Pauly's equation was 0.09 for females and 0.11 for males, respectively. According to the catch length–frequency distribution and mesh size, L_c was assumed at 260 mm TL for capture, and obtained $\acute{L}$=334.0 mm in the lower reaches. Total mortality (Z) was estimated 0.28 per year in the lower reaches. The Z of Hoenig's estimate in the upper and middle reaches was close to the M of Pauly's estimate because of the unexploited or lightly exploited stocks (Table 2). Therefore, fishing mortality F was 0.17 per year in the lower reaches.

Discussion

According to the ageing result, about 5–10 years age difference in approximate TL group was observed

Table 2 Estimations of the growth and mortality parameters for *P. dipogon* of the upper-middle and lower stocks in the Lhasa River

Stocks	Habitat conditions			Life-history parameters							
	Altitude (m)	Temperature (°C)	Harvest	L_∞ (mm)	k (year^{-1})	t_0 (year)	r^2	Z	M	F	n
The upper-middle reaches	>3,900	<7.5	Little	563.7	0.1084	−0.8902	0.95	0.11	0.11	0	127
The lower reaches	<3,900	>7.5	Much	540.6	0.1172	−0.4090	0.89	0.28	0.11	0.17	257

Table 3 Comparison of growth characters of schizothoracine fishes in different water systems

Species	Region	Age material	Body length (mm)	Age (year)	Sex	Growth parameters				Sources
						L_∞ (mm)	k (year^{-1})	t_0 (year)	Ø	
Schizothorax prenanti	Western Sichuan	Anal scale	–	1–14	♂	623.59	0.1151	−0.0919	4.6509	Tsao and Wu (1962)[a]
Diptychus pachycheilus	Western Sichuan	Anal scale	–	1–13	♀	2,713.67	0.0162	−0.8465	5.0726	Tsao and Wu (1962)[a]
			–	1–10	♂	837.02	0.0698	−0.1783	4.6893	
Gymnocypris eckloni	Western Sichuan	Anal scale	–	1–12	♀	565.17	0.0937	−0.6452	4.4761	Tsao and Wu (1962)[a]
Schizopygopsis malacanthus	Western Sichuan	Anal scale	–	1–9	♀	317.71	0.1305	−0.1467	4.1197	Tsao and Wu (1962)[a]
Platypharodon extremus	Western Sichuan	Anal scale	–	1–14	♀	637.69	0.1110	0.2156	4.6545	Tsao and Wu (1962)[a]
			–	1–14	♂	753.92	0.0864	−0.3426	4.6912	
Gymnocypris przewalskii	Qinghai Lake	Anal scale	70–348	1–14$^+$	♀+♂	1,103.2	0.0287	−0.0189	4.5432	Zhao et al. (1975)[a]
Gymnocypris Selincuoensis	Selincuo Lake	otolith	34–430	1–29	♀	485.33	0.0710	0.5679	4.2233	Chen et al. (2002b)
			34–405	1–26	♂	484.19	0.0684	0.6028	4.2051	
Gymnocypris cuoensis	Cuoe Lake	otolith	182–460	7–29	♀+♂	639.71	0.0291	4.6745	4.0759	Yang et al. (2002)
Ptychobarbus dipogon	Lhasa River	otolith	70–490	1$^+$–44$^+$	♀	598.66	0.0898	−0.7261	4.5076	Present study[b]
			70–593	1$^+$–23$^+$	♂	494.23	0.1197	−0.7296	4.4659	

[a] Growth parameters were calculated based on the origin data of Tsao and Wu (1962) and Zhao et al. (1975)

[b] Total length

between the upper-middle stock and lower one in the Lhasa River. In other words, there was a great overlapping in total length among the different age groups. For example, the total length of adult *P. dipogon* at 10 years ranged from 295.0 to 478.0 mm. Tsao and Wu (1962) and Zhao et al. (1975) also demonstrated this phenomenon for other schizothoracine species. This marked difference could be attributable to different growth or mortality rates. The growth dispersant of *P. dipogon* resulted from the lower growth rate in the cold water of the high altitude area. In addition, the low fishing effort probably played an important role.

The estimated oldest ages of *P. dipogon* were 23$^+$ for males and 44$^+$ years for females, suggesting that there were much older females than males. The maximum age of 44$^+$ years estimated for female is the oldest reported for schizothoracinae fishes. Tsao and Wu (1962) indicated the maximum age (A_{max}) 9$^+$–14$^+$ years for five schizothoracine fishes from Western Sichuan by counting annuli in anal scales. Zhao et al. (1975) also recorded 25 years for *Gymnocypris przewalskii* in the Qinghai Lake. Chen et al. (2002a) reported *G. selincuoensis* in the Selincuo Lake attained 29 years of age based on otolith determination. These analyses have revealed that the subfamily Schizothoracinae are a long-lived species. Delayed senescence is a common trait of all populations of schizothoracine fishes, particularly for the underexploitation population. Genetic and environment factors, such as low temperature, poor food availability and slight fishing pressure have likely driven plasticity adaptation in longevity of *P. dipogon*.

In theory, the maximum observed TL could be exceeded the L_∞ from VBGF in unexploided populations. Tracey and Lyle (2005) considered that the standard von Bertalanffy function provided a poor representation of growth in older individuals, resulting in an unrealistically low L_∞. The L_∞ obtained in this study were 598.6 mm for females and 494.2 mm for males, and close to maximum observed TL in the upper-middle reaches (593.0 and 490.0 mm for females and males, respectively). The observed TL of species of the subfamily Schizothoracinae appeared to be close to their physiological limits. In addition, the results also showed that adult females grew larger than adult males, while the k value for males was higher than that for females (0.0898 for females and 0.1197 for males).

In general, the k had been used as an index of the vulnerability of a stock subjected to excessive mortality and was useful for comparing life history strategies (Pratt and Casey 1990; Musick 1999). Branstetter (1987) categorized the k values as 0.05–0.10/year for slow growth species, 0.10–0.20/year for species with moderate growth, and 0.20–0.50/year for rapid growth. Most schizothoracine fishes have the k value of around 0.1 (Table 3). Therefore, *P. dipogon* is particularly vulnerable for to excessive exploitation. The growth traits of *P. dipogon* are adaptations to cold water temperature and food deficiency, particularly thc short growth seasons, approximately 5–6 months (Tsao and Wu 1962; Chen et al. 2002a). The values of Ø found in the present study are similar to the average value of 4.49, calculated form 9 species of the subfamily schizothoracinae.

The age structure of a population and the parameter k of the VBGF are key factors to understand the responses of fish population in exploited condition. The population structure could also affect the growth pattern, so that growth is usually favoured by capture (Donald and Alger 1989). Healey (1980) observed a stimulation of growth in exploited populations of *Coregonus clupeaformis*. Brańa et al. (1992) also found that brown trout exhibited faster growth in some exploited sections of northern Spanish rivers. In the present study, our findings indicated that capture reduced the life span of *P. dipogon* downstream. The lower stocks was characterized by the relative faster growth rate and high Z (0.28), whereas in the upper-middle stocks the estimated Z was close to the M. Tsao and Wu (1962) found the same effects of fishing on age structure in the drainages of Western Sichuan.

The effects of fishery management on population dynamics depend on the environmental and biological characteristics of fish. Our analysis indicates that fishing mortality was slightly higher than natural mortality in the lower reach. Therefore, fishing regulations for *P. dipogon* should be adapted to the river-specific ecological conditions. As aforementioned, protracted longevity, slow growth in later life and low natural mortality, are typical characteristics of *P. dipogon*. The more restrictive harvest regulations, like catch size limits, may prevent overexploitation of stocks, maintaining the abundance of large *P. dipogon* and improving the natural recruitment of populations. Because of these stocks embody thousands of years of evolutionary adaptations to local rigorous environments, the complete study of the life-history strategies and better understanding of the carrying capacity of the rivers are required. Finally, a long-term monitoring of populations and restrictive fishing activity are to be carried out.

Acknowledgements This study was supported by the Knowledge Innovation Project of the Chinese Academy of Sciences (grant no. KSCX2-SW-125; KSCX1-SW-13–04) and the National Natural Science Foundation of China (grant no. 30471345). We are grateful to Fuhua Hao and Feng Chen for assistance in sample collection. We particularly thank Dekui He for his assistance in the field and valuable guidance. Additional sincerely thanks extended to several reviewers for providing helpful comments on earlier drafts of this manuscript.

References

Beverton RJH, Holt SJ (1956) A review of methods for estimating mortality rates in exploited fish populations, with special reference to sources of bias in catch sampling. Rapp P-V Réun Cons Int Explor Mer 140:67–83

Brańa F, Nicieza AG, Toledo MM (1992) Effects of angling on population structure of brown trout, *Salmo trutta* L., in mountain streams of Northern Spain. Hydrobiologia 237:61–66 doi:10.1007/BF00008428

Branstetter S (1987) Age, growth and reproductive biology of the silky shark, *Carcharhinus falciformis*, and the scalloped hammerhead, *Sphyrna lewini*, from the northwestern Gulf of Mexico. Environ Biol Fishes 19:161–173 doi:10.1007/BF00005346

Campana SE (2001) Accuracy, precision and quality control in age determination, including a review of the use and abuse of age validation methods. J Fish Biol 59:197–242 doi:10.1111/j.1095-8649.2001.tb00127.x

Campana SE, Thorrold SR (2001) Otoliths, increments, and elements: keys to a comprehensive understanding of fish populations? Can J Fish Aquat Sci 58:30–38 doi:10.1139/cjfas-58-1-30

Cao WX, Chen YY, Wu YF, Zhu SQ (1981) Origin and evolution of Schizothorscine fishes in relation to the upheaval of the Qinghai-Xizang Plateau. In: The comprehensive scientific expedition to the Qinghai-Xizang Plateau, CAS (ed) Studies on the period, amplitude and type of the uplift of the Qinghai-Xizang Plateau. Science Press, Beijng, pp 118–130 (in Chinese)

Chen YF, Cao WX (2000) Schizothoracinae. In: Yue PQ (ed) Fauna sinica osteicthtyes cypriniformes III. Science Press, Beijing, pp 273–388 (in Chinese)

Chen Y, Jackson DA, Harvey HH (1992) A comparison of von Bertalanffy and polynomial functions in moldelling fish growth data. Can J Fish Aquat Sci 49:1228–1235

Chen YF, He DK, Cai B (2001) Status and sustainable utilization of fishery resources of Selincuo Lake, northern Tibet. Biod Sci 9(1):85–89 in Chinese, with English abstract

Chen YF, He DK, Chen YY (2002a) Age discrimination of Selincuo schizothoracini (*Gymnocypris selincuoensi*) in Selincuo Lake. Tibetan Plateau. Acta Zool Sin 48(4):527–533 in Chinese, with English abstract

Chen YF, He DK, Cao WX, Duan ZH (2002b) Growth of Selincuo schizothoracini (*Gymnocypris selincuoensis*) in Selincuo Lake. Tibet. Acta Zool Sin 48(5):667–676 in Chinese, with English abstract

Chen YF, He DK, Cai B, Chen ZM (2004) The Reproductive Strategies of an Endemic Tibetan Fish, *Gymnocypris selincuoens*. J Freshw Ecol 19(2):255–262

Donald DB, Alger DJ (1989) Evaluation of exploitation as a means of improving growth in a stunted population of brook trout. N Am J Fish Manage 9:77–183 doi:10.1577/1548-8675(1989)009<0177:EOEAAM>2.3.CO;2

Fei HN, He BQ (1983) Ecology parameters and types of life history of the fishes in continental shelf in Guangdong Province. In: Corpus of Fisheries Science and Technology II. Agriculture Press, Beijing, 6–16 (in Chinese)

Healey MC (1980) Growth and recruitment in experimentally exploited lake whitefish (*Coregonus clupeaformis*) populations. Can J Fish Aquat Sci 37:255–267

Hoenig JM (1983) Empirical use of longevity data to estimate mortality rates. Fish Bull (Wash D C) 82:898–902

Hu A, Tang SS, Gong SX (1975) Reproductive biology of *Gymnocypris przewalskii przewalskii* (Kessler). In: Institute of Biology, Qinghai Province (ed) The Fish Fauna of Qinghai Lake Region and Biology of *Gymnocypris przewalskii przewalskii* (Kessler). Science Press, Beijing, pp 49–62 (in Chinese)

Li XQ, Chen YF, Chen F, He DK (2008) Otolith characteristics and age determination of an endemic *Ptychobarbus dipogon* (Regan, 1905) (Cyprinidae: Schizothoracinae) in the Yarlung Zangbo River, Tibet. Environ Biol Fishes (in press)

Munro JL, Pauly D (1983) A simple method for comparing the growth of fishes and invertebrates. Iclarm Fishbyte 1(1):5–6

Musick JA (1999) Life in the slow lane: ecology and conservation of long-lived marine animals. Am Fish Soc Symp 23. Bethesda, Maryland

Pauly D (1980) On the interrelationships between natural mortality, growth parameters and mean environmental temperature in 175 fish stocks. J Cons Int Explor Mer, 39(3): 179–192

Pauly D (1984) Fish population dynamics in tropical waters: a manual for use with programmable calculators. ICLARM Stud Rev 8:325

Pratt HL, Casey JG (1990) Shark reproductive strategies as a limiting factor in directed fisheries, with a review of Holden's method of estimating growth parameters. NOAA Tech Rep NMFS 90:97–109

Ren ML, Sun L (1982) An investigation of the fish resource and exploitation problems in Namucuo Lake, in Tebit. Freshwat Fish Sin 2:1–10 in Chinese

Tracey SR, Lyle JM (2005) Age validation, growth modeling, and mortality estimates for striped trumpeter (*Latris lineata*) from southeastern Australia: making the most of patchy data. Fish Bull (Wash D C) 103:169–182

Tsao WX, Wu XW (1962) An investigation of the fish biology and fishery problems in Ganze-Apa region of western Szechwan Province. Acta Hydrobiol Sin 2:79–111

von Bertalanffy L (1938) A quantitative theory of organic growth (Inquires on growth laws. II). Hum Biol 10:181–213

Wu YF, Wu CZ (1992) The fishes of the Qinghai-Xizang Plateau. Sichuan Publishing House of Science & Technology, Chengdu (in Chinese)

Xiong Y, Tang QC (1998) The hydrology of river in cold area of the Tibetan Plateau. In: Xiong Y, Tang QC (eds) The hydrology of river in China. Science Press, Beijng, pp 118–130 (in Chinese)

Yang JS, Chen YF, He DK, Chen ZM (2002) Studies on age determination and growth characteristics of *Gymnocypris cuoens*. Acta Hydrobiol Sin 26(4):378–387 in Chinese, with English abstract

Zhao LH, Wang SH, Zhao TQ (1975) The age and growth of *Gymnocypris przewalskii przewalskii* (Kessler). In: Institute of Biology, Qinghai Province (ed) The Fish Fauna of Qinghai Lake Region and Biology of *Gymnocypris przewalskii przewalskii* (Kessler). Science Press, Beijing, pp 37–45 (in Chinese)

Environ Biol Fish (2009) 86:107–108
DOI 10.1007/s10641-007-9284-8

Threatened fishes of the world: *Gobiocypris rarus* Ye and Fu, 1983 (Cypinidae)

Dongmei Xiong · Congxin Xie · Lu Xia

Received: 13 April 2007 / Accepted: 3 July 2007 / Published online: 14 September 2007

Common name: Rare gudgeon or rare minnow.

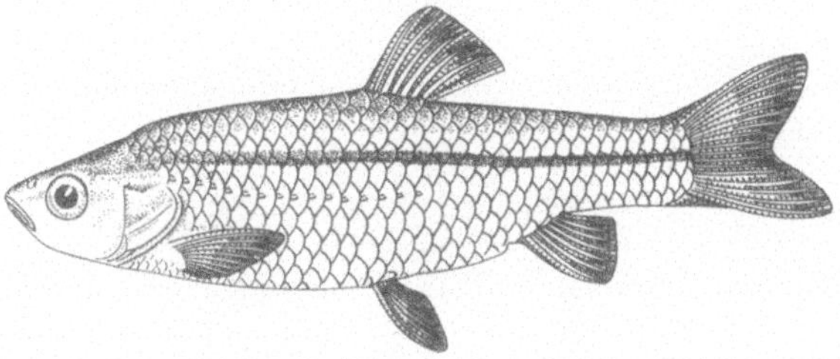

Conservation status: Endangered, China Red Data book of endangered animals (Yue and Chen 1998). *Identification*: A small freshwater Cypinidae (maximum 83 mm), slightly compressed body, round belly, short snout, incomplete lateral line. Lateral line scales 31–34, D: iii, 7–8; A: iii, 6–7; P: i, 11–12; V: ii, 7–8, C: 17, two rows pharyngeal teeth, barbel absent, small forked caudal fin, no keel on abdomen (Ye and Fu 1983). Gray dorsum, white belly, a longitudinal lateral yellowish zone. This zone becomes brighter, and small nuptial organs on operculum and pectoral fins in spawning males (Wang 1992). *Distribution*: Endemic to southwest China, in Liushahe River and some rivulets in Hanyuan county and Chengdu city of Sichuan province. *Habitat and ecology*: Gregarious in small water systems with a half-mud, half-stone bottom such as paddyfields, ditches, and rivulets, especially in weedy areas with low water flow. The species eats some small invertebrates, such as the aquatic earthworms (Oligochaeta). It is eurythermal, tolerates 1 to 35°C, and 0.50 $mg{\cdot}l^{-1}$ oxygen and 70 $mg{\cdot}l^{-1}$ carbon dioxide at 25°C (Wang 1995). Easy to be fed and reproduced, so it is a good experimental animal in China. *Reproduction*: Minimum age of sexual maturity 4 months. Females spawn between March and November, when water temperature above 14°C (Wang 1999). Fertilized eggs are ball, transparent, sticky, heavier than water and 1.25–1.70 mm in diameter (Wang 1992). *Threats*: Narrow habitat range and limited natural population, pesticides and ecological instability have caused rapid declines (Yue and Chen 1998). *Conservation action*: No specific action has been taken, not IUCN listed. *Conservation recommendation*: On-site conservation and restriction of fishing during its spawning season. Studies to estimate population size and structure, life history and genetic diversity, are needed for management and protection.

D. Xiong · C. Xie (✉) · L. Xia
College of Fisheries, Key Laboratory of Agricultural Animal Genetics, Breeding and Reproduction of Ministry of Education, Huazhong Agricultural University,
Wuhan 430070, China
e-mail: xiecongxin@mail.hzau.edu.cn

References

Wang JW (1992) Reproductive biology of *Gobiocypris rarus*. Acta Hydrobiologica Sinica 16:165–174 (in Chinese with English abstract)

Wang JW (1995) Acute effects of high concentration of dissolved free carbon dioxide and low dissolved oxygen on rare minnow. Acta Hydrobiologica Sinica 19:84–88 (in Chinese with English abstract)

Wang JW (1999) Spawning performance and development of oocytes in *Gobiocypris rarus*. Acta Hydrobiologica Sinica 23:161–166 (in Chinese with English abstract)

Ye MR, Fu TY (1983) Description of a new genus and species of Danioninae from China (Cypriniformes: Cyprinidae). Acta Zootaxonomica Sinica 8:434–437 (in Chinese with English abstract)

Yue PQ, Chen YY (1998) China Red Data Book of Endangered Animals – Pisces (pp 74–75). Beijing: Science Press

Environ Biol Fish (2009) 86:109–117
DOI 10.1007/s10641-008-9372-4

Degradation of icefishes (Salangidae) in the Yangtze River basin of China: threats and strategies

Zhongsuo Wang · Jieqiang Shi · Chongren Xu · Guangchun Lei

Received: 1 August 2007 / Accepted: 26 May 2008 / Published online: 22 July 2008

Abstract The Yangtze River basin has been a major area in terms of icefish diversity conservation and fisheries. In the past two to five decades, however, the wild icefish resources have degraded continuously in both the brackish estuary and freshwater portions of the basin. Stocks have declined in all icefish fisheries, resulting in an annual fishing yield decrease from the maximum of 4,000 metric tons to less than 1,300 metric tons in this river basin. The stocks decline significantly ($t=-6.74$, df=1, $p<0.001$) correlated with icefish fishing stress in the estuary. Meanwhile, species composition changed markedly with smaller sized icefish species increased relative to the number of larger ones in terms of their relative abundances. Following the stocks decline and size reduction, icefish commercial value decreased to the point that 91.9% of the traditional icefish-producing waters (57 out of 62) terminated their icefish fisheries and some species became locally extinct. Based on the degradation process, we suggest that these changes in stock size and species diversity are the result of overfishing and habitat loss and degradation. We propose a number of measures aimed at the conservation and restoration of the wild icefish resources in the Yangtze River basin.

Z. Wang (✉) · J. Shi
College of Life Sciences, Capital Normal University,
Beijing 100037, People's Republic of China
e-mail: zhongsuo@yahoo.com.cn

Z. Wang · C. Xu · G. Lei
College of Life Sciences, Peking University,
Beijing 100871, People's Republic of China

Keywords Stock decline · Local extinction · Economic extinction · Habitat · Diversity · Conservation

Introduction

Icefishes (Salangidae) are small fishes, ranging from 45 to 250 mm in standard length (SL), found in both freshwater and marine areas in eastern Asia, ranging from Japan to Korea, China and Vietnam (Nelson 2006). This family is comprised of two subfamilies, six genera and 17 species (Fang 1934; Zhang 1993; Zhu 1995). Freshwater icefishes primarily occur in the lentic bays, gulfs and estuaries of open lakes and reservoirs, and marine species are distributed in estuaries and nearby costal areas (Sun 1982; Chen and Zhang 1986).

As annual fishes, icefishes reproduce in discrete annual pulses with parental individuals dying shortly after spawning (Wang et al. 2002). So, in feeding seasons, the sampled icefish individuals in our investigation were the same cohort at the same age, and this provided a great convenience for our population dynamics study. Moreover, the short life span and simple life history allow the rapid response of population size to environmental changes, and thus icefish species are very sensitive to habitat degradation and environmental changes (Wang et al. 2004, 2005; Islam et al. 2006).

Despite their small body sizes, icefishes have high commercial value and thus have been traditionally exploited in China (Chen 1956). In the Yangtze River basin, icefishes have been commercially fished since early 1900s in the mid-lower reaches (from the estuary to Gezhouba Reservoir near Yichang City), and icefish fishery was virtually an essential component of historical fisheries in this river basin (Zhu 1985; Wang et al. 2002).

In the last few decades, however, the wild icefish resources have degraded markedly and continuously in the Yangtze River basin as well as in other icefish-inhabited river basins (Takita 1996; Xie and Chen 1999; Tang et al. 2000). Stocks decline and yields reduction of wild icefish resources in both the freshwaters and estuaries have already been reported (Sun 1982; Zhu 1985), but few papers have focused on the spatial dynamics and diversity change of icefish assemblages so far (Wang et al. 2004).

This paper is mostly based on data collected from an icefish-specific survey in the Yangtze River basin during 2001–2003. Our objective is to document icefish status in the Yangtze River basin and to discuss the threats on the wild icefishes while suggesting some basic strategies for wild icefish conservation and restoration in the Yangtze River basin.

Materials and methods

Study area

Our study area was the original distribution region of wild icefishes in the Yangtze River basin, including 86 associated waters (85 lakes and one reservoir) in the mid-lower reaches and three fisheries in the Yangtze estuary (Fig. 1). These 86 inland waters generally represented the natural distribution region of the freshwater icefish in this river basin, and the three sampling stations in the estuary were the most important fishing grounds for brackish icefishes in history (Fig. 1). Sampling sites in the freshwaters were the same as those in our previous work (Wang et al. 2005). Our investigation mainly focused on the freshwater icefishes and their habitats.

Sampling methodology

In the freshwaters, icefish sampling was conducted from July to September in 2001–2003, with a specially designed icefish trawl. Sampling methodology in the 86 freshwaters was the same as that in Wang et al. (2005). For the Yangtze estuary, we randomly sampled icefishes from fishers' catch in commercial fishing season from August to December 2003 (Jackson and Harvey 1997). Icefish specimens were kept in 5% formalin solution for further species composition analysis in our laboratory.

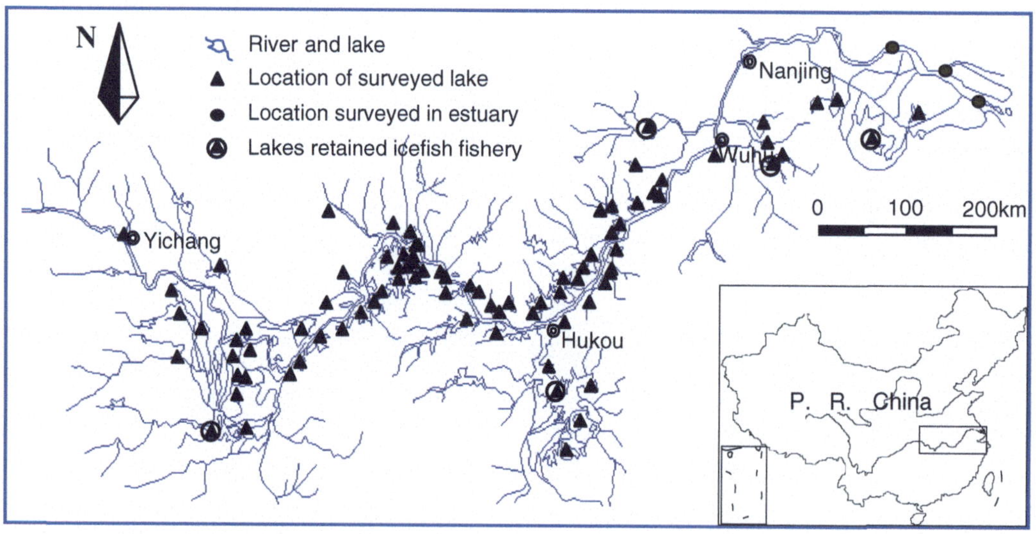

Fig. 1 Map of river–lake network system in the Yangtze River basin with the central locations of surveyed waters illustrated

Historical data on icefish yield and species composition were partly provided by local fishery bureaus and fishery corporations, and partly collected from documents and study reports (Zhu 1985; Chu 1998).

Data analysis

Economic viability for commercial exploitation depends on stock abundance, fishing cost and valuation of target species. Only when fishing is economically viable is icefish exploitation feasible (Wootton 1998). Hereby we divided the economic viability of icefish species into three groups (Table 1): high value species (>500 metric tons per year), medium value species (100–500 metric tons per year) and low value species (<100 metric tons per year). If a species is found in too low densities, its exploitation is not economically viable and therefore the commercial fishing activity should be terminated. At this critical point, we defined the species status as economic extinction (no commercial value; Jiang and Fan 2003). At the same time, we defined local ecological extinction of icefish species as no records for that species in at least 10 years in a specific location.

Relative abundance (P_i) is defined by the individual proportion of a species out of the overall icefish catch. The historical changes of relative abundances of icefish species in a specific locality is epitomized by changes in species composition of the icefish assemblage. To analyze historical changes of icefish species composition, we sorted icefish species into large, medium and small species by their body sizes of matured individuals. The large species included *Protosalanx hyalocranius* and *Hemisalanx brachyrostralis*, with SL over 100 mm. The medium species included *Neosalanx taihuensis*, with SL over 70 mm. *Neosalanx oligodontis* was the small species with SL less than 50 mm (Chen 1956; Wang et al. 2004). When the proportion of larger body-sized species is reduced, the body-size composition changed and icefish catch devalued, which was defined as size reduction of target assemblage by Cao et al. (1991).

The historical data used in size reduction analysis were based on document records, which was only available for four lakes of Taihu, Nanyi, Chaohu and Poyang (Zhu 1985; Diao 1986; Wang 1995; Chu 1998). Lack of more historical data prevented us from a more complete representation on icefish stock size reduction. In calculating the relative abundance, we omitted rare species with low population size (generally their $P_i < 0.0001$).

Given a fixed stock, fishing impact can be generally positively correlated with the number of fishing boats. The more fishing boats, the higher impact on the icefish stocks, and the lower the catch per boat. Thus, the mean catch per boat in a fishing year was a good indication of the impact of icefish fishing stress, which was used in simple linear regression between icefish stocks decline and the fishing stress changes (with Minitab 14.0 package). Other statistic analyses were simply conducted with Microsoft Excel 2003 for Windows.

Table 1 Distribution range and economic value of the nine icefish species in the Yangtze River ecosystem

Species	Distribution area	Commercial value	
		History	2003
Hemisalanx brachyrostralis	Mid-lower reaches	++	+
Hemisalanx prognathus	Estuary	+++	–
Neosalanx taihuensis	Mid-lower reaches	+++	+++
Neosalanx oligodontis	Mid-lower reaches	+	++
Neosalanx tangkahkeii	Lake Nanyi only	–	–
Neosalanx anderssoni	Estuary	–	–
Protosalanx hyalocranius	Lower reaches, estuary	+++	++
Salanx ariakensis	Estuary	–	–
Salanx cuvieri	Estuary	–	–

+++ Means high value, ++ means medium value, + means low value, and – means no value for commercial exploitation

Results

Species composition and economic value of icefish species

There are nine species of icefishes in the Yangtze River basin. Four of them are distributed in freshwater lakes and reservoirs in the mid-lower reaches, another four in the brackish estuary, and the other one distributed in both fresh and brackish waters (Table 1).

Among the five species in the estuary, only one species of *Hemisalanx prognathus* had been commercially fished before 1987, when it was found in high densities. The other four species existed in too low stocks to exploit, and showed no or low value for commercial production (Table 1). In the inland waters, however, four out of the five species were part of the

Table 2 Historical change of fished populations, fishery ranges and relative abundances (P_i) of the four economic icefish species in the freshwaters of the Yangtze River basin

Species	Population		P_i (%)		Fishery range	
	History	2003	History	2003	History	2003
H. brachyrostralis	18	2	5	1.3	Gezhouba to Nanyi	Dongting/Poyang
N. oligodontis	39	3	5	2.4	Dongting to Nanyi	Dongting/Poyang/Nanyi
N. taihuensis	60	5	62	80.6	Dongting to Taihu	Five lakes[a]
P. hyalocranius	9	2	28	15.7	Dongting to Taihu	Chaohu/Taihu

[a] Five lakes: the lakes of Dongting, Poyang, Nanyi, Chaohu and Taihu

icefish fishery. *Neosalanx taihuensis* dominated in stocks and species densities, and the other three economic icefishes *P. hyalocranius*, *N. oligodontis* and *H. brachyrostralis* constituted a much lower proportion of the annual icefish yield (Table 2). *N. tangkahkeii*, which is rare in Lake Nanyi, showed no economic potential for commercial exploitation (Table 1).

Stocks decline at the river basin scale

In the estuary, the maximum yield of *H. prognathus* reached 944.6 metric tons in 1960, but the yield decreased gradually in the following years falling to 24.3 metric tons in 1987 (Table 3). In the following 3 years (1988–1990), icefish fishing was banned in this area, but no icefish fishery resumed thereafter. Simple linear regression analysis showed that a significant decline of icefish yield (t=–6.74, df=1, p<0.001) was correlated to the increase of icefish-specific fishing intensity during 1960–1987 (Fig. 2).

Similarly, there was a simultaneous decline of icefish stocks in freshwater areas. Icefish yield declined from the peak of 2,170 to 990 metric tons in Lake Taihu and from 260 to 20 metric tons in Lake Nanyi in the past two decades (Fig. 3). Meanwhile, icefishes stocks in some other lakes, with maximum yield over 100 metric tons, all declined in more than 50% (Table 3). As a result, the total icefish yield declined dramatically at the river basin scale. In early 1980s, gross yield of freshwater icefishes averaged 4,000 metric tons per year, but in the last 10 years this value decreased to no more than 1,300 metric tons, a loss of 67.5% of icefish stocks in the year of 1990 (Table 3).

Size reduction of icefish assemblage in freshwater area

Decrease of gross icefish stocks was accompanied by a decline in species composition (relative abundance, P_i) in almost all localities. Relative abundance of smaller body-sized species increased, whereas larger species diminished (Fig. 4). This process of overall size reduction was well illustrated in the four large lakes with traditional icefish fisheries.

Table 3 Historical change of historical peak yield and present yield of icefishes (metric ton) in the five icefish-producing lakes and the estuary, with maximum yield over 100 metric tons respectively

Water area	Maximum yield in history		Present yield	Reference to historical record
	Yield	Year		
Freshwater areas				
Chaohu Lake	400	1983	200	Diao 1986
Dongting Lake	200	1958	5	Wang 1995
Nanyi Lake	260	1983	20	Chu 1998; Wang et al. 2004
Poyang Lake	600	1967	50	Wang et al. 2006
Taihu Lake	2,170	1985	990	Zhu 1985
Estuary	945	1960	0	Zhang 1992
Total	4,000	Early 1980s	1,300	Zeng 1990

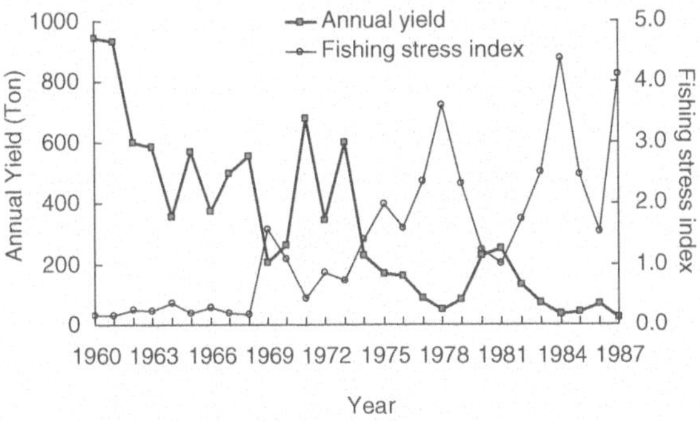

Fig. 2 Decline tendency of icefish yield in relation to the increasing fishing stress in the Yangtze estuary

In Lake Taihu, the large body-sized species of *P. hyalocranius* decreased its P_i from 39.4% in 1974 to 14.6% in 2003, while the medium body-sized species of *N. taihuensis* (smaller one) increased from 60.6% to 85.4% (TH in Fig. 4). A similar process occurred in Lake Chaohu. The proportion of *P. hyalocranius* (larger species) decreased from 20.0% in 1986 to 2.0% in 2003, and the smaller species of *N. taihuensis* increased from 80.0% to 98.0% in the same period of time (CH in Fig. 4).

In the Poyang and Nanyi lakes, the economic icefishes can be divided into three body-sized groups: the large one is *H. brachyrostralis*, medium one is *N. taihuensis* and small one is *N. oligodontis*. In the past two to three decades, species abundance changed with the small species increasing from 12.0% and 5% to 71.5% and 91.0% in Lake Poyang and Nanyi respectively. Meanwhile, the large and medium species decreased dramatically and *N. taihuensis* lost its dominant status (PY & NY in Fig. 4).

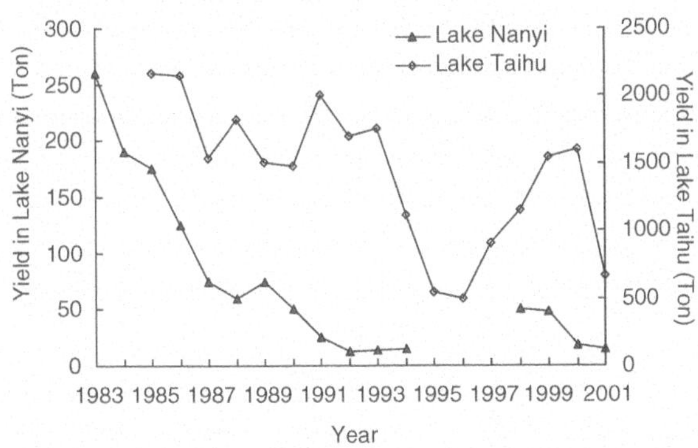

Fig. 3 Decline of icefish yield in Lake Nanyi and Taihu from the year with maximum yield to 2001 (commercial icefish fishing was forbidden in Lake Nanyi during 1995–1997)

Economic extinction in local waters

In freshwater areas, the number of lakes in which *N. taihuensis*, *N. oligodontis* and *H. brachyrostralis* were traditionally fished was 60, 39 and 18 respectively, but then these numbers reduced to 5, 3 and 2 respectively in the year of 2003 (Fig. 1, Table 2). In the estuary, icefishes decreased much more severely and icefish fishery diminished drastically after 1987, following the severe decline of *H. prognathus* in this area (Fig. 2).

Taking icefish assemblage in consideration as a whole, an icefish fishery once existed in 71.3% (62 out of 87) of the studied areas (A and D in Fig. 5). But 91.9% (57 out of 62, D in Fig. 5) of them lost an icefish-targeting fishery. Currently, icefish fisheries occur in only five large lakes, e.g. lakes Taihu, Chaohu, Nanyi, Poyang and Dongting (Fig. 1, Table 2), which occupy only 5.7% of the 87 surveyed water areas (A in Fig. 5).

Local extinction and spatial decline of key species

P. hyalocranius was once distributed broadly in the mid-lower reaches, ranging from the estuary to the upstream of Dongting plain, but then disappeared from the large area upriver from Wuhu river section before 1990. In 2003 this species was found only in Lake Chaohu, Lake Taihu and the estuary, and was commercially fished in these two lakes only (Table 2).

Discussion

Historically the Yangtze River ecosystem was an important area for both icefish diversity and icefish fisheries. Among all icefish inhabiting river basins, this river system showed the highest species richness, with nine out of 17 species of icefishes occurring in its river basin (Sun 1982; Zhang 1993; Nelson 2006). Besides this, the distribution range of icefishes in this river basin also turns out to be larger than any other river basins (Takita et al. 1988; Islam et al. 2006). In the broad mid-lower reaches (1,800 km along the Yangtze River course), icefishes occurred in most of the associated waters (Zhang 1992; Wang et al 2005). More importantly, most of these populations were historically large enough for commercial exploitation, and the icefish fishery in this area was not only the one that started the earliest but also the largest worldwide (Chen 1956;

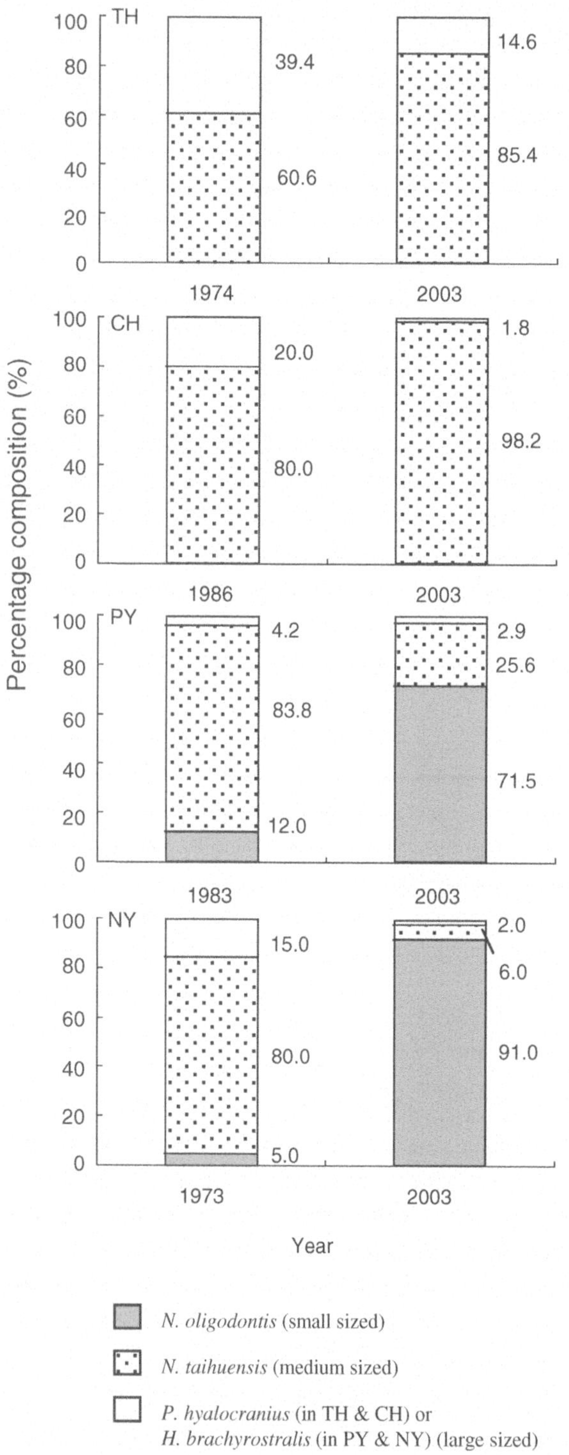

Fig. 4 Historical change of species composition in relation to their body sizes. The four representative lakes are: *TH* Taihu, *CH* Chaohu, *PY* Poyang and *NY* Nanyi

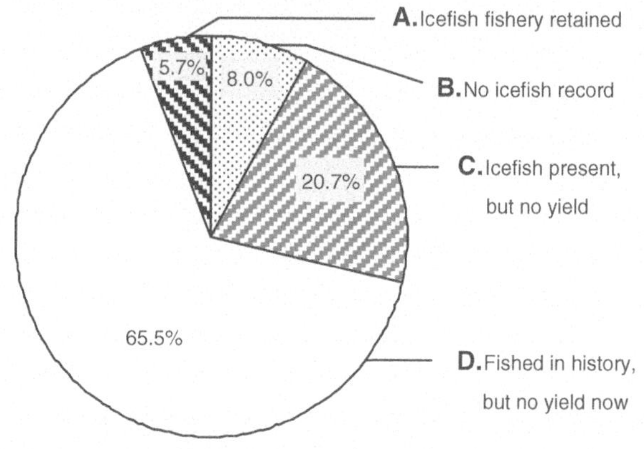

Fig. 5 Allocation of studied waters related to changing history of icefish fishery. Data partly referred to Wang et al. 2005

Zeng 1990; Takita 1996). Moreover, *N. taihuensis* and *P. hyalocranius* in the Taihu, Poyang, Dongting and Chaohu lakes became an important source for icefish introduction and stock enhancement in China (Hu et al. 2001). So the wild icefishes in the Yangtze River basin played an important role in world icefish fisheries, and therefore the decline of the icefish in this area was of serious concern.

Icefish degradation is in fact a worldwide problem. Icefish species are all critically endangered in Japan (Takita 1996; Kawanabe et al. 2001; Islam et al. 2006). And in China, Wild icefish resources also degraded in almost all icefish-inhabiting river basins. Traditional icefish fisheries have already diminished since late 1980s in the river basins of Yellow and Pearl (Zheng 1989; Dou and Chen 1994), and icefish stocks in the Yalu River basin fluctuated dramatically from year to year in the mid 1990s and lost commercial viability recently (Tang et al. 2000). In our study area, icefish resources have been reduced severely throughout the whole Yangtze River basin. The first record of icefish decline was in the early 1960s in the Yangtze estuary (Sun 1982), and then the stocks in this estuary continued to decline and dropped to so low a level that icefish sampling became quite hard in this area and icefish exploitation was not economically viable to the year of 1990 (Zeng 1990; Zhang 1992). In freshwaters, icefish depletion was already evident in some localities in early 1980s, and then more and more water areas lost their traditional economic value (Chen and Zhang 1986; Chu 1998). By the late 1990s, icefish fishery was reduced to small numbers in the Yangtze River basin as mentioned above.

Besides the stocks decline, the above-mentioned changes of species composition and body size reduction of icefish assemblages as well as the local extinction of *P. hyalocranius* further suggested the severe degradation of icefish resources at the river basin level.

Causes of icefish depletion

Factors influencing icefish dynamics are fairly complicated and key factors impacting on icefish degradation are quite different in different river basins. Kim and Park (2002) believed that water pollution had acted as a key factor leading to the heavy decline of *Salanx ariakensis* in Ariake Bay, Japan. Islam et al. (2006) similarly ascribed the depletion of *Neosalanx reganius* to environmental degradation in the upper Ariake Bay. Dou and Chen (1994) and Wang et al. (2004) reported a deep impact of historical overfishing on icefish assemblages in the Yellow River estuary and Lake Nanyi. And recently, the impacts of habitat fragmentation from irrigation works and reclamation on icefishes distribution and population dynamics were revealed at both a lake area and the large river basin scale in the Yangtze River basin (Wang et al. 2005, 2006). According to the declining process, the icefish depletion in the Yangtze River basin can be attributed to several factors mostly related to overfishing and habitat degradation.

Overfishing

Increasing fishing usually leads to decrease in fish stocks. This in turn leads to loss of economic value of fishing resources (Pauly 1998) and further to economic extinction (Myers et al. 1995). If a fish stock can resume in terms of fishing ban, the previous fishing press inflicted on the population was generally thought to be overstressed (Wootton 1998). Wang et al. (2004) studied the relationships between stocks decline of icefish assemblage and the long period of exploitation in Lake Nanyi. Our present paper further illustrates that the declined icefish stocks in this lake have once rebounded from 15 to 50 metric tons after a 3-year fishing ban (1995–1997). Severe decline of icefish yield and its subsequent loss of economic value in the large amount of water areas as well as the stocks rebound with fishing ban in Lake Nanyi are all good illustrations of the impacts of icefish overfishing.

In the late 1970s, powerboat was used in icefish fishing in the Yangtze River basin which directly improved the fishing efficiency. When the number of boats used in icefish fishing kept increasing, icefishes suffered heavier and heavier fishing press after 1980, which necessarily led to the long time depletion of wild icefish resources. The significant relationships between the serial stock decline and the increasing fishing stress in the Yangtze estuary provides another good illustration of the impacts of overfishing on icefish resources.

Fishing can also be size selective, and overfishing generally results in the reduction of the average fish size (tropicalization) of the target fish assemblage (Cao et al. 1991; Pauly 1998). The size reduction of icefish assemblages in the four above-mentioned lakes provides evidence for overfishing in these waters. Exploited with specially designed nets, larger icefishes were the most targeted by this kind of fishery decreasing the mean size of the individuals of the species than smaller ones (Wang et al. 2004). Thus, species composition was affected. The average size reduction and the changes of species composition reported in this paper further support the suggestion that overfishing is a major factor in the reduction of these fishes.

Habitat loss and degradation

Habitat loss and degradation can directly influence fish stocks by decreasing reproduction rate and increasing larvae mortality (Taylor 1996; Kapuscinski et al. 2007). Differing from overfishing, the influence of habitat degradation is generally impossible to recover just through fishing ban (Myers et al. 1995). Icefishes are very sensitive to habitat change, especially to spawning ground degradation (Chen and Zhang 1986; Wang et al. 2006). Water pollution and environmental degradation have affected the diet, fecundity and survival of icefishes and contributed to their stocks decline in the Yellow River estuary and the river basins upstream to Ariake Bay (Dou and Chen 1994; Islam et al. 2006). Similarly, habitat fragmentation and reclamation have also showed significant relationships with icefish depletion in Lake Poyang and Ariake Bay (Wang et al. 2006; Kim et al. 2006). In our study area, most of the lakes associated to the Yangtze River lost their original linkages with the main river course due to the large amount of irrigation works in the past three to five decades. As result, original river-lake networks in the Yangtze basin were artificially destroyed and habitat fragmentation occurred on icefish-inhabited waters (Wang et al. 2005).

Moreover, the simultaneous massive reclamation of shoal lake areas directly destroyed 50–80% of icefish spawning grounds in our studied lakes (Chen and Zhang 1986; Wang and Dou 1998). These changes must have directly contributed to the serious decline of icefish stocks and the subsequent economic extinction of economic species, and even led to the local extinction of *P. hyalocranius* reported in this paper. Although icefish fishing has been banned in the years of 1988–1990 in the Yangtze estuary, icefish fishery hasn't been resumed in this area. This is most probably due to habitat or environment changes, whereas not due to overfishing.

Considering the different sensitivity of icefish species to habitat change, their response to habitat degradation must be different (Magnuson et al. 1998; Olden et al. 2001). Among the studied icefish species, *P. hyalocranius* and *H. brachyrostralis* (the two larger body-sized species), were much stricter in habitat selection and were more sensitive to habitat degradation (Wang et al. 2004). Thus, the above-mentioned habitat loss and degradation in our studied waters must have had much larger impacts on them resulting in the mean body-size reduction of icefish individuals.

Suggestions to icefish protection

To protect and restore the natural icefishes in the Yangtze River basin, the following measures should be carried out:

1) Intensify basic research on icefish dynamics: Appropriate management needs better understanding of the basic characteristics of the wild icefishes in both the separate water area and the large river basin scale. Population dynamics, interspecific relationships and environmental determinants to icefishes are all important, particularly the relationships between icefish degradation and the threats from overfishing and habitat degradation.
2) Improve icefish fishery management: Fishing season and duration, fishing methods and intensity are key factors leading to overfishing. So fishing activities must be forbidden in icefishes' spawning seasons, and the fishing gears and fishing duration must also be under effective management in fishing season so as to decrease fishing stress for sustaining icefish fishery. Sometimes, icefish stock enhancement is necessary when a population size is too small to naturally restore.
3) Protect and restore icefish habitat: The number, area and quality of suitable habitats have a direct impact on the reproduction and development of icefishes. Taking effective steps to reduce spawning disturbance and improve habitat conditions will directly contribute to icefish conservation. Meanwhile, keeping and restoring the connection between spawning grounds, feeding grounds and wintering grounds will greatly facilitate the appropriate use of natural resources and increasing icefish stocks. Furthermore, reconnecting the isolated water bodies will be beneficial for inter-population recruitment contributing to icefish population stability at broad river basin scale.

Acknowledgements This research was financially supported by the National Natural Science Foundation of China (30570290), Beijing Municipal Commission of Education (KM200610028009) and WWFChina (CN008804). We are indebted to the local authorities and fishermen for their kind and consistent help in the long time fieldwork, to Aldemaro Romero for his kind editorial help and to three anonymous reviewers for their valuable comments on this manuscript.

References

Cao WX, Zhang G, Ma J, Yu D (1991) Preliminary studies on the phenomenon of size diminution of the fish resources in Lake Honghu. In: Honghu research group (ed) Studies on comprehensive exploitation of aquatic biological productivity and improvement of ecological environment in Lake Honghu. China Ocean, Beijing

Chen NS (1956) Preliminary study on *Neosalanx tangkahkeii taihuensis* Chen. Acta Hydrobiol Sin 2:324–335. in Chinese

Chen GH, Zhang B (1986) Reproducing field survey of Salangid fishes in Poyang Lake. Freshwat Fish 5:13–16. in Chinese

Chu YX (1998) Degradation of icefishes resources in Nanyi Lake and conserving strategies. Agric Sci Anhui Prov 26:84–86. in Chinese

Diao ZS (1986) Population composition of icefishes in Lake Chaohu and fishery suggestions. Fish Resour Chaohu 3:24–26. in Chinese

Dou SZ, Chen DG (1994) Taxonomy, biology and abundance of icefishes, or noodlefishes (Salangidae), in the Yellow River estuary of the Bohai Sea, China. J Fish Biol 45:737–748. doi:10.1111/j.1095-8649.1994.tb00940.x

Fang PW (1934) Study on the fishes referring to Slangidae of China. Sinensia 4:231–268

Hu CL, Liu JS, Peng FH, Yu FH (2001) Salangids in China and their ecological management principles for transplantation. J Lake Sci 13(3):204–210

Islam MS, Hibino M, Ohta T, Nakayama K, Tanaka M (2006) Environmental effect on diet, fecundity and condition of an endangered fish *Neosalanx reganius* (Osmeriformes) in the Chikugo Estuary, in the upper Ariake Bay, Japan. Aquat Living Resour 19:59–68. doi:10.1051/alr:2006005

Jackson DA, Harvey HH (1997) Qualitative and quantitative sampling of lake fish communities. Can J Fish Aquat Sci 54:2807–2813. doi:10.1139/cjfas-54-12-2807

Jiang ZG, Fan EY (2003) Exploring the endangered species criteria: rethinking the IUCN Red List Criteria. Biodiv Sci 11:383–392

Kapuscinski KL, Belonger BJ, Fajfer S, Lychwick TJ (2007) Population dynamics of muskellunge in Wisconsin waters of Green Bay, Lake Michigan, 1989–2005. Environ Biol Fishes 79:27–36. doi:10.1007/s10641-006-9132-2

Kawanabe H, Mizuno N, Hosoya K (2001) Freshwater fishes of Japan. Yama-kei, Tokyo (in Japanese)

Kim IS, Park JY (2002) Freshwater fishes of Korea. Kyohak, Seoul (in Korean)

Kim JK, Doiuchi R, Nakabo T (2006) Molecular and morphological differences between two geographic populations of Salanx ariakensis (Salangdidae) from Korea and Japan. Ichthyol Res 53:52–62. doi:10.1007/s10228-005-0315-1

Magnuson JJ, Tonn WM, Banerjee A, Toivonen J, Sanchez Q, Rask M (1998) Isolation vs. extinction in the assembly of fishes in small northern lakes. Ecology 79:2941–2956

Myers RA, Barrowman NJ, Hutchings JA, Rosenberg AA (1995) Population dynamics of exploited fish stocks at low population levels. Science 269:1106–1108. doi:10.1126/science.269.5227.1106

Nelson JS (2006) Fishes of the world. Wiley, Hoboken, New Jersey

Olden JD, Jackson DA, Peres-Neto PR (2001) Spatial isolation and fish communities in drainage lakes. Oecologia 127:572–585. doi:10.1007/s004420000620

Pauly D (1998) Beyond our original horizons: the tropicalization of Beverton and Holt. Rev Fish Biol Fish 8:307–334. doi:10.1023/A:1008863215253

Sun GY (1982) Study on Salangidae of the Yangzte River Delta and its adjacent waters. J East China Nor University 1:111–119

Takita T (1996) Threatened fishes of the world, *Neosalanx reganius* Wakiya and Takahashi, 1937 (Salangidae). Environ Biol Fishes 47:100. doi:10.1007/BF00002383

Takita T, Kawaguchi K, Masutani H (1988) Distribution and morphology of the salangid fish, *Neosalanx reganius*. Jpn J Ichthyol 34:497–503

Tang ZP, Xie H, Li B, Xie YH, Zhang SD, Yu F (2000) The biological aspects of Ariake icefish (*Salanx ariakensis* Kshinouye) in Yalujiang River. J Dalian Fish Univ 15:113–118

Taylor CM (1996) Abundance and distribution within a guild of benthic stream fishes: local processes and regional patterns. Freshw Biol 36:385–396. doi:10.1046/j.1365-2427.1996.00104.x

Wang GP (1995) Survey study on the environmental resources of *Neosalanx taihuensis* in Lake Dongting. Inland Fish 3:11. in Chinese

Wang SM, Dou HZ (1998) Lakes in China. Science Press of China, Beijing. (in Chinese)

Wang ZS, Fu CZ, Lei GC (2002) Biodiversity of Chinese icefishes (Salangidae) and their conserving strategies. Biodiv Sci 10:416–424. in Chinese

Wang ZS, Lu C, Hu HJ, Xu CR, Lei GC (2004) Dynamics of icefish (Salangidae) stocks in Nanyi Lake, eastern China: degradation and overfishing. J Freshw Ecol 19:271–278

Wang ZS, Lu C, Hu HJ, Zhou Y, Xu CR, Lei GC (2005) Freshwater icefishes (Salangidae) in the Yangtze River basin of China: spatial distribution patterns and environmental determinants. Environ Biol Fishes 73:253–262. doi:10.1007/s10641-005-2146-3

Wang ZS, Chen MH, Lu C, Xu CR, Lei GC (2006) Species diversity and spatio-temporal distribution patterns of icefishes (Salangidae) in Poyang Lake. Acta Ecol Sin 26:1337–1344

Wootton RJ (1998) Ecology of teleost fishes. Kluwer Academic, London

Xie P, Chen YY (1999) Threats to biodiversity in Chinese inland waters. Ambio 28:54–61

Zeng XC (1990) Fishery resources in the Yangtze River basin. Marine Press, Beijing. (in Chinese)

Zhang GQ (1992) Preliminary study on stock variation of icefish *Hemisalanx Prognathus* Regan. Oceanol Limnol Sin 23(5):517–526. in Chinese

Zhang YL (1993) The phylogenetic biogeography of Salangidae. In: Zhang GX (ed) Scientific Treatise on Systematic and Evolutionary zoology, Volume 2. Science and Technology Press of China, Beijing

Zheng CY (1989) Fishes in Zhu River. Science Press of China, Beijing. (in Chinese)

Zhu CD (1985) A preliminary study on growth and feeding habits of icefish in Taihu Lake. J Fish China 9:275–287. (in Chinese)

Zhu SQ (1995) Synopsis of freshwater fishes of China. Jiangsu Science and Technology, Nanjing. (in Chinese)

Environ Biol Fish (2009) 86:119–120
DOI 10.1007/s10641-008-9356-4

Threatened fishes of the world: *Yunnanilus discoloris* Zhou & He 1989 (Cobitidae)

Yan-E Jiang · Xiao-Yong Chen · Jun-Xing Yang

Received: 4 September 2007 / Accepted: 1 April 2008 / Published online: 30 May 2008

Keywords IUCN red list · Loach · Cypriniformes · Dianchi

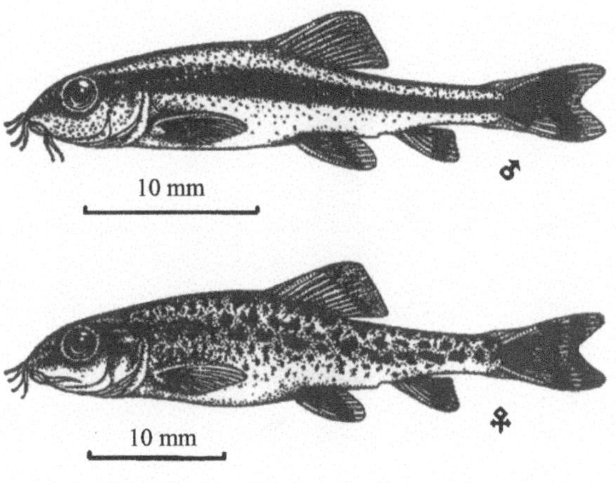

Common name: Yi se yun nan qiu. *Conservation status*: This species is currently not listed. *Identification*: Body naked, origin of dorsal fin at middle of body, branched caudal rays 14, branched ventral rays seven, dwarfism (Zhou and He 1989; Figure). *Distribution*: *Yunnanilus discoloris* is one of the endangered fish in Lake Dianchi (24°28′–25°28′N, 102°30′–103°00′E, 1,886 m above sea level) and White Dragon spring (Li 2001), which connect with effluent from the lake, Yunnan, China. This species was not found in Lake Dianchi from the 1950–60s, few around Lake Dianchi from the 1970–80s. Now it is found in the White Dragon spring at Chenggong, Kunming which flows into Lake Dianchi. *Abundance*: The number of the total population is estimated about 500 individuals in White Dragon spring ponds. *Y. discoloris* is considered to be rare in Lake Dianchi now. *Habitat and ecology*: Lives in clear springs, at a depth of 80 cm, with sand, stones and some macrophytes on the bottom. A difference of the color pattern between male and females in mature individuals (Zhou and He 1989). *Reproduction*: The White Dragon spring has been protected, so these fish are protected too. *Threats*: Its loss from the lake is likely due to introduced fish species ("the four cultured carp", Black carp, *Mylopharyngodon piceus*, Grass carp, Silver carp, *Hypophthalmichthys molitrix* and Big head carp, *Hypophthalmichthys nobilis*) (He et al. 1983; Gao et al. 1990), water pollution and loss of macrophytes. Current threats to the springs are introduced species and modification of spring structure. *Conservation actions*: Beginning in 2003, a GEF/World Bank funded project was established in Kunming, Yunnan to conserve and try to restore the aquatic biodiversity of Lake Dianchi. *Conservation recom-*

Xiao-Yong Chen and Jun-Xing Yang contributed equally to this paper

Y.-E. Jiang (✉) · X.-Y. Chen · J.-X. Yang
Department of systematic Zoology,
Kunming Institute of Zoology,
Chinese Academy of Sciences,
32 Jiao Chang Dong Road, Kunming,
Yunnan 650223, China
e-mail: jiangyane03@yahoo.com

X.-Y. Chen
e-mail: chenxy@mail.kiz.ac.cn

J.-X. Yang
e-mail: yangjx@mail.kiz.ac.cn

Y.-E. Jiang
Graduate School of Chinese Academy of Sciences,
Beijing 100039, China

mendations: *Y. discoloris* should be listed as a province protected animal. Continuing monitoring and surveys are required for survival of this endangered fish.

References

Gao LC, Zhuang DD, Guo QZ, Wang YH (1990) Fishery resources of lakes in Yunnan plateau. Jiangsu Science and Technology, Nanjing, China, p 78

He JC, Liu ZH, Li CJ (1983) A primary investigation on the impact of pollution on fishes in Dianchi Basin. In: Pollution and Aquatic life in Lake Dianchi. Yunnan, pp 38–47

Li WX (2001) The distribution of surviving *Sinocyclocheilus grahami* and other endemic fishes in Dianchi Lake Valley. Journal of Jishou University (Natural Science Edition) 22 (4):72–74

Zhou W, He JC (1989) A new species of dwarfish in *Yunnanilus* (Cypriniformes: Cobitiae). Acta Zootaxonomica Sinica 14(3):380–384

Environ Biol Fish (2009) 86:121–130
DOI 10.1007/s10641-008-9412-0

Feeding, morphological changes and allometric growth during starvation in miiuy croaker larvae

Xiu-Juan Shan · Liang Cao · Wei Huang · Shuo-Zeng Dou

Received: 22 November 2007 / Accepted: 31 July 2008 / Published online: 30 August 2008

Abstract We investigated the effects of the timing of first feeding (larvae in F0, F1, F2, F3 and S were first fed on day 3, 4, 5, 6 days after hatching (DAH) and unfed, respectively) on feeding, morphological changes, survival and growth in miiuy croaker larvae at 24°C. The fed larvae initiated feeding on 3 DAH and reached point of no return (PNR) on 6 DAH. Larvae in F0 and F1 groups survived apparently better than F2 group at the end of the experiment on 36 DAH. High larval mortality occurred from 3 to 7 DAH in all feeding groups, accounting for 40% (F0, F1 and F2 groups) to 90% (F3 and S groups) of the total mortality. Larvae in F0 and F1 groups grew better than F2 group throughout the experiment. Eye diameter, body height, head height and mouth gape of the first feeding larvae were more sensitive to starvation than other morphometrics and could be used as indicators for evaluating their nutritional status. Results indicated that delayed first feeding over 1 day after yolk exhaustion could lead to poor larval survival and growth. To avoid starvation and obtain good growth in culturing, larvae feeding should be initiated within 1 day after yolk exhaustion at 24°C.

Keywords *Miichthys miiuy* larvae · Delayed feeding · Morphometrics · Growth and survival

X.-J. Shan · L. Cao · W. Huang · S.-Z. Dou (✉)
Institute of Oceanology, Chinese Academy of Sciences,
7 Nan-hai Lu Road,
Qingdao 266071, People's Republic of China
e-mail: szdou@ms.qdio.ac.cn

X.-J. Shan · L. Cao · W. Huang
Graduate School, Chinese Academy of Sciences,
Beijing 100049, People's Republic of China

Introduction

Starvation is commonly considered to be a critical cause for mortality and poor growth performance in fish during their larval and juvenile stages both in captivity and in the wild (Blaxter and Hempel 1963; May 1971; Leggett and Deblois 1994; Dou et al. 2002; Yúfera and Darias 2007). It occurs when fish larvae fail to establish successful first feeding for various reasons such as low feeding ability, shortage of food or environmental perturbations (Fogarty et al. 1991; Iguchi and Mizuno 1999). Many studies indicate that starvation at first feeding stage could result in drastic morphological deformity and thus lead to low feeding ability and high mortality during the larval development (McGurk 1984; Kjørsvik et al. 1991; Gisbert et al. 2004). Effects of starvation on the larval development are usually evidenced in morphological changes and allometric growth because the organs related to feeding, swimming and digesting mainly develop and function at larval stage (Yúfera et al. 1993; Bisbal and Bengtson 1995; Osse et al. 1997). In many fish species, even a short period of food deprivation at first feeding could induce severe developmental, behavioral and nutritional problems in the larvae and adversely affect their ability to withstand starvation, feeding, growth and survival (Yin and Blaxter 1987; Gisbert and Williot 1997; Peña and Dumas 2005).

The miiuy croaker, *Miichthys miiuy*, is an important commercial species that inhabits the coastal waters of China, Japan and Korea (Zhang and Hong 2000; Choi et al. 2002). Due to overexploitation, wild populations of this fish have drastically decreased in recent years. On the other hand, morphological deformity and high mortality in the artificially bred fingerlings have hindered their commercial mass production. So far, few studies have dealt with the morphological development (Zhong et al. 2005; Luo et al. 2006), growth, survival and digestive enzymes of miiuy croaker during their early life stage (Shan and Dou 2008; Shan et al. 2008a,b). Little information is available at present to understand how starvation may affect the development, growth and survival of the larvae. This study is to investigate the effects of delayed first feeding on these biological parameters in order to better understand their larval growth and survival strategies during starvation and provide information for feeding management in larval rearing.

Materials and methods

Egg incubation and larval rearing

We obtained fertilized eggs from adult miiuy croakers that were stocked in a 100 m^3 indoor concrete pond and spontaneously spawned at 24°C under natural photo-period from September to October in Sanduao bay, Fujian, China. We collected the eggs within 6 h of fertilization with a screen placed inside a rectangular flow-through tank and transferred them into cylindrical incubators (500 l) at an estimated density of 800 eggs per liter. Temperature of the hatching tanks was maintained at 24°C. A light regime of 12L (800–1,000 lx at 0600–1800 hours):12D (<1 lx at 1800–0600 hours) was provided by using fluorescent light during the egg incubation. Gentle aeration was provided by air diffusers to keep the water constantly oxygen-saturated. The time at which 90% of the viable eggs (i.e. hatched eggs plus unhatched viable eggs) were hatched was defined as hatching time.

We transferred the newly hatched larvae with a 500 ml beaker from the incubators into the 50 l experimental tanks. They were supplied with overflowing filtered sea water. The light regime in the experimental tanks was the same as that of the egg incubators. Water exchange rate was about 1 l min^{-1}. Gentle aeration was provided to maintain dissolved oxygen level and promote a homogeneous distribution of microalgae and live food after first feeding commenced. Water temperature was controlled to 24°C via water bath throughout the experiment. Larvae started to take food on 3 days after hatching (DAH). The feeding schedule was designed as following (adjusted according to the desired timing of first feeding for each treatment; Fig. 1): rotifers *Brachionus rotundiformis* were given to the larvae at a density of about 10 ind ml^{-1} (twice a day at 1000 and 1600 hours, respectively) from mouth opening (3 DAH) to 9 DAH. They were fed with the microalgae *Nannochloropsis oculata* and enriched with docosahexaenoic acid (DHA; Kangke B, Qingdao) for 12 h prior to being given to the larvae. Meanwhile, microalgae were added into the experimental tanks as food for the rotifers. *Artemia* nauplii (2–5 ind ml^{-1}) enriched with DHA were offered to the larvae from 8 to 15 DAH. The larvae were provided freshly-collected, live copepods from 14 to 26 DAH at a daily ration of 20–40 mg l^{-1}. Compound diet (Tianbang, Shanghai, 1.0–1.5 mm in diameter) was offered to the larvae at a daily ration of 30–60 mg l^{-1} from 24 to 36 DAH. Uneaten food and dead larvae were removed from the experimental tanks every day.

To provide experimental fish for the point of no return (PNR) experiment, we stocked 2,000 larvae in a 50 l unfed holding tank where the larvae were deprived of food from hatching until no larvae survived. We held another 2,000 larvae in a separate 50 l holding tank and offered rotifers to them from 3 DAH up to the end of the experiment on 7 DAH. Other rearing conditions in the holding tanks were identical to those in the experimental tanks.

Experiment I: PNR and effects of delayed first feeding on larval growth

To determine the PNR of miiuy croaker larvae, we set up two feeding treatments in triplicates: starting from 3 DAH, 20 larvae from both the unfed and fed

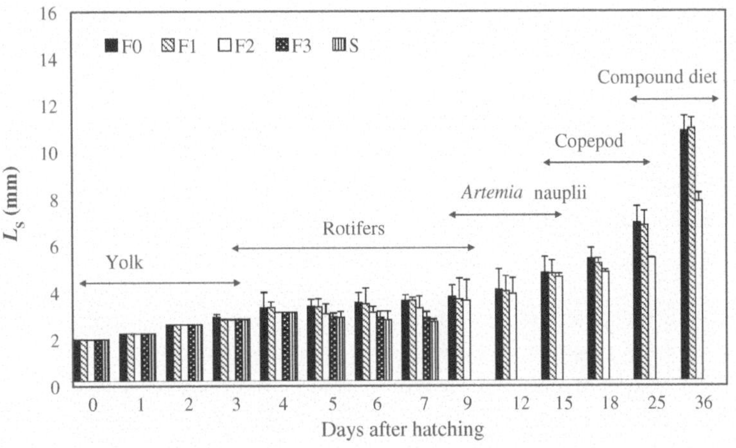

Fig. 1 Feeding schedule and growth (L_s, mean±SD) of miiuy croaker larvae at different ages of delayed first feeding. Codes for feeding treatments: F0, F1, F2, F3 and S were first fed on 3, 4, 5, 6 DAH and unfed, respectively

holding tanks were introduced into each of the 500 ml experimental tanks every day, respectively. They were provided rotifers at a density of about 10 ind ml^{-1} for 4 h, then sampled and examined under a binocular microscope (Nikon SMZ-1000) with ACT-2U image system to identify the presence of rotifers in their guts. Feeding rate was the percentage of larvae that took food to the total number of larvae initially stocked. PNR was defined as the time after hatching when the first feeding rate of the starved larvae reached less than half of the first highest feeding rate when food became available (Dou et al. 2002). The guts of the larvae were dissected and the number of the distinguishable rotifers was counted under a binocular microscope to determine the feeding intensity, which was described as the number of rotifers taken by a larva per hour.

To investigate the effects of delayed first feeding on larval growth, we set up five feeding treatments in triplicates: larvae in F0, F1, F2 and F3 treatments were first fed on 0 (3 DAH), 1 (4 DAH), 2 (5 DAH) and 3 (6 DAH) days after mouth opening, respectively, while those in S treatment were never offered food and kept as starved controls. All first feedings were started during the light period because miiuy croaker larvae are visual feeders. Experimental tanks were 50 l and larval stocking density was 50 ind l^{-1}. Lighting (800–1,000 lx at 0600–1800 hours) was provided using fluorescent light. All experimental tanks were enclosed within black plastic sheets to prevent light escaping.

Starting from hatching, we sampled 20 larvae at random from each experimental tank every day up to 7 DAH and then on 9, 12, 15, 18, 25, and 36 DAH, respectively, to investigate morphological changes and growth of the larvae during starvation. The sampled larvae were anaesthetized and sacrificed with tricaine methanesulphonate (MS-222), preserved in 10% formalin and then stored in the dark at 4°C for morphological measurement at laboratory. Morphometrics included standard length (L_S), tail length (L_T), head length (L_H), head height (H_H), body height (H_B), eyes diameter (D_E) and mouth gape (G_M) (Fig. 2). They were measured to the nearest 0.1 mm under a microscope. Specific growth rate (SGR) was calculated as: $SGR = 100 \times (\ln L_t - \ln L_0)/t$, where L_0 and L_t were the initial and final L_s (mm) of the larvae, respectively, and t is time duration of the experiment in days. Coefficient of variation (CV_L) of L_s ($CV_L = 100 \times SD/\text{mean } L_s$) in each feeding treatment was

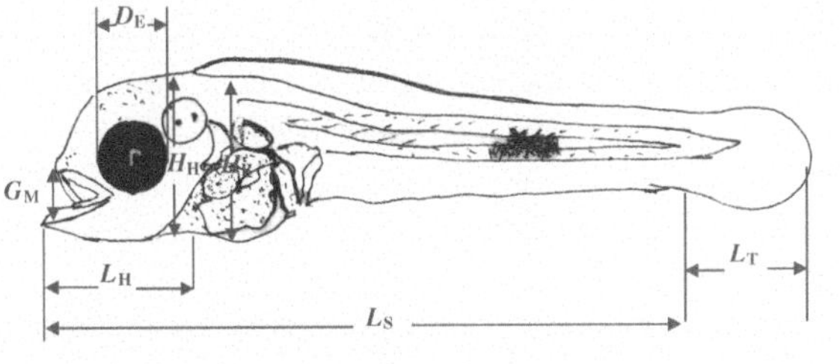

Fig. 2 Morphological measurements of miiuy croaker larvae. L_S Standard length, L_T tail length, L_H head length, H_H head height, H_B body height, D_E eyes diameter, G_M mouth gape

analyzed to investigate the effects of delayed feeding on individual size variation. The allometric growth was evaluated by the following ratios of morphometrics: L_T/L_S, L_H/L_S, H_H/L_S, H_B/L_S, D_E/L_S, G_M/L_S, D_E/L_H and H_H/L_H.

Experiment II: Effects of delayed first feeding on larval survival

Experimental protocol was identical to the growth experiment except that no samplings were conducted throughout the experiment. Experimental tanks were cleaned and dead larvae were removed and counted twice every day by siphoning.

Data analysis

Each pair of the morphometrics between fed (F0) and starved (S) larvae was compared using a t-test. A one-way ANOVA was performed to test the effects of delayed first feeding on larval growth (L_S and SGR) between delayed feeding groups, followed by Tukey's test to compare the differences between treatments wherever a significant effect ($P<0.05$) was detected.

Results

Onset of first feeding and PNR

The earliest first feeding was detected on 3 DAH, one day earlier than yolk exhaustion. At first feeding, most of the larvae had a functional mouth but only 10% of them started to take food (Fig. 3a). Feeding rate of the fed larvae continuously increased from 3 DAH and reached over 90% at the end of the experiment on 7 DAH. Feeding rate of the starved survivors when first offered food increased with the age of delayed feeding up to 5 DAH (85%) but decreased sharply to 25% and 4% on 6 and 7 DAH, respectively (Fig. 3a). PNR was on 6 DAH, when the feeding rate of the starved larvae started to decrease to less than half of the initial peak (85%) on 5 DAH. As starvation progressed, larvae tended to become less active or sensitive to external stimulus of food.

Feeding intensity in the fed larvae continuously increased from less than one rotifer per larva per on 3 DAH to three rotifers per larva per hour on 7 DAH. In the starved controls, it increased slightly from 3 to 5 DAH and then gradually decreased until the end of the experiment on 7 DAH, without reaching one rotifer per larva per hour during the experiment (Fig. 3b).

Morphological changes and larval growth

All the morphological measurements of the fed larvae (F0) increased gradually as they developed. Except for D_E and H_H, no morphological measurement was significantly different from that of the starved larvae (S) before PNR on 6 DAH ($P>0.05$ for each comparison; Fig. 4). Compared to the fed controls, starved larvae started to show significantly slow morphological development in D_E and H_H on 5 DAH, in H_B and G_M on 6 DAH, in L_H and L_T on 7 DAH, respectively ($P<0.05$ for each comparison; Fig. 4). H_B/L_S, D_E/L_S and D_E/L_H in the fed larvae started to significantly differ from the starved ones on 6 DAH ($P<0.05$ for each comparison), while no significant difference was observed between these two groups in L_H/L_S, L_T/L_S, H_H/L_S, G_M/L_S or H_H/L_H ($P>0.05$ for each comparison; Fig. 5).

Larval growth strongly depended on the timing of first feeding. L_S and SGR of the 7 DAH larvae in F0 (3.6 mm, 9.9%), F1 (3.3 mm, 8.4%) and F2 groups

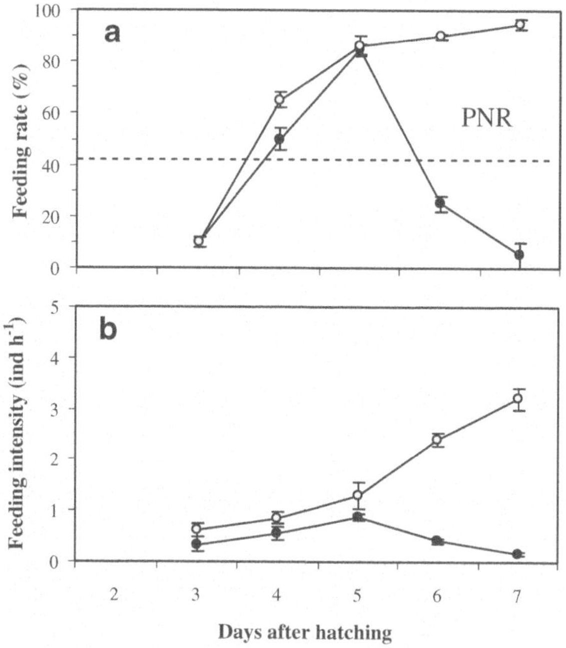

Fig. 3 Feeding rate (**a**, mean±SD) and feeding intensity (**b** mean±SD) of fed (*open circles*) and starved (*solid circles*) miiuy croaker larvae when food was first offered

Fig. 4 Morphometrics (mean±SD) of fed and starved miiuy croaker larvae at first feeding stage. Codes for the morphometrics as in Fig. 2

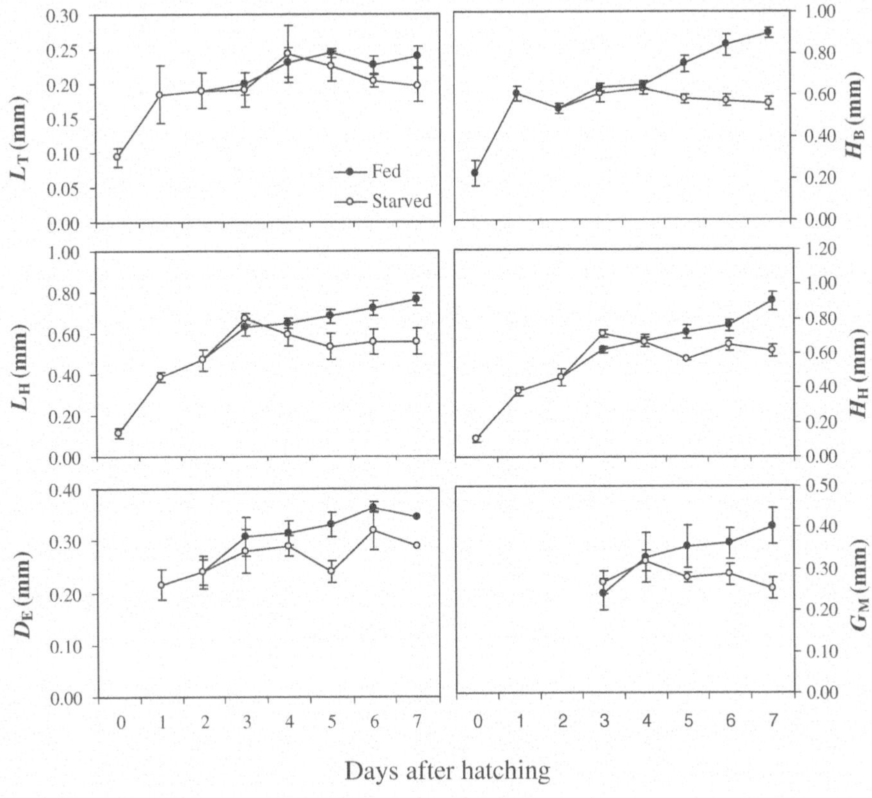

(3.2 mm, 7.5%) were significantly larger than in F3 (2.6 mm, 5.0%) and S (2.5 mm, 4.6%) groups ($P<0.05$ for each comparison) (Figs. 1 and 6a). Neither L_S nor SGR of the 36 DAH larvae in F0 (9.7 mm, 4.8%) and F1 (9.5 mm, 4.8%) groups were significantly different ($P>0.05$ for each comparison), but they were both significantly larger than in F2 (7.1 mm, 4.1%) group ($P<0.05$ for each comparison; Figs. 1 and 6b). CV_L of the first feeding larvae (7 DAH) was 5.0, 6.8, 6.0, 4.5 and 4.5 in F0, F1, F2, F3, S groups, respectively, in contrast with 15, 13 and 11 in F0, F1 and F2, respectively, in the 36 DAH larvae.

Larval survival

Larval survival was 71%, 70%, 61%, 4% and 3% on 7 DAH in F0, F1, F2, F3 and S groups, respectively. No larvae survived up to 8 DAH in F3 and S groups (Fig. 7). In these two groups, most of the larvae died on 3–7 DAH with an average daily mortality rate of 18.5%. On 36 DAH, it was 16% and 15% in F0 and F1 groups, respectively, which was apparently higher than 9% in F2 group (Fig. 7). Mortality in F2 group mainly occurred on 3–7 DAH averaging a daily rate of 6.5% but kept at a relatively low level (1.8%) after 7 DAH till the end of the experiment on 36 DAH. Similarly, high mortality was also observed on 3–7 DAH in F1 and F0 groups with a daily rate of 5%, which then decreased gradually (1.6%) afterwards (Fig. 7).

Discussion

Onset of first feeding and feeding ability of fish larvae

The timing of first feeding in fish larvae largely depended on egg quality, yolk absorption rate and temperature. Shan and Dou (2008) reported that, among the 51 fish species in previous studies, a majority of the species could initiate first feeding within 1–3 DAH at 22–32°C, in contrast with 4–8 DAH at 6–21°C in other species. High temperature generally promoted fish larvae to initiate feeding earlier. For instance, first feeding in Japanese flounder, *Paralichthys olivaceus*, larvae advanced from 7 to 2 DAH as temperature

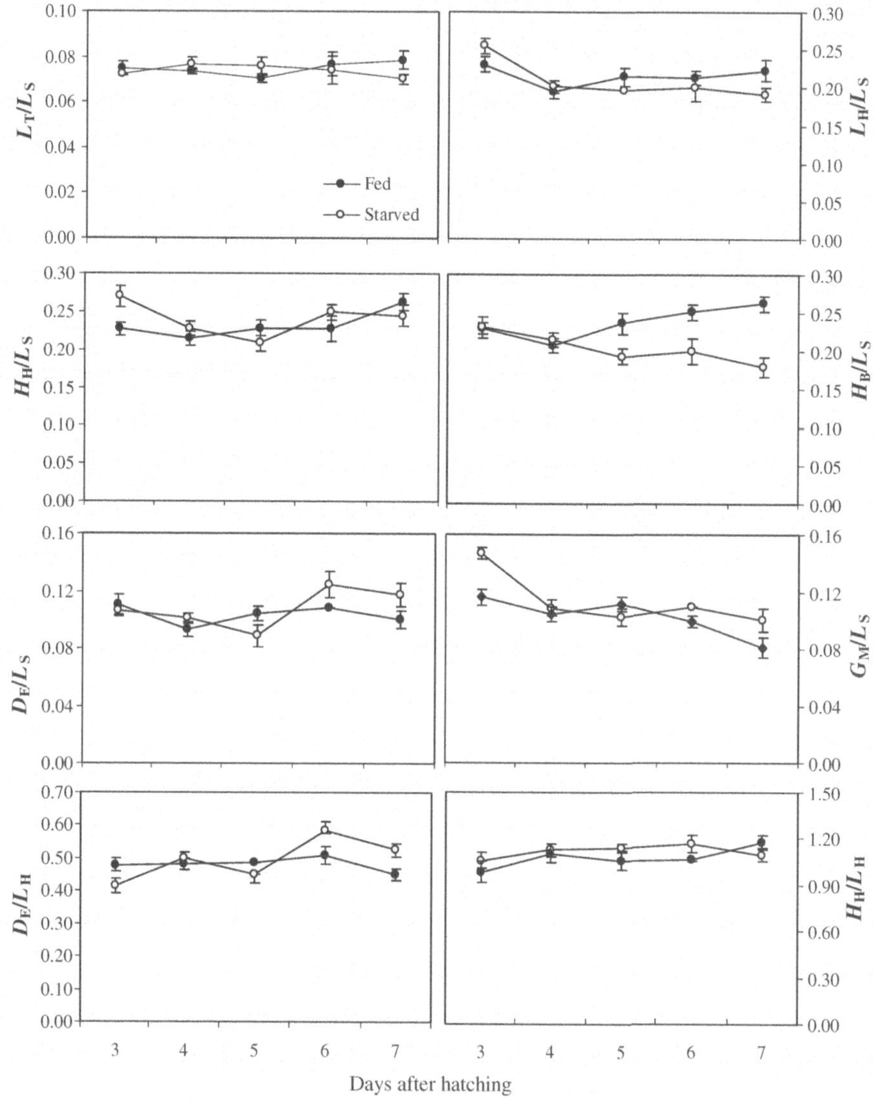

Fig. 5 Ratios of morphometrics (mean±SD) of fed and starved miiuy croaker larvae at first feeding stage. Codes for the morphometrics as in Fig. 2

increased from 12°C to 21°C (Seikai et al. 1986; Dou et al. 2002, 2005), while it advanced 3 days in Atlantic herring *Clupea harengus* larvae as temperature increased from 7.5°C to 9.2°C (Yin and Blaxter 1987). However, Atlantic salmon, *Salmo salar*, and tilapia, *Oreochromis mossambicus*, larvae did not start to feed until 41 DAH at 12.4°C (Hansen and Møller 1985) and 7 DAH at 28°C (Rana 1985), respectively. The long-delayed first feeding in these fish larvae might be due to the facts that both the salmon and tilapia had large eggs (5.5 and 2.0 mm in diameter, respectively) and their yolk sacs took a relatively long time to complete absorption (on 70 and 15 DAH, respectively) (Hansen and Møller 1985; Rana 1985). In the present study, miiuy croaker larvae started to take food on 3 DAH at 24°C, 1 day earlier than at 22°C (Luo et al. 2006). Moreover, the yolk absorption was completed on 4 DAH, suggesting that there existed a 1-day mixed nutrition period in the first feeding larvae. Similar findings were also obtained for other sciaenidae fishes such as *Nibea chui* (on 2 DAH at 21–23°C; Huang et al. 2005) and large yellow croaker *Pseudosciaena crocea* (on 3 DAH at 23–25°C; Yu et al. 2003).

First feeding of the miiuy croaker larvae coincided with the occurrence of eye pigmentation. Moreover,

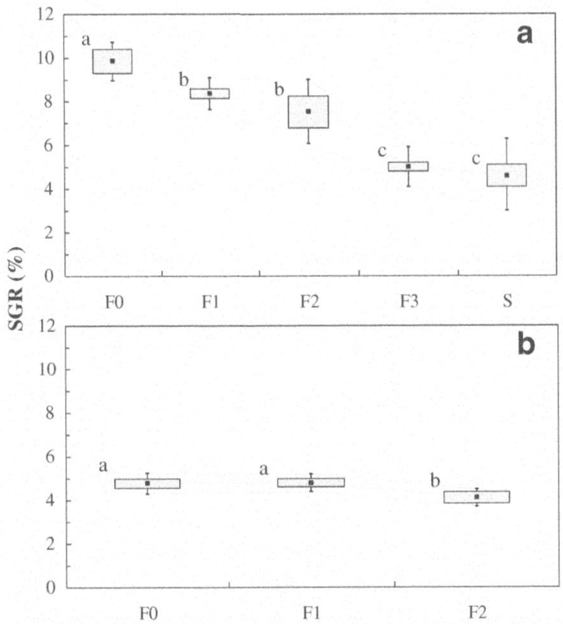

Fig. 6 Specific growth rate (SGR) (mean±SD) of miiuy croaker larvae at different ages of delayed first feeding on 7 DAH (**a**) and 36 DAH (**b**). Different superscripts indicated significant differences at $P<0.05$. Codes for feeding treatments as in Fig. 1

active swimming and foraging were frequently observed in the first feeding larvae. Similar results were also obtained in other visual feeders such as Japanese flounder (Dou et al. 2005) and spotted sand bass, *Paralabrax maculatofasciatus* (Peña and Dumas 2005). Therefore, occurrence of eye pigmentation and active swimming behavior could be used as potential indicators to determine the time of initially offering food to visual feeders in aquaculture.

Feeding ability of first feeding larvae was influenced by the functional development of searching, cruising and feeding organs (Yin and Blaxter 1989). Early studies indicated that feeding rate was generally low (10–50%) in most of the first feeding larvae (Yúfera et al. 1993; Brinkmeyer and Holt 1998; Olsen et al. 1999; Hamlin and Kling 2001; Puvanendran and Brown 2002; Huang et al. 2005; Qiao et al. 2007). Since starvation could lead to malformed development of these organs, delayed first feeding plays a critical role in determining the feeding ability of fish larvae. In the present study, feeding rate of the miiuy croaker larvae in F0 was 10% on 3 DAH and reached 85% on 5 DAH. However, the larvae that were starved over 5 DAH could not establish successful first feeding even when food was available, suggesting that their ability to feed decreased as starvation was prolonged. Larvae first fed on 3 DAH showed the ability to take food, but 20–40% of the larvae could not successfully establish active feeding until 4 DAH. Therefore, even a short time period of starvation (e.g. 1 day after yolk exhaustion) might have adverse impacts on the feeding ability of the survivors. This was also supported by the finding that feeding intensity of the starved larvae started to decrease drastically after 5 DAH.

In many fish species, larvae show a mixed feeding period, during which they start to take food before yolk exhaustion (Yúfera et al. 1993; Mookerji and Rao 1999; Dou et al. 2005). Duration of this period varies with species and is considerably influenced by elemental cues such as temperature (Seikai et al. 1986; Yin and Blaxter 1987; Dou et al. 2005). To minimize the discrepancies caused by temperature and yolk absorption rate, Shan and Dou (2008) followed Kamler (1992) and employed the R ratio ($R=t_y/t_f$, t_y is the time from first feeding to yolk exhaustion, t_f is the time from hatching to first feeding) to evaluate and compare the mixed feeding period of 51 fish species. Among them, some fish larvae had a relatively long mixed feeding time ($R>1$) such as Japanese pufferfish, *Takifugu rubripes* ($R=$ 2:1; Jiang et al. 2002) and rohu, *Labeo rohita* ($R=$4:1; Mookerji and Rao 1999). They generally had sufficient time to learn to feed and improve feeding ability before yolk exhaustion. However, many of the fish larvae had a very short mixed feeding period ($R<1$) such as in Peruvian anchoveta, *Engraulis ringens* ($R=$

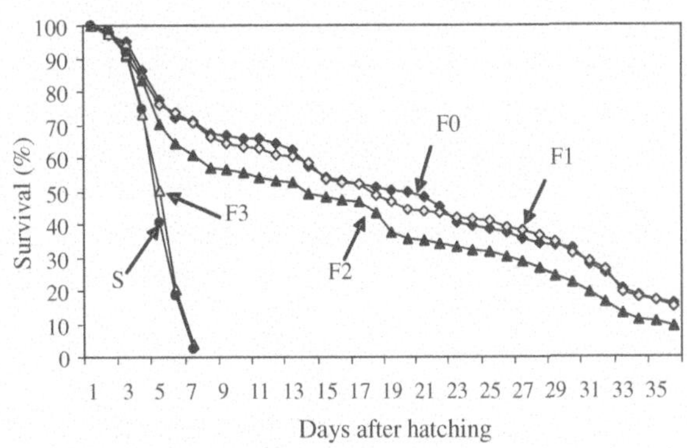

Fig. 7 Survivorship of miiuy croaker larvae at different ages of delayed first feeding. Codes for feeding treatments as in Fig. 1

1:5; Ware et al. 1981) and flounder, *Platichthys flesus* (R=1:6; Yin and Blaxter 1987). Similar to these species, miiuy croaker larvae had only 1 day mixed feeding period (R=1:3) and exhibited low feeding ability at first feeding, suggesting that they were very likely to suffer starvation during this critical period. Starvation commonly led to high larval mortality in these fish species both in wild populations and at laboratory larval rearing.

Larval survival and growth

The ability to withstand starvation in fish larvae is species specific and generally deteriorates with increased temperature. Shan and Dou (2008) compared the PNR of 51 fish species and found that many larvae reached PNR within 2 days or less after yolk exhaustion at 22–32°C such as in lined sole, *Achirus lineatus* (0.5 days at 24–28°C) and anchovy, *Anchoa mitchilli*, (0.6–0.9 days at 28–32°C) (Houde 1974), in contrast with 3–8 days at 6–21°C in other species. However, a few species such as Bagrid catfish, *Pelteobagrus vachelli*, and sturgeon (*Huso huso*♀ × *Acipenser baerii*♂) exhibited PNR as long as 11.5 days at 24–25°C (Ma et al. 2006) and 16 days at 16–17°C (Gao 2001), respectively. The relative large yolk sac (1.7 and 4.8 mm in egg diameter, respectively) and oil globules in this fish might provide a relatively large nutrition reserve for the larvae and make contributions to their ability to withstand starvation. PNR of the miiuy croaker larvae at 24°C was 6 DAH (i.e. 2 d after yolk exhaustion). This meant that they had only 3 days after mouth opening on 3 DAH to establish successful first feeding. Otherwise, they would suffer starvation-induced mortality. Small egg size (<1.0 mm) of miiuy croaker and relatively high rearing temperature (24°C) in this experiment might promote their early development and thus prompt energy consumption, which led to their reaching PNR in a short time after yolk exhaustion.

In this study, larval mortality mainly occurred during 3–7 DAH in F3 and S groups, accounting for over 90% of the total mortality throughout the experiment. In F0, F1 and F2 groups, larval mortality also concentrated on 3–7 DAH (accounting for 40% of the total mortality) but decreased gradually afterwards. Dead larvae were characterized with starvation-induced morphological deformity such as concavity of head, gut collapse and body malformation. These showed that miiuy croaker larvae suffered a course of starvation mortality around the first feeding period even if initial feeding was not delayed. This might be due to the fact that the time in which the larvae had to learn to take food and establish successful first feeding before reaching PNR was too limited. Moreover, although the survival curve pattern in F2 group was similar to those in F0 and F1 groups, it was apparently lower than the latter two groups during the experiment. This indicated that starvation effect on the first feeding larvae might retain and function on the survival of the fish throughout their whole larval stage.

On 7 DAH, SGR of the fed larvae (F0) was apparently higher than all the delayed feeding larvae. Moreover, it was significantly larger in F1 and F2 groups than in F3 and S groups. This suggested that the larval growth was sensitive to starvation at first feeding stage and even a short time period of starvation could adversely affect their growth. However, on 36 DAH, SGR of the larvae in F0 and F1 was not significantly different, although it was significantly larger than in F2. This suggested that 1 day-delayed feeding larvae (F1) might resume growth through a certain mechanism of compensatory growth during the later larval development. Similar results were obtained in grunion, *Leuresthes tenuis* (May 1971); striped bass, *Morone saxatilis* (Rogers and Westin 1981) and *P. maculatofasciatus* larvae (Peña and Dumas 2005). Delayed feeding for 2 days after yolk exhaustion (F2) would have adverse impacts on growth throughout the larval stage in miiuy croaker at 24°C. Moreover, individual variation in larval growth decreased with the age of delayed first feeding. This was in agreement with the findings for other species such as summer flounder, *Paralichthys dentatus* (Bisbal and Bengtson 1995); California halibut, *Paralichthys californicus* (Gisbert et al. 2004) and Japanese flounder (Dou et al. 2002, 2005). Larvae fed before PNR exhibited variable responses to food and thus varied considerably in individual feeding ability, leading to high individual growth variation. As starvation progressed, a large number of larvae died and only the strong survivors could establish successful feeding when food was provided, possibly resulting in relatively high individual growth uniformity.

Morphological measurements are often used as important indicators to evaluate starvation and nutritional status in fish larvae because of its ease and practicality of application (Ehrlich et al 1976; Theilacker 1978; Bisbal and Bengtson 1995; Theilacker and Porter 1995; Dou et al. 2002; Gisbert et al. 2004). It is particularly useful and convenient to assess larval starvation in the wild when morphological development, age determination and growth of the larvae have been well studied. However, the disadvantage of this method is that individual morphological development of fish larvae varies considerably even when initial feeding is not delayed and therefore, it is a difficult task to deriver valid criteria at laboratory for application in the wild. From this study, we concluded that D_E, H_B, H_H, G_M, H_B/L_S, D_E/L_S and D_E/L_H in miiuy croaker larvae were more sensitive to starvation than other morphometrics and thus could be used as potential indicators for evaluating nutritional status of the first feeding larvae. However, given the fact that knowledge about the larval development of this fish is still lacking, we should be cautious to extrapolate the experimental data to investigate larval starvation in nature. A combination with other methods such as histological and biochemical analysis should be applied to validate the effectiveness of the morphological indicators inferred from this study in the future.

Acknowledgements This study was supported by the Ministry of Science and Technology of P. R. China under grant contracts No. 2007CB407305 and No. 2006BAD09A02. We thank DC Zhang, ZZ Xiao, MY Chen and F Gao for their advice on experimental methods. We greatly appreciate the editor and reviewers' constructive comments on the manuscript. We are grateful to MQ Zhang for his support in the rearing of rotifers.

References

Bisbal GA, Bengtson DA (1995) Effects of delayed feeding on survival and growth of summer flounder *Paralichthys dentatus* larvae. Mar Ecol Prog Ser 121:301–306 doi:10.3354/meps121301

Blaxter JHS, Hempel G (1963) The influence of egg size on herring larvae (*Clupea harengus* L.). J Cons Int Explor Mer 28:211–240

Brinkmeyer RL, Holt GJ (1998) Highly unsaturated fatty acids in diets for red drum (*Sciaenops ocellatus*) larvae. Aquaculture 161:253–268 doi:10.1016/S0044-8486(97)00274-3

Choi Y, Kim JH, Park JY (2002) Marine fishes of Korea. Kyohaksa Publ Co Ltd, Seoul

Dou S, Masuda R, Tanaka M, Tsukamoto K (2002) Feeding resumption, morphological changes and mortality during starvation in Japanese flounder larvae. J Fish Biol 60:1363–1380 doi:10.1111/j.1095-8649.2002.tb02432.x

Dou S, Masuda R, Tanaka M, Tsukamoto K (2005) Effects of temperature and delayed initial feeding on the survival and growth of Japanese flounder larvae. J Fish Biol 66:362–377 doi:10.1111/j.0022-1112.2005.00601.x

Ehrlich KF, Blaxter JHS, Pemberton R (1976) Morphological and histological changes during the growth and starvation of herring and plaice larvae. Mar Biol (Berl) 35:105–118 doi:10.1007/BF00390932

Fogarty MJ, Sissenwine MP, Cohen EB (1991) Recruitment variability and the dynamics of exploited marine populations. Trends Ecol Evol 6:241–246 doi:10.1016/0169-5347(91)90069-A

Gao LJ (2001) Studies on development, growth, physiology and lipid nutrition of sturgeon larvae and juvenile. Dissertation, Huadong Normal University

Gisbert E, Williot P (1997) Larval behaviour and effect of the timing of initial feeding on growth and survival of Siberian sturgeon (*Acipenser baeri*) larvae under small scale hatchery production. Aquaculture 156:63–76 doi:10.1016/S0044-8486(97)00086-0

Gisbert E, Conklin DB, Piedrahita RH (2004) Effects of delayed first feeding on the nutritional condition and mortality of California halibut larvae. J Fish Biol 64:116–132 doi:10.1111/j.1095-8649.2004.00289.x

Hamlin HJ, Kling LJ (2001) The culture and early weaning of larval haddock (*Melanogrammus aeglefinus*) using a microparticulate diet. Aquaculture 201:61–72 doi:10.1016/S0044-8486(01)00557-9

Hansen TJ, Møller D (1985) Yolk absorption, yolk sac constrictions, mortality, and growth during first-feeding of Atlantic salmon (*Salmo salar*) incubated on Astro-turf. Can J Fish Aquat Sci 42:1073–1078

Houde ED (1974) Effects of temperature and delayed feeding on growth and survival of larvae of three species of subtropical marine fishes. Mar Biol (Berl) 26:271–285 doi:10.1007/BF00389257

Huang LM, Xie YJ, Zhang GH, Hu JC (2005) Effect of delayed initial feeding on foraging, growth and survival of Chu's drum *Nibea chui* larvae. J Dalian Fish Univ 20:300–304 (in Chinese with English abstract)

Iguchi K, Mizuno N (1999) Early starvation limits survival in amphidromous fishes. J Fish Biol 54:705–712 doi:10.1111/j.1095-8649.1999.tb02027.x

Jiang ZQ, Jiang GJ, Zhang B (2002) Studies on feeding and growth of larval tiger puffer *Takifugu rubripes*. J Dalian Fish Univ 17:20–24 (in Chinese with English abstract)

Kamler E (1992) Early life history of fish: an energetics approach. Chapman & Hall, London

Kjørsvik E, van der Meeren T, Kryvi H, Arnfinnson J, Kvenseth PG (1991) Early development of the digestive tract of cod larvae, *Gadus morhua* L., during start-feeding and starvation. J Fish Biol 38:1–15 doi:10.1111/j.1095-8649.1991.tb03086.x

Leggett WC, Deblois E (1994) Recruitment in marine fishes: is it regulated by starvation and predation in the egg and larval stages? Neth J Sea Res 32:119–134 doi:10.1016/0077-7579(94)90036-1

Luo HZ, Fu RB, Chen B, Luo HJ, Li CJ (2006) Embryos and early larvae occurrence of *Miichthys miiuy* off Zhoushan and the relationship to salinity. J Zhejiang Ocean Univ 25:15–23 (in Chinese with English abstract)

Ma XZ, Wang W, Gan L, Yuan Y, Zhang WB (2006) Effects of delayed feeding on survival, feeding and growth of *Pelteobagrus vachelli* larvae. J Fish China 30:323–328 (in Chinese with English abstract)

May RC (1971) Effects of delayed initial feeding on larvae of the grunion, *Leuresthes tenuis* (Ayres). Fish Bull (Wash D C) 69:411–425

McGurk MD (1984) Effects and delayed feeding and temperature on the age of irreversible starvation and on the rates of growth and mortality of Pacific herring larvae. Mar Biol (Berl) 84:13–26 doi:10.1007/BF00394522

Mookerji N, Rao TR (1999) Rates of yolk utilization and effects of delayed initial feeding in the larvae of the freshwater fishes rohu and singhi. Aquacult Int 7:45–56 doi:10.1023/A:1009244819835

Olsen Y, Evjemo JO, Olsen A (1999) Status of the cultivation technology for production of Atlantic halibut (*Hippoglossus hippoglossus*) juveniles in Norway/Europe. Aquaculture 176:3–13 doi:10.1016/S0044-8486(99)00045-9

Osse JWM, van den Boogaart JGM, van Snik GMJ, van der Sluys L (1997) Priorities during early growth of fish larvae. Aquaculture 155:249–258 doi:10.1016/S0044-8486(97)00126-9

Peña R, Dumas S (2005) Effect of delayed first feeding on development and feeding ability of *Paralabrax maculatofasciatus* larvae. J Fish Biol 67:640–651 doi:10.1111/j.0022-1112.2005.00765.x

Puvanendran V, Brown JA (2002) Foraging, growth and survival of Atlantic cod larvae reared in different light intensities and photoperiods. Aquaculture 214:131–151 doi:10.1016/S0044-8486(02)00045-5

Qiao ZG, Chang GL, Zhang JP, Niu JY, Wang W (2007) Effect of delayed feeding on feeding ability, survival and growth of *Silurus asotus* larvae. J Shanghai Fish Univ 16:130–134 (in Chinese with English abstract)

Rana KJ (1985) Influence of egg size on the growth, onset of feeding, point-of-no-return, and survival of unfed *Oreochromis mossambicus* fry. Aquaculture 46:119–131 doi:10.1016/0044-8486(85)90196-6

Rogers BA, Westin DT (1981) Laboratory studies on effects of temperature and delayed initial feeding on development of striped bass larvae. Trans Am Fish Soc 110:100–110 doi:10.1577/1548-8659(1981)110<>2.0.CO;2

Seikai T, Tanangonan JB, Tanaka M (1986) Temperature influence on larval growth and metamorphosis of the Japanese flounder *Paralichthys olivaceus* in the laboratory. Bull Jpn Soc Sci Fish 52:977–982

Shan XJ, Dou SZ (2008) Effects of delayed first feeding on growth, survival and biochemical composition of croaker *Miichthys miiuy* larvae. Oceanol Limnol Sin 39:14–23 (in Chinese with English abstract)

Shan XJ, Quan HF, Dou SZ (2008a) Effects of delayed first feeding on growth and survival of rock bream *Oplegnathus fasciatus* larvae. Aquaculture 277:14–23 doi:10.1016/j.aquaculture.2008.01.044

Shan XJ, Xiao ZZ, Huang W, Dou SZ (2008b) Effects of photoperiod on growth, mortality and digestive enzymes in miiuy croaker larvae and juveniles. Aquaculture 281:70–76 doi:10.1016/j.aquaculture.2008.05.034

Theilacker GH (1978) Effect of starvation on the histological and morphological characteristics of jack mackerel, *Trachurus symmetricus*, larvae. Fish Bull (Wash D C) 76:403–414

Theilacker GH, Porter SM (1995) Condition of larval walleye Pollock, *Theragra chalcogramma*, in the western Gulf of Alaska assessed with histological and shrinkage indices. Fish Bull (Wash D C) 93:333–344

Ware DM, de Mendiola BR, Newhouse DS (1981) Behavior of first feeding Peruvian anchoveta, *Engraulis ringens* J. Rapp P-V Reun Cons Int Explor Mer 178:467–474

Yin MC, Blaxter JHS (1987) Feeding ability and survival during starvation of marine fish larvae reared in the laboratory. J Exp Mar Biol Ecol 105:73–83 doi:10.1016/S0022-0981(87)80030-8

Yin MC, Blaxter JHS (1989) Cruising speeds during early development and starvation of marine fish larvae. Oceanol Limnol Sin 20:1–9 (in Chinese with English abstract)

Yu HR, Mai KS, Duan QY, Ma HM, Liufu ZG, Tan BP et al (2003) Feeding habits and growth performance of larvae and juveniles of *Pseudosciaena crocea* under artificial rearing conditions. J Fish Sci China 10:495–501 (in Chinese with English abstract)

Yúfera M, Darias MJ (2007) The onset of exogenous feeding in marine fish larvae. Aquaculture 268:53–63 doi:10.1016/j.aquaculture.2007.04.050

Yúfera M, Pascual E, Polo A, Sarasquete MC (1993) Effect of starvation on the feeding ability of gilthead sea bream (*Sparus aurata* L.) larvae at first feeding. J Exp Mar Biol Ecol 169:259–272 doi:10.1016/0022-0981(93)90196-U

Zhang QY, Hong WS (2000) Status and prospects of artificial propagation and breeding technique of marine fish in China in the 1990s. Mod Fish Info 15:3–6 (in Chinese)

Zhong JS, Lou B, Yuan JF (2005) Study on the early development in larvae and juveniles of *Miichthys miiuy*. J Shanghai Fish Univ 14:231–237

Environ Biol Fish (2009) 86:131–135
DOI 10.1007/s10641-008-9384-0

Isolation of cellulose—producing microbes from the intestine of grass carp (*Ctenopharyngodon idellus*)

He Li · Zhang Zheng · Xie Cong-xin · Hao Bo ·
Wang Chao-yuan · He Gang

Received: 23 July 2007 / Accepted: 25 March 2008 / Published online: 5 September 2008

Abstract The cellulase activities of bacterial strains in the intestine of grass carp were analyzed, using filter paper and absorbent cotton as substrates and measuring the concentration of glucose by calorimetry. Six strains were isolated and determined high cellulase activity in all grass carp. Strains showed different abilities to produce cellulase, which suggests that they interact in the grass carp intestine to digest cellulose. The presence of cellulose activity suggests that grass carp have the ability to digest cellulose in the diet. The cellulase enzymatic activity increased dramatically after 6 days of culture and reached its peak at the 7th day. Microbes are probably the main source of cellulase in grass carp diets.

Keywords *Ctenopharyngodon idellus* · Bacterial strain · Cellulase activity

H. Li · X. Cong-xin (✉) · H. Bo · H. Gang
Huazhong Agriculture University,
Wuhan, Hubei 430074, China
e-mail: xiecongxin@mail.hzau.edu.cn

H. Li · Z. Zheng
Yangtze River Fisheries Research Institute,
Jingzhou, Hubei 434000, China

W. Chao-yuan
South-central University for Nationalities,
Wuhan 430000, China

Introduction

Cellulose is produced by plants and is recycled by microbes. It is a sustainable energy resource, as one of the main components of plants and also the most abundant organic materials in the world. The structure of cellulose in plants is very complicated and difficult to degrade because of its unique molecular structure. Cellulose can not be degraded directly by animals, but only by some microbes such as bacteria and fungi (Xiao et al 2002). Cellulose is thought to be indigestible and therefore of little nutritional value in formulated fish feeds. It is vitally important to develop methods to improve enzymatic hydrolysis of cellulose, since developments in fishery nutrition and aquaculture technology have encouraged the use of cheaper feed ingredients, including cellulose. Enzymatic hydrolysis of cellulose has been extensively reported since the utilization of cellulosic biomass as a renewable resource has great potential for reducing emissions of carbon dioxide to help prevent global warming (Claassen et al 1999; Li et al 2004a, 2004b, 2005). However, despite the fact that many studies have been conducted worldwide, the significance of cellulose in fish diets is not yet fully elucidated. Fishes cannot produce cellulase and thus, they can not utilize cellulose directly.

Some reports have suggested that soluble, high molecular weight non-starch polysaccharides such as those common in cellulose increase digestive viscosity and reduce digestive enzyme access to other nutrients (Castanon et al. 1997; Bedford 2000; Francis et al. 2001). In fishes this can result in poor growth and low feed efficiency, depending on ingredient type and proportion, in cultured fishes (Watanabe 2002).

Inclusion of exogenous enzymes as additives in plant-based feeds has greatly improved feed utilization in terrestrial animals (Bedford 1995; Castanon et al. 1997; Bedford 2000). Cellulase isolated from grass carp must be adapted to the internal environment of the fish intestine and have the ability to effectively digest cellulose. In this study, bacterial strains with cellulase-producing ability were identified from the intestine of grass carp. The capacity of cellulase to hydrolyze chemically defined cellulosic substrates was evaluated and it was determined that it might play a significant role in the utilization of plant materials in fish diets.

Materials and methods

Isolation and screening of microbes

Sample collection: Ten 10 grass carp were purchased from a retail market, having been fed mainly grass. Their individual live weights were 3.91, 3.94, 3.97, 3.97, 3.99, 4.00, 4.02, 4.03, 4.05, and 4.08 kg, respectively. The full length of the intestine from each fish was removed and food collected. Intestine lengths ranged from 79 to 144 cm.

In the first round of screening (Horikoshi et al. 1984), 1 gm of food from the intestine was ground and 9 ml of de-ionized (DI) water added to dissolve the material. This solution was diluted by 10 times to 10^{-1}, 10^{-2}, 10^{-3}, 10^{-4}, 10^{-5}, 10^{-6} and then 0.1 ml of the diluted solutions was inoculated onto plates and cultured at 37°C. Na-CMC was used to detect endo-β-1,4-glucanase activity. 0.1 ml of the screening strain suspension was added to the cellulose medium and mixed it at 50°C. The clearance area around colonies of strains of cellulase producing microbes was recorded as white medium.

In the second round of screening (Qi 2003), the diameter of the clear area was measured around each circle. The ratio of the diameter of the circle and the clearance area for each strain were calculated. Those with high clearance cycle/strain cycle ratios were selected for further testing.

Cellulase enzymatic activity measurement

The glucose concentrations in enzyme assays (see *Wave length for the production after cellobiose, cellotriose and glucose oxidized by DNS*) was determined relative to a standard curve for glucose in sodium acetate buffer (Liu 2002). Selected strains obtained were inoculated into 10 ml of Na-CMC medium and cultured them at 37°C for 5, 6 and 7 days. 1 ml of the supernatant was used to measure β-1,4-glucosidase activity, 0.5 ml of which was used to measure exo-β-1,4-glucanase and endo-β-1,4-glucanasc activity. Control tubes were heated in boiling water for 5 min to inactivate them. After allowing the samples to cool, 2 ml of acetic acid/sodium acetate buffer and 2 ml of Na-CMC were added to the β-1,4-glucosidase activity assay. 2 ml of acetic acid/sodium acetate buffer and 2 ml of absorbent cotton were added to the assay for exo-β-1,4-glucanase activity. 2 ml of acetic acid/sodium acetate buffer was added to a sample of 1 cm×3 cm filter paper to assay for endo-β-1,4-glucanase activity. Tubes were incubated at 50°C for 5 min and then cooled with ambient temperature water. 2 ml of 3,5-dinitrosalicylic acid (DNS) was added to the tubes, which were then placed tubes into 100°C bath for 5 min, allowed to cool with tap water and made the final volume to 10 ml. Optical density (OD) was measured at 490 nm.

$$\beta\!-\!1,4 - \text{glucosidase} = \frac{B}{1.0}\,\mu\text{g/ml}\times 5\text{min}\times 50^{\circ}\text{C}$$

$$\text{endo}\!-\!\beta\!-\!1,4 - \text{glucanase} = \frac{B}{0.5}\,\mu\text{g/ml}\times 1\text{h}\times 50^{\circ}\text{C}$$

$$\text{exo}\!-\!\beta\!-\!1,4 - \text{glucase} = \frac{B}{0.5}\,\mu\text{g/ml}\times 24\text{h}\times 50^{\circ}\text{C}$$

where B is the weight (μg) obtained from the standard curve above.

Wave length for the production after cellobiose, cellotriose and glucose oxidized by DNS

The most commonly-used wavelengths for detecting oxidized and reduced DNS by cellobiose, cellotriose and glucose are 550 nm, 540 nm 520 nm and 490 nm, respectively. The OD value obtained was different from wavelengths for the detection of reduced DNS and the resultant enzymatic activity was measured.

The absorption properties of the deoxidized DNS by cellobiose, cellotriose and glucose were studied. Brown-red products had strong absorption between the wave lengths of 540–470 nm with a peak at 490 nm. Thus a wave length of 490 nm was used to detect cellulase activity.

Results

Screening of cellulase producing microbes

After the first round of screening with cellulase selective medium, positive strains were transferred to plates for the second round of screening at 37°C to culture for 48 h. Strains were picked out by colony morphology observation. Strains obtained were transferred from the second round screening to new plates to purify them. 31 strains were screened and 2 strains with cellulase activity were obtained.

The strains obtained were further inoculated to a liquid fermentation medium. The strains were cultured and levels of β-1,4-glucosidase, endo-β-1,4-glucanase and exo-β-1,4-glucanase activity were measured. 6 strains of microbes with relatively high cellulase activity were collected and designated as: X1, X2, X3, X4, X5 and X6 (shown in Table 1).

The effect of pH on absorbance

Two milliliters of acetic acid/sodium acetate buffer with different pH was added to the samples. 1 ml of enzymatic solution with 0.5% CMC-Na Two milliliters was then added, after complete hydrolysis at 50° C, DNS was then added and the absorbance was measured. The absorbance was the highest at pH 4.8 which was why the acetic acid/sodium acetate solution was chosen for the experiment.

The effect of culturing time on cellulase production

The X1 microbe was chosen to study the effects of culturing time on cellulase production. Cultures on filter paper fermentation medium were obtained and then analyzed for enzymatic activity. From the 3rd day on, enzymatic activity was measured every 24 h (Fig. 1).

Enzymatic activity increased dramatically after 6 days of culture and reached its peak at the 7th day (Fig. 1). The enzymatic activity of the other microbe strains obtained in this study was also studied. Similar results for the enzymatic activity were obtained, with the highest values around the 7th day. Therefore cellulase activity was measured during the period between 5 and 8 days.

Table 1 Cellulose enzymatic activity after different fermentation period (OD_{490})

Strain number	Fermentation time (days)	β-1,4-glucosidase activity (OD_{490})	Endo-β-1,4-glucanase activity (OD490)	Exo-β-1,4-glucanase activity (OD_{490})
X1	5	0.004	0.007	0.1
	6	0.021	0.027	0.13
	7	0.056	0.096	0.11
X2	5	0.019	0.178	0.041
	6	0.025	0.075	0.095
	7	0.062	0.126	0.12
X3	5	0.097	0.001	0.107
	6	0.15	0.025	0.145
	7	0.14	0.074	0.137
X4	5	0.041	0.035	0.013
	6	0.059	0.072	0.083
	7	0.055	0.13	0.092
X5	5	0.083	0.028	0.063
	6	0.164	0.016	0.019
	7	0.159	0.149	0.056
X6	5	0.054	0.104	0.114
	6	0.135	0.071	0.098
	7	0.142	0.096	0.091

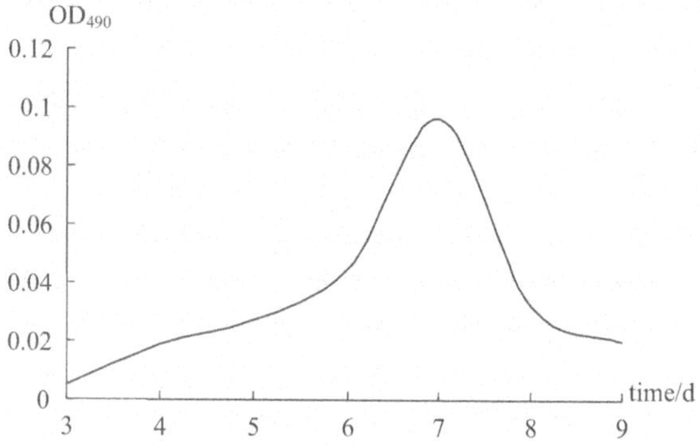

Fig. 1 The effect of culturing time on the enzymatic activity of cellulose

Cellulase activities and X-ray dispersion

According to the results obtained for the OD490/glucose concentration above, the highest OD490, which represented the cellulase activity, was measured between 5 and 7 days culture of all the microbes to calculate the enzymatic activity (Table 1).

As shown in Table 2 and Table 3, the enzymatic activity of the same strain of microbe was different on different substrates. There were also differences in enzymatic activity among different strains. Among all the 6 strains, CMC enzymatic activity was higher in X3, X5 and X6, with X5 obtaining the highest level of activity. This indicates that this strain has a strong ability to decompose water-soluble substrates. For filter paper enzymatic activity, X2, X4 and X5 demonstrated high activity, with X2 obtaining the highest. X2 had the greatest ability to decompose multi-component cellulose substrates. Higher exo-β-1,4-glucanase activity appeared in X1, X2 and X3, with the highest level of activity being observed in X3.

Table 2 Cellulase activities of the 6 bacterial strains

Strain	β-1,4-glucosidase concentration (u/ml)	Endo-β-1,4-glucanase concentration (u/ml)	Exo-β-1,4-glucanase concentration (u/ml)
X1	181	314.33	427.67
X2	201	587.67	394.33
X3	494.33	241	477.67
X4	191	427.67	301
X5	541	491	204.33
X6	467.67	341	374.33

Table 3 X-ray dispersion of different cellulose materials (marked as crystallinity)

Cellulose materials	In-between area percentage of cellulose	Crystallized percentage in cellulose crystallized area
Absorbent cotton	12.4	87.6
Filter paper	38.9	61.1

Discussion

There are a few reports concerning microbial flora and microbial cellulase production in the intestinal tracts of fishes (Stickney and Shumway 1974; Lesel et al. 1986; Das and Tripathi 1991; Stellwag et al. 1995; Saha and Ray 1998; Bairagi et al. 2002). Grass carp is a fish that feeds only on hydrophytes in natural waters. Therefore, there must be some kind of mechanism for them to use plant resources effectively, which suggests that cellulase plays an important role in their digestive system. Because grass carp can not produce cellulase by itself, there should be some cellulase-producing microbes in the grass carp's intestine.

Grass carp with high intestinal microbial activity may be supplied with additional carbohydrate energy through alternative routes. Cellulase activity has been observed in several fish species indicating that fish may be able to utilize cellulose (Chakrabarti et al. 1995). Our experiment has demonstrated that cellulase-producing microbes exist in the intestine of grass carp.

Appropriate amounts of dietary fibre are important for digestion in fishes because fibre can enhance the peristaltic movements of the intestine, stimulate the secretion of digestive enzymes and enhance the contact surface between food and enzymes. Previous reports have demonstrated that microbes in the fish intestine could secrete cellulase, which could help fish to digest cellulose into cellobiose, cellotriose and other oligosaccharides and eventually into glucose to be utilized by the fish. The grass carp gut is three times the length of body. Because cellulose is difficult to digest and the fish intestine is relatively long, food passes through the intestine relatively slowly. Therefore, it is necessary to add cellulase to fish feed to increase gut absorption. Until now, few reports are available concerning the application of cellulase in fish feed.

All of the six strains with high cellulase-producing ability were obtained from the middle or rear part of the intestine, while the strains screened from the front part of the intestine had relatively low cellulase activity. This suggests that all these high cellulase-producing strains were established and stable strains. The cellulase activities for different strains were found to be quite different. Some of them had high CMC activity while some of them had high endo-β-1,4-glucanase. This indicates that these strains might work together to decompose cellulose in the grass carp intestine.

As the grass carp do not have a stomach to digest cellulose, it can only digest cellulose in its intestine. Shcherbina and Kazlawlene (1971) suggested that some portion of dietary cellulose is digested in the anterior portion of the gut while the remaining portion of the cellulose is digested in the posterior portion of the digestive tract, indicating the probable presence of microbial cellulase in the posterior region. Analysis of the residual food obtained from grass carp intestine showed that the ability for cellulose digestion was very limited. Large amounts of cellulose were still found even in the rear part of its intestine.

Bacteria in the diet of the fish may adapt themselves to the environment of the gastrointestinal tract and form a symbiotic association. In our study, a considerable population of bacterial symbionts was isolated from the alimentary tracts of grass carp and some of the strains were shown to exhibit significant cellulolytic activity.

Acknowledgements This study was funded by an Excellent Project for Scholars Returned from Abroad award, Ministry of Human Resource (2005) and the Key Laboratory of Freshwater Fish Germplasm Resource and Biotechnology, Chinese Academy of Fish Sciences (CAFS).

References

Bairagi A, Sarkar GK, Sen SK, Ray AK (2002) Enzyme producing bacterial flora isolated from fish digestive tracts. Aquacult Int 10:109–121. doi:10.1023/A:1021355406412

Bedford MR (1995) Mechanism of action and potential environmental benefits from the use of feed enzymes. Anim Feed Sci Technol 53:145–155 doi:10.1016/0377-8401(95)02018-U

Bedford MR (2000) Exogenous enzymes in monogastric nutrition—their current value and future benefits. Anim Feed Sci Technol 86:1–13. doi:10.1016/S0377-8401(00)00155-3

Castanon JIR, Flores MP, Pettersson D (1997) Mode of degradation of non-starch polysaccharides by feed enzyme preparations. Anim Feed Sci Technol 68:361–365. doi:10.1016/S0377-8401(97)00046-1

Chakrabarti I, Gani MA, Chaki KK, Sur R, Misra KK (1995) Digestive enzymes in 11 freshwater teleost fish species in relation to food habit and niche segregation. Comp Biochem Physiol A 112:167–177. doi:10.1016/0300-9629(95)00072-F

Claassen PAM, Van Lier JB, Contreas AML (1999) Utilization of biomass for the supply of energy carriers. Appl Microbiol Biotechnol 52:741–755. doi:10.1007/s002530051586

Das KM, Tripathi SD (1991) Studies on the digestive enzymes of grass carp, *Ctenopharyngodon idella* (Val.). Aquaculture 92:21–32. doi:10.1016/0044-8486(91)90005-R

Francis G, Makkar HPS, Becker K (2001) Antinutritional factors present in plant-derived alternate fish feed ingredients and their effects in fish. Aquaculture 199:197–227. doi:10.1016/S0044-8486(01)00526-9

Horikoshi K, Nakao M, Kurono Y (1984) Cellulase of an alkalophilic Bacillus strain isolated from soil. Can J Microbiol 30:774–779

Lesel R, Fromageot C, Lesel M (1986) Cellulose digestibility in grass carp, Ctenopharyngodon idella and in goldfish, *Carassius auratus*. Aquaculture 54:11–17. doi:10.1016/0044-8486(86)90249-8

Li C, Yoshimoto M, Fukunaga K, Nakao K (2004a) Preparation and characterization of cellulase-containing liposomes for their immobilization suitable for enzymatic hydrolysis of cellulose. J Chem Eng of Jpn 37:680–684. doi:10.1252/jcej.37.680

Li C, Seki K, Matsunaga T, Yoshimoto M, Fukunaga K, Nakao K (2004b) Enzymatic hydrolysis of waste paper in an external loop airlift bubble column with continuous ultrasonic irradiation. J Chem Eng of Jpn 37:1041–1049. doi:10.1252/jcej.37.1041

Li C, Yoshimoto M, Ogata H, Tsukuda N, Fukunaga K, Nakao K (2005) Effects of ultrasonic intensity and reactor scale on kinetics of enzymatic saccharification of various waste papers in continuous irradiated stirred tanks. Ultrason Sonochem 12:373–384. doi:10.1016/j.ultsonch.2004.02.004

Liu DH (2002) Methods for cellulose activity measurement. Chin Feed 17:27–28

Qi Y (2003) A study on isolation and characterization of cellulase producing microbes. Nat Prod Res Dev 15:510–513

Saha AK, Ray AK (1998) Cellulase activity in rohu fingerlings. Aquacult Int 6:281–291. doi:10.1023/A:1009210929594

Shcherbina MA, Kazlawlene OP (1971) The reaction of the medium and the rate of absorption of nutrients in the intestine of carp. J Ichthyol 11:81–85

Stellwag EJ, Smith TD, Luczkovich JJ (1995) Characterization and ecology of carboxymethylcellulase-producing anaerobic bacterial communities associated with the intestinal tract of the pinfish, *Logodon rhomboides*. Appl Environ Microbiol 61:813–816

Stickney RR, Shumway SE (1974) Occurrence of cellulase activity in the stomachs of fish. J Fish Biol 6:779–790. doi:10.1111/j.1095-8649.1974.tb05120.x

Watanabe T (2002) Strategies for further development of aquatic feeds. Fish Sci 68:242–252. doi:10.1046/j.1444-2906.2002.00418.x

Xiao CL, Xu CX (2002) A study on the application of cellulose from microbes. J Microbiol 22:33–35

Environ Biol Fish (2009) 86:137–142
DOI 10.1007/s10641-008-9344-8

Sinocyclocheilus guilinensis, a new species from an endemic cavefish group (Cypriniformes: Cyprinidae) in China

Yahui Zhao · Chunguang Zhang · Jie Zhou

Received: 22 September 2007 / Accepted: 26 March 2008 / Published online: 13 May 2008

Abstract The endemic Chinese cavefish genus *Sinocyclocheilus* possesses very rich species diversity. Specimens of this genus collected from Guilin, Guangxi, China, were originally described in an unpublished document in 1982. Later, they were recognized as *Sinocyclocheilus jii*. Comparison to the type specimens of *S. jii* leads to the conclusion that, though similar, they represent two different species. Herein we describe them as a new species, *Sinocyclocheilus guilinensis* sp. nov. It can be distinguished from all congeners by its possession of soft, unbranched dorsal ray without serrations, 19–20 (11–12) scale rows above (below) the lateral line, 34–36 circumpeduncular scales and 8–9 pre-dorsal vertebrae. *Sinocyclocheilus guilinensis* sp. nov. is most similar to *S. jii*. In addition to the scale count differences, a morphometrics-based principal component analysis (PCA) also supports designation of *S. guilinensis* and *S. jii* as two distinct species.

Keywords China · *Sinocyclocheilus* · Cavefish · New species

Y. Zhao · C. Zhang (✉)
Institute of Zoology, Chinese Academy of Sciences,
Beijing 100101, China
e-mail: fish@ioz.ac.cn

J. Zhou
Fishery Institute of Guangxi Zhuang Autonomous Region,
Nanning 530021, China

Introduction

The cyprinid genus *Sinocyclocheilus* is endemic to the karstric caves of the eastern Yunnan, middle-southern Guizhou and notrthwestern Guangxi, China (Zhao and Zhang 2006). Almost all species of the genus *Sinocyclocheilus* are obligatory or facultative cave dwellers. Rapid speciation due to isolation of small populations in separate caves has contributed to the genus' speciation (Zhao and Zhang 2006).

In 1982, Mr. Ji, a Chinese ichthyologist, described a species based on specimens collected in Guilin, which he named *Sinocyclocheilus guilinensis*. This original description of *S. guilinensis* was in an unpublished governmental report[1]. The original description is extremely succinct and lacks any comparison to other species. In 1984, Zhou mentioned this species along with a very simple description in another internal report[2]. Since the original reference describing the species was not formally published, this species description is not valid according to the rules (Article 8.1) of the International Code of Zoological Nomenclature (International Commission on Zoological Nomenclature 1999). Therefore, Chu

[1] An internal report for the government: Fishery Institute of Guangxi Zhuang Autonomous Region (1982) Influences on fishery resources from pollution in the Lijiang River

[2] Fishery Institute of Guangxi Zhuang Autonomous Region (1984) Investigative Report on the fishery stocks of the inland waters in Guangxi Zhuang Autonomous Region, China, 487 pp.

and Cui (1985) suggested that *S. guilinensis* should be regarded as an unrecognized species.

Zhang and Dai collected two specimens of *Sinocyclocheilus* in 1984 from Fuchuan, Guangxi, near Guilin. Fuchuan and Guilin are located along different rivers, the Hejiang and Guijiang rivers, respectively, both of which belong to the Pearl River drainage. Zhang and Dai (1992) compared the specimens from Fuchuan with the unpublished descriptions of *S. guilinensis*. They thought the main characters of the specimens they collected were similar to the description of *S. guilinensis* and described the specimens from Fuchuan as a new species, *Sinocyclocheilus jii* (Zhang and Dai 1992). The specific name "*jii*" was proposed to recognize the contribution of the original researcher, Ji. After that, both specimens from Guilin and Fuchuan were recognized as *S. jii* by all other researchers (Shan et al. 2000).

During a phylogenetic study on the genus, we found that specimens from Guilin actually have some stably-different characteristics compared with the type specimens of *Sinocyclocheilus jii*. Therefore, we described it herein as a new species. Out of respect for Mr. Ji, who first reported the species, we keep his original scientific name for the species, *S. guilinensis*.

Materials and methods

The holotype and paratypes and comparative specimens are deposited in different collections. Their abbreviations are as listed in Leviton et al. (1985) with the following exceptions: IHB (Institute of Hydrobiology, Chinese Academy of Sciences), GXIF (Fishery Institute of Guangxi Zhuang Autonomous Region, China), and Lan (personal collection of Mr. J.H. Lan). Measurements were taken using a digital caliper to the nearest 0.1mm. Morphometric and meristic characters were selected according to the methods described in Zhao et al. (2006). Osteological characters were observed using X-ray photographs following the methods described in Zhao et al. (2002). Vertebrae counts excluded the Weberian complex. Systat Version 10 (Wilkinson 2001) software was used in the statistical analysis (principal component analysis).

Sinocyclocheilus guilinensis sp. nov. (Fig. 1)

Sinocyclocheilus jii, Zhang and Dai 1992, Acta Zootaxon Sin 17: 377 (Fuchuan, Guangxi Zhuang Autonomous Region, China); Shan et al. 2000, Fauna Sinica, Osteichthyes, Cypriniformes III: 59 (Guilin, Guangxi Zhuang Autonomous Region, China).

Holotype ASIZB 113753, 107.3mm standard length (SL), from Zhujiang River Basin: Lijiang River, Feifengshan Hill, Guilin City, Guangxi Zhuang Autonomous Region, China; collected in 1982 by CS Ji and J Zhou.

Paratypes ASIZB 113754, 101.2mm SL; GXIF 82389, 115.9mm SL; GXIF 82374, 110.7mm SL; GXIF 82375, 108.1mm SL; data as for holotype. IHB 82IV123, 136.5mm SL, collected in April, 1982 from Guilin, collector unknown. KIZ 845005, 60.9mm SL, collected in May, 1984 from Guilin, collector unknown.

Diagnosis The new species is distinguished from all congeners by the following combination of characteristics: body completely scaled, eyes well developed; last unbranched ray of dorsal fin soft, without serration; scale row counts above and below lateral line 19–20 and 11–12, respectively; circumpeduncular scale count 34–36; gill rakers 8–11; pre-dorsal vertebrae 8–9.

Fig. 1 *Sinocyclocheilus guilinensis* sp. nov. Holotype, ASIZB 113753, 107.3 mm SL, scale bar 10 mm

Description General body features are shown in Fig. 1. Meristics and proportional measurements as percentages of standard length are listed in Table 1.

Body compressed. Dorsal profile convex, not forming a humped back, greatest body depth exactly at dorsal fin insertion; ventral profile slightly concave from snout tip to pelvic insertion, then slightly convex to anal-fin end and straight toward caudal fin.

Head compressed. Snout round. Nostrils midway between snout tip and anterior margin of orbit. Anterior

Table 1 Meristics and morphmetrics of *Sinocyclocheilus guilinensis* sp. nov. and *S. jii*

	Sinocyclocheilus guilinensis sp. nov						*Sinocyclocheilus jii*				
	N	Min	Max	Mean	SD	Holotype	*N*	Min	Max	Mean	SD
Dorsal fin rays	7	7	7	7		7	5	7	7	7	
Anal fin rays	7	5	5	5		5	5	5	5	5	
Pectoral fin rays	7	13	15	14		13	5	14	15	15	
Pelvic fin rays	7	8	8	8		8	5	8	8	8	
Total vertebrae	3	36	37	37		37	2	36	38	37	
Predorsal vertebrae	3	8	9	8		8	2	8	9		
Caudal vertebrae	3	17	18	18		18	2	17	17		
Gill rakers	7	8	11	9		9	5	6	7	7	
Lateral line scales	7	43	53	48		49	5	47	52	49	
Scale rows above lateral line	7	19	20	20		19	5	27	29	28	
Scale rows below lateral line	7	11	12	12		12	5	15	17	16	
Circumpeduncular scales	7	34	36	36		34	5	46	50	47	
Standard length	7	60.9	136.5	105.8	22.8	107.3	5	100.5	123.6	112.2	8.3
In % of standard length											
Body depth	7	24.6	29.0	27.0	1.9	27.3	5	27.0	28.8	28.1	0.7
Predorsal length	7	56.1	58.0	56.8	0.6	56.1	5	54.5	57.1	55.6	1.2
Dorsal-fin base length	7	10.5	12.9	11.9	0.8	11.7	5	12.2	13.6	13.0	0.6
Dorsal fin length	7	20.8	22.4	21.6	0.7	21.0	5	19.8	23.9	22.2	1.6
Preanal length	7	74.6	79.1	76.7	1.7	76.8	5	74.1	76.7	74.8	1.1
Anal-fin base length	7	7.5	9.7	8.7	0.8	8.4	5	7.7	8.7	8.3	0.5
Anal fin length	7	16.6	19.8	18.0	1.1	16.8	5	17.4	19.5	18.7	1.0
Prepectoral length	7	25.7	30.0	27.9	1.5	27.7	5	25.7	27.5	26.6	0.6
Pectoral-fin base length	7	3.5	4.8	4.2	0.5	4.5	5	3.7	4.3	4.0	0.2
Pectoral fin length	7	19.0	24.9	21.5	1.8	20.6	5	20.3	23.9	21.7	1.7
Prepelvic length	7	49.8	55.1	52.2	1.7	52.1	5	50.7	55.5	52.8	2.0
Pelvic-fin base length	7	4.0	5.4	4.5	0.4	4.4	5	4.0	4.9	4.3	0.4
Pelvic fin length	7	15.4	20.7	18.1	1.7	18.1	5	16.1	18.5	17.5	1.0
Caudal peduncle length	7	16.4	19.2	18.0	0.9	17.3	5	17.9	19.6	18.9	0.7
Caudal peduncle depth	7	11.4	14.1	12.7	0.8	13.1	5	11.8	13.5	12.8	0.7
Head length	7	26.6	31.4	28.7	1.5	28.0	5	25.3	27.4	26.5	0.8
Head depth	7	15.3	17.9	16.8	1.0	17.1	5	14.4	16.0	15.2	0.7
Head width	7	12.4	15.1	14.1	1.0	14.3	5	13.3	14.2	13.7	0.4
Snout length	7	7.6	10.3	8.9	0.9	9.3	5	6.9	7.7	7.2	0.4
Eye diameter	7	7.1	9.1	8.1	0.7	8.1	5	7.2	8.6	7.8	0.5
Interorbital width	7	6.4	7.8	7.3	0.6	7.4	5	6.1	7.1	6.5	0.4
Prenostril length	7	5.1	6.7	5.8	0.5	5.7	5	5.3	5.8	5.5	0.2
Width between posterior nostrils	7	5.2	7.0	6.4	0.7	5.2	5	5.8	6.8	6.3	0.4
Upper jaw length	7	7.3	9.3	8.3	0.8	7.4	5	6.1	7.7	6.8	0.7
Lower jaw length	7	6.8	8.7	7.8	0.8	6.8	5	5.9	7.1	6.4	0.4
Mouth width	7	7.9	8.4	8.1	0.2	8.0	5	7.7	8.4	8.1	0.3
Maxilla barbel length	7	8.4	13.6	10.8	1.7	11.2	5	10.5	14.1	12.4	1.7
Rictal barbel length	7	9.3	15.5	12.4	2.1	12.8	5	10.3	17.8	13.3	3.0

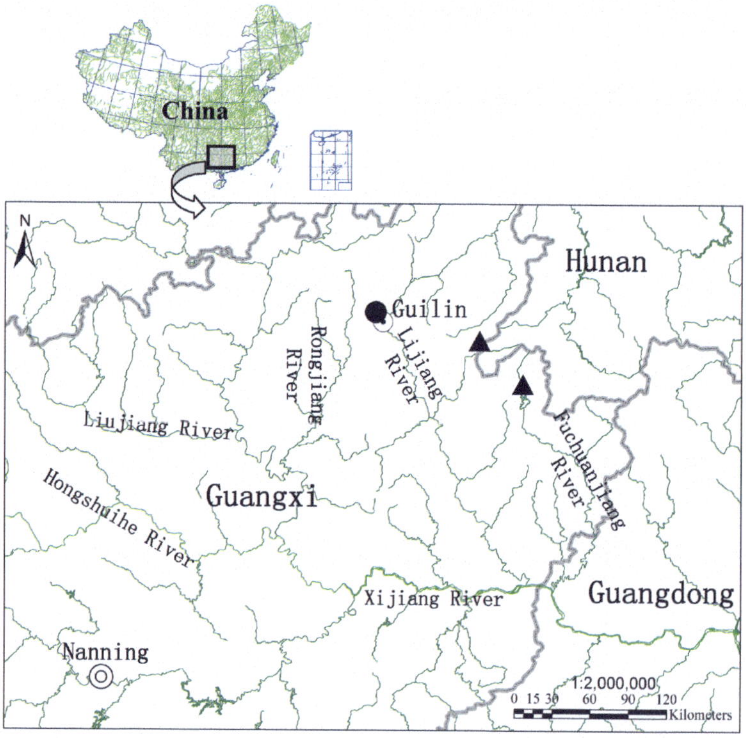

Fig. 2 Sampling sites of *Sinocyclocheilus guilinensis* sp. nov. (*circle*) and *S. jii* (*triangle*)

nostrils round, possessing a rim with a fleshy flap forming a complete tube. Posterior nostrils elliptical and opened. Mouth terminal and curved; lips simple, upper lip base covered by rostral cap, lower lip closely adnate to lower jaw, upper and lower jaws connected at rictus. Two pairs of barbels: maxillary barbel inserted in front of anterior nostril, barbels extending beyond anterior margin of eye; rictal barbels a little longer, extending to posterior margin of eye, but not reaching posterior margin of preopercle. Eye round, normal, and moderate in size, eye diameter one fourth to one third of head length. Cranial sensory canal developed, superaorbital and infraorbital canals connected, canals below eye radially arranged. Gill opening large, operculum membrane not connected at isthmus. Joints of dentary-angulars not close to each other at isthmus. Gill rakers triangular, well-developed, 8–9 in first gill arch, epibranchial with 2 (five specimens) or 3 (2); ceratobranchial with 6 (2), 7 (2) or 8 (2). Pharyngeal teeth in three rows with counts of 2, 3, 4–4, 3, 2.

Pectoral fin insertion under posterior margin of operculum; pectoral fin not reaching pelvic fin insertion. Pelvic fin insertion anterior to dorsal fin insertion, midway between pectoral and anal fin insertions; pelvic fin stretching two-thirds of distance between pelvic and anal fin insertions, not reaching anus. Dorsal fin origin

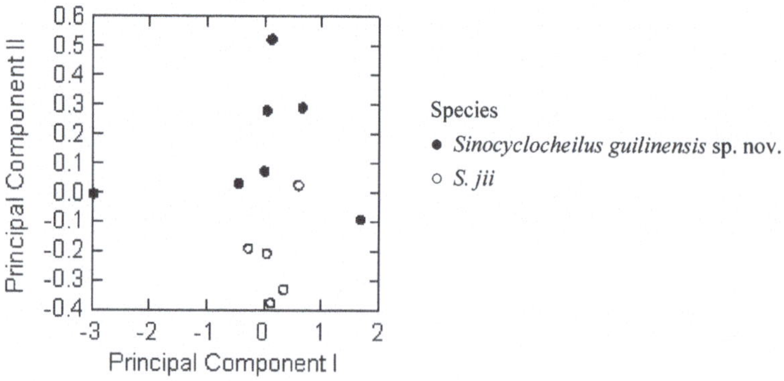

Fig. 3 Principal component analysis based on morphometric characters of *Sinocyclocheilus guilinensis* sp. nov. and *S. jii*

Table 2 Character loadings on principal components I–V for measurements from specimens of *Sinocyclocheilus guilinensis* sp. nov. (seven specimens) and *S. jii* (five specimens)

	PC I	PC II	PC III	PC IV	PC V
Standard length	0.1933	−0.0019	0.0245	−0.0131	0.0009
Body depth	0.2264	−0.0054	0.0351	−0.0257	0.0059
Predorsal length	0.1884	0.0050	0.0221	−0.0035	−0.0035
Distance from head end to dorsal fin insertion (dorsal view)	0.2118	0.0053	0.0225	0.0034	0.0048
Dorsal-fin base length	0.2195	−0.0513	0.0023	−0.0456	−0.0138
Dorsal fin length	0.1873	−0.0347	0.0284	−0.0038	0.0160
Preanal length	0.1851	0.0133	0.0272	−0.0026	−0.0002
Anal-fin base length	0.1766	−0.0021	−0.0058	0.0275	−0.0102
Anal fin length	0.1995	−0.0423	0.0216	0.0087	−0.0130
Prepectoral length	0.1773	0.0057	−0.0086	0.0053	−0.0133
Pectoral-fin base length	0.1648	−0.0119	−0.0183	0.0445	0.0096
Pectoral fin length	0.2220	−0.0405	0.0125	0.0333	0.0062
Prepelvic length	0.1804	0.0038	0.0181	−0.0178	−0.0170
Pelvic-fin base length	0.2035	−0.0168	−0.0320	0.0080	0.0394
Pelvic fin length	0.2078	−0.0372	0.0050	0.0299	0.0363
Caudal peduncle length	0.2171	−0.0200	0.0429	−0.0367	0.0133
Caudal peduncle depth	0.2146	−0.0316	0.0069	0.0053	0.0296
Head length	0.1688	0.0196	−0.0124	0.0083	−0.0118
Head depth	0.1832	0.0463	0.0008	0.0145	0.0164
Head width	0.1569	0.0202	0.0270	−0.0156	0.0056
Snout length	0.1865	**0.1146**	0.0052	0.0248	0.0120
Eye diameter	0.1497	−0.0000	−0.0019	−0.0326	0.0171
Interorbital width	0.2290	0.0602	−0.0115	0.0026	0.0267
Prenostril length	0.2206	0.0432	0.0207	0.0042	−0.0199
Width between posterior nostrils	0.2005	0.0197	−0.0019	−0.0107	−0.0564
Upper jaw length	0.1867	**0.1008**	−0.0525	0.0014	−0.0189
Lower jaw length	0.1923	0.0918	−0.0552	−0.0029	−0.0125
Mouth width	0.2033	0.0102	0.0231	−0.0239	−0.0096
Maxilla barbel length	0.2493	**−0.1184**	−0.0086	0.0483	−0.0455
Rictal barbel length	0.2190	−0.0798	−0.1335	−0.0376	0.0107

The three highest loadings on PC II are shown in boldface.

approximately midway between snout tip and caudal-fin base, and posterior to pelvic fin insertion; last unbranched ray of dorsal fin soft, without serrations along posterior edge. Anal fin 17–20% of SL, with insertion approximately midway between pelvic fin origin and caudal fin base. Caudal fin bifurcate.

Lateral line complete, descending to point above pectoral fin rays from posterior margin of operculum, then ascending to body midline above anal fin base, extending to end of caudal peduncle. Scales small. Lateral line scales larger than other scales. Lateral line scale counts 43–51, scale row counts above lateral line 19–20, below lateral line 11–13; circumpeduncular scale counts 34–36. Predorsal scales irregularly arranged and difficult to accurately count. Pelvic fin axillary scales (generally 2) present.

Coloration in alcohol Specimens were fixed in 10% formalin and then preserved in 75% alcohol. Body yellow-brownish, dorsum darker and abdomen lighter, no spots or blotches. All fins light yellowish.

Distribution Known only from subterranean rivers in suburbs of Guilin City, tributaries of the Lijiang River, which belongs to the Xijiang River drainage, the largest tributary of the Zhujiang River basin (Fig. 2). From an unpublished report[3], this species may be found in caves or subterranean rivers within

[3] Fishery Institute of Guangxi Zhuang Autonomous Region, 1984. Report of the survey on inland fishery resources of Guangxi Zhuang Autonomous Region.

approximately 30km north-east, 20km south-east, 25km south and 15km north-west of Guilin City.

Etymology The name of the new species, *guilinensis*, is derived from the name of the collection locality, Guilin City.

Remarks Only a few species of *Sinocyclocheilus*, including *S. tianlinensis* (Zhou et al. 2004), *S. macrolepis* (Wang 1989), *S. yishanensis* (Li and Lan 1992), *S. jii* and this new species, have soft unbranched dorsal fin rays without serrations. *S. tianlinensis* is completely blind. *S. macrolepis* is distinct in that the lateral line scales are not larger than scales located just above or below the lateral line. *S. yishanensis* has fewer pre-dorsal vertebrae than *S. jii* and the new species (6–7 vs. 8–9).

Sinocyclocheilus jii is most similar in body shape to the new species. This is why these two species were mistaken with each other in the past. In fact, *S. jii* has higher scale row counts above and below the lateral line (27–29 vs. 19–20 and 15–18 vs. 11–12, respectively) and circumpeduncular scale counts (46–50 vs. 34–36) than the new species. *S. jii* also possesses fewer of gill rakers than the new species (6–7 vs. 8–11). In addition, a principal component analysis (PCA) based on morphometric characteristics shows marked differences between *S. jii* and *S. guilinensis* sp. nov. (Fig. 3). The features with the highest PCA loading scores (Table 2) were snout length, upper jaw length, and maxillary barbel length.

Comparative materials Sinocyclocheilus jii: ASIZB 62726 (Holotype), 123.6 mm SL, ASIZB 62727 (Paratype), 100.5 mm SL, Fuchuan county, Guangxi, China, 1984, C. G. Zhang and D. Y. Dai; Lan 02081410, 114.4 mm SL, Lan 02081388, 112.7 mm SL, Lan 02081407, 109.9 mm SL, Guanyin Town, Gongcheng County, Guangxi, China, October 2002, J. H. Lan.

Acknowledgments The work was supported by the projects of National Natural Science Foundation of China (contract number NSFC-30470251 and NSFC-J0030092). Many thanks go to Dr. Junxing Yang (KIZ), Mrs. Guihua Cui (KIZ) and Dr. Shunping He (IHB) for allowing to access specimens. Thanks to Dr. Aldemaro Romero (Arkansas State University, USA) for revising the paper.

References

Chu XL, Cui GH (1985) A revision of Chinese cyprinid genus *Sinocyclocheilus* with reference to its relationship (in Chinese with English abstract). Acta Zootaxon Sin 10:435–441

International Commission on Zoological Nomenclature (1999) International code of zoological nomenclature, 4th edn. International Trust for Zoological Nomenclature, London

Leviton AE, Gibbs RH, Heal E et al (1985) Standards in herpetology and ichthyology: part I. Standard symbolic codes for institutional resource collections in herpetology and ichthyology. Copeia 1985:802–832

Li WX, Lan JH (1992) A new genus and three new species of Cyprinidae from Guangxi, China (in Chinese with English abstract). J Zhanjiang Fish Coll 12(2):46–51

Shan XH, Lin RM, Yue PQ et al (2000) Barbinae. In: Yue PQ (ed) Fauna Sinica (Osteichthyes: Cypriniformes III). Science Press, Beijing, pp 52–84

Wang DZ (1989) Descriptions of three new species of Cyprinidae from Guizhou Province, China (Cypriniformes: Cyprinidae) (in Chinese with English abstract). Acta Acad Med Zunyi 12(4):29–34

Wilkinson L (2001) SYSTAT: the System for Statistics, Version 10.0. SPSS, Chicago, IL, USA

Zhang CG, Dai DY (1992) A new species of *Sinocyclocheilus* from Guangxi, China (in Chinese with English abstract). Acta Zootaxon Sin 17:377–380

Zhao YH, Zhang CG (2006) Past research and future development on endemic Chinese cavefish of the genus *Sinocyclocheilus* (Cypriniformes, Cyprinidae) (in Chinese with English abstract). Acta Zootaxon Sin 31:769–777

Zhao YH, Zhang CG, Peng JX (2002) Methods of making transparent skeletons and X-ray photography of fish. In: Lin Q, Zhang CG, Jin HN (eds) Proceedings of the Conference, China Biological Collections for the 21st Century. China Science and Technology Press, Beijing, pp 298–301

Zhao YH, Watanabe K, Zhang CG (2006) *Sinocyclocheilus donglanensis*, a new cavefish (Teleostei: Cypriniformes) from Guangxi China. Ichthyol Res 53:121–128

Zhou J, Zhang CG, He AY (2004) A new species of the genus *Sinocyclocheilus* (Cypriniformes, Cyprinidae) (in Chinese with English abstract). Acta Zootaxon Sin 29:594–598

Environ Biol Fish (2009) 86:143–144
DOI 10.1007/s10641-008-9355-5

Threatened fishes of the world: *Squalidus iijimae* (Oshima, 1919) (Cyprinidae)

I-Shiung Chen

Received: 24 September 2007 / Accepted: 26 March 2008 / Published online: 27 May 2008

Common names: Iijima's gudgeon; Tai-Yin-Ju (Mandarin); Fan-Dao-Shih-Yin-Ju (Mandarin); Larm-Hon-Hi (Taiwanese). *Conservation status*: Endangered pending Taiwanese Red-list, but not listed IUCN Red-list (Chen and Chang 2005). *Identification*: D 3+7; P 1+13–14; V 1+8; A 3+6; LL 34–35; PreD11–13; Ph 5.3–3.5. Body elongate, compressed. Eye large. Mouth terminal. Lips thin. Two barbels present about half length of orbit. Body light greenish-silver dorsally, silver-white ventrally. Scale pockets with posterior gray margin. Dorsal, caudal fins with small black spots. A longitudinal shiny-green stripe with six to seven rounded grey spots. Size 5–7 cm, maximum 10 cm (SL) (Tzeng 1986; Shen 1993; Chen and Fang 1999; Yue 2000; Chen and Chang 2005). *Distribution*: Endemic to the To-Chien (Oshima 1919) and Hou-Long River basins (Chen and Chang 2007), no collection record in To-Chien since Oshima (1919). Abundance: An estimated five to ten adult individuals per 10 m^2 of stream in the densest populations. *Habitat and ecology*: Freshwater benthopelagic (0.7–3.0 m) species prefers weakly alkaline (pH 7.8–8.5), well-oxygenated (DO> 75%), clear (0–80 NTU), slowly flow (<40 $cm \cdot s^{-1}$) waters in lower and middle stream reaches between 10 and 300 m of altitude (Chen and Chang 2005). *Reproduction*: Ascending reproductive migration during summer (May to July), spawns eggs on hard substrates May to August (Chen and Chang 2005; Han et al. 2007). *Threats*: (1) Riparia habitat destruction and vegetation degradation; (2) illegal (poaching) electro-fishing; (3) dam construction in middle reaches blocked seasonal migration; (4) water pollution; and (5) severe competition for ecological niches with invasive species like *Oreochromis* spp. (Chen 2001). *Conservation action*: Application to Taiwanese Red-list as "Endangered". *Conservation recommendation*: Avoid introduction of exotic species, especially ichthyophagus and direct competitors, protect natural river habitats, restore critical riverine habitats critical for the species, prohibit illegal fishing, avoid new dam construction or habitat fragmentation. *Remarks*: Establishing population by back-up artificial breeding is important.

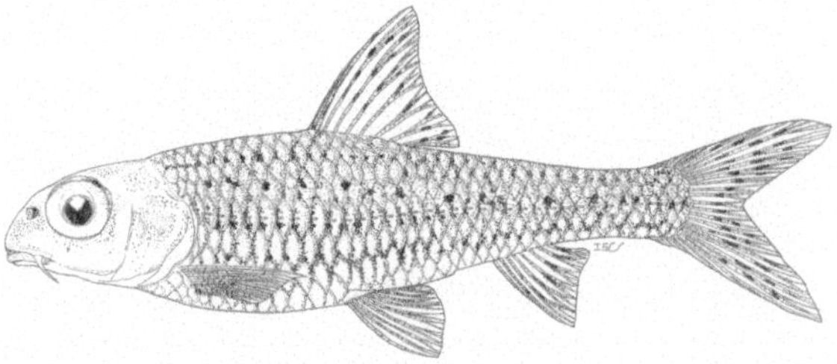

I-S. Chen (✉)
Institute of Marine Biology and Department of Life Science, National Taiwan Ocean University,
Keelung 202 Taiwan, Republic of China
e-mail: isc@mail.ntou.edu.tw

References

Chen I-S (2001) Research history, geographical distribution, conservation strategy and future perspective of the freshwater fishes in Taiwan. Taiwan Ref 52:45–60 (in Chinese)

Chen I-S, Chang YC (2005) A photographic guide to the inland-water fishes of Taiwan. Vol. I. Cypriniformes. Sueichan, Keelung

Chen I-S, Chang YC (2007) Taxonomic revision and mitochondrial sequence evolution of the cyprinid genus *Squalidus* (Teleostei: Cyprinidae) in Taiwan with description of a new species. Raffles Bull Zool S14:69–76

Chen I-S, Fang LS (1999) The freshwater and estuarine fishes of Taiwan. National Museum of Marine Biology/Aquarium Press, Pingtung (in Chinese)

Han CC, Tew KS, Fang LS (2007) Spatial and temporal variations of two cyprinids in a subtropical mountain reserve—a result of habitat disturbance. Ecol Freshw Fish 16:395–403

Oshima M (1919) Contributions to the study of the freshwater fishes of the Island of Formosa. Ann Carnegie Mus 12:169–328

Shen SC (1993) Fishes of Taiwan. National Taiwan University Press, Taipei (in Chinese)

Tzeng CS (1986) Distribution of the freshwater fishes of Taiwan. J Taiwan Mus 39:127–146

Yue P (ed) (2000) Fauna Sinica. Osteichthyes. Cypriniformes III. Science, Beijing (in Chinese)

Environ Biol Fish (2009) 86:145–153
DOI 10.1007/s10641-007-9283-9

Comparative and evolutionary analysis in natural diploid and tetraploid weather loach *Misgurnus anguillicaudatus* based on cytochrome *b* sequence data in central China

Chengtai Yang · Ling Cao · Weimin Wang · Yi Yang · Khalid Abbas · Binlun Yan · Hanping Wang · Li Su · Yuhua Sun · Huanling Wang

Received: 12 March 2007 / Accepted: 10 May 2007 / Published online: 20 September 2007

Abstract To obtain the phylogenetic relationship between diploid and tetraploid *Misgurnus anguillicaudatus*, the mitochondrial cyt *b* gene in the diploid and tetraploid weather loach were isolated and sequenced. The DNA sequences were analyzed using MEGA 3.0 software to determine the phylogenetic relationship. Forty-five variable sites among cyt *b* gene sequences and 18 amino acid substitutions occurred within the diploid and tetraploid loaches as deduced from the nucleotide sequences analysis of the cyt *b* gene. The nucleotide pairwise distance between diploid and tetraploid loach ranged from 0.001 to 0.025. Phylogenetic analysis revealed evolutionary relationships between diploid and tetraploid loach. Our results indicated a significant difference between diploid and tetraploid loach about the cyt *b* gene. AMOVA analysis indicated that there were no significant genetic variations within diploid loaches (Fst=0.2529, $P>0.05$) and within tetraploid loaches (Fst=0.0564, $P>0.05$), neither. However, significant genetic differences were found between diploid and tetraploid loaches (Fst=0.7634, $P<0.05$). Thus, it is concluded that no reproductive isolation was found within the same cytotypes of different localities, but there was reproductive isolation between these two cytotypes. The diploid loach existed before the tetraploid loach in nature. The present study is the first to describe the phylogenetic relationships of natural polyploidy weather loach using mtDNA cyt *b* gene.

Keywords *Misgurnus anguillicaudatus* · Cytochrome *b* · Ploidy · Genetic variation · Genetic distance · Evolution

C. Yang · L. Cao · W. Wang (✉) · K. Abbas · Y. Sun · H. Wang
College of Fisheries, Key Lab of Agricultural Animal Genetics, Breeding and Reproduction of Ministry of Education, Huazhong Agricultural University, Wuhan, Hubei 430070, People's Republic of China
e-mail: wangwm@mail.hzau.edu.cn

Y. Yang
Aquaculture and Aquatic Resources Management, School of Environment, Resources and Development, Asian Institute of Technology, P.O. Box 4, Klong Luang, Pathum Thani 12120, Thailand

B. Yan
Jiangsu Key Lab of Marine Biotechnology, Huaihai Institute of Technology, Lianyungang, Jiangsu 222005, People's Republic of China

H. Wang
Aquaculture Genetics and Breeding Laboratory, Ohio State University South Centers, 1864 Shyville Road, Piketon, OH 45661, USA

L. Su
Huangshi Institute of Technology, Huangshi, Hubei 435003, People's Republic of China

Introduction

The weather loach, *Misgurnus anguillicaudatus* (Cobitidae, Cypriniformes), is an important freshwater fish species with both high economic and nutritional value throughout Asia. This omnivorous species inhabits the bottom of the water column and at times can be found in the sediment, feeding upon algae, bottom dwelling animals, detritus, and cladocerans. The weather loach has long been used in traditional Chinese medicine as a folk remedy for the treatment of hepatitis, osteomyeitis, carbuncles, inflammations and cancers. It has also been used to restore health from debilities caused by various pathogens and aging. Its attractive characteristics such as moderate size (10–20 g of body weight in marketable size), easy to culture, soft egg chorion, transparent embryos, fast embryonic development (24 h at 25°C), short generation time (3–4 months), and year-round spawning under controlled conditions make this loach suitable as a candidate for field and laboratory studies (Shao et al. 2005). It has been successfully used for several genetic manipulations, including ploidy manipulation, hybridization, and sex control.

Polyploidization is well known as an important evolutionary force in plants and indications of its importance in the evolution of animals are constantly accumulating (Soltis and Soltis 1995, 1999; Le Comber and Smith 2004; Slechtova et al. 2006). Examples of fish groups, in which changes in ploidy level have been already identified as key events in their evolution include Acipenseridae (Ludwig et al. 2001), Cyprinidae (Alves et al. 2001; David et al. 2003), Catostomidae (Ueno et al. 1988), and Salmonidae (Phillips and Ráb 2001; Crespi and Fulton 2004). Multiple origins of polyploidy were demonstrated in African barbs (Tsigenopoulous et al. 2002) and Cobitidae (Janko et al. 2003). Weather loaches represent an interesting model to study the role of polyploidization in vertebrate evolution. Although the taxonomic relationship among these loaches has not yet been clarified, presence of taxonomically or genetically distinct loaches has been suggested. Cytogenetic studies have shown that the natural loaches in China are usually diploid (2n=50 chromosomes) and tetraploid (2n=100 chromosomes). To evaluate the origin of polyploidisation within weather loach and its contribution to its evolution, it is necessary to understand the phylogenetic relationships between the diploid and tetraploid species.

The mtDNA is being widely used across many groups of fishes, especially Cypriniformes, as a genetic marker in the analysis of genetic variation and for establishing phylogenetic relationships (Xiao et al. 2001). As a conserved protein-coding gene with relatively slow evolutionary rate, mitochondrial cytochrome *b* (cyt *b*) has been demonstrated to be useful for phylogenetic studies among closely related species (Tang et al. 2006). Many population genetic and phylogenetic studies have been conducted using cyt *b* gene sequences in fishes (Briolay et al. 1998; Song et al. 1998; He et al.1999; Matsuo et al. 2001; Peng et al. 2004). It has been reported that restriction fragment length polymorphism analysis of mitochondrial DNA should be used to analyze genetic variation of Japanese loach (Khan et al. 2005). However, little work on the cyt *b* gene in the natural polyploid loach has been reported. There has been no research related to genetic differences and phylogenetic relationships between diploid and tetraploid loach.

The objectives of this study were to investigate the ploidy level distribution of weather loach collected from nine selected sites in central China and to analyze genetic diversity and phylogenetic relationships between diploid and tetraploid using cyt *b* gene sequence analyses to estimate the evolution of weather loach. To our knowledge, the present study is the first report to describe the phylogenetic relationships of natural polyploidy weather loach using mtDNA cyt *b* gene.

Material and methods

Specimen examination and ploidy determination

To survey the distribution and frequency of natural loach polypoids, six samples were collected from nine different geographic locations in central China (Fig. 1). The chromosome specimens belonging to *M. anguillicaudatus* were obtained from metaphase of kidney cells by in vivo injection of PHA and colchicin, hypatoic-air drying technique, and Giemsa staining (Quan et al. 2000; Zheng et al. 2005). The chromosome results were analyzed according to Levan et al (1964).

Total DNA extraction

The total DNA was extracted using scales or underlying muscle tissue following the description found in

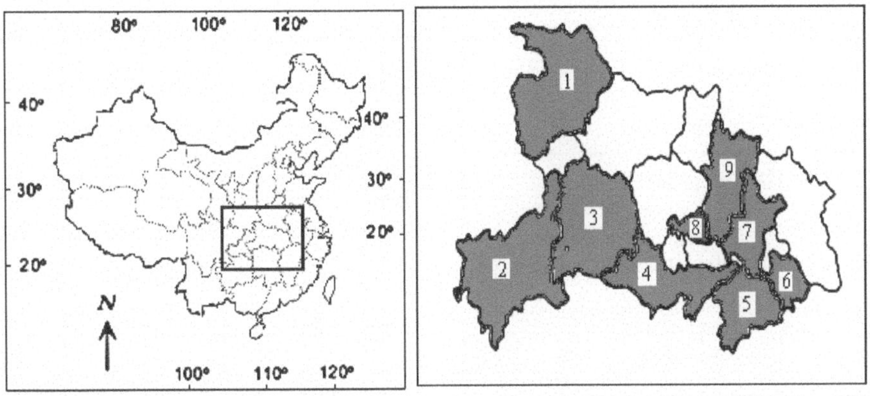

Fig. 1 Sampling localities of weather loach *M. anguillicaudatus* in central China (*nine black areas* indicating nine sampling sites): *1* Shiyan City (SY); *2* Enshi City (ES); *3* Zigui City (ZG); *4* Shashi City (SS); *5* Chibi City (CB); *6* Yangxin City (YX); *7* Wuhan City (WH); *8* Tianmen City (TM); *9* Xiaogan City (XG)

Kocher et al. (1989) with some simplifications. The tissues from the sample fish were ground to powder in liquid nitrogen. Approximately 0.1 g of ground tissue was added to 700 μl of lysis buffer (50 mM Tris–HCl, pH 7.5, 50 mM EDTA, and 3% SDS) and digested by adding proteinase K (final concentration 100 μg ml^{-1}) and incubated at 50°C overnight. The supernatant was extracted once with chloroform/phenol (1:1). The nucleic acids were precipitated in a solution containing 10 μl of 3 M sodium acetate (pH 5.2) and 1 ml of absolute isopropanol. The resulting pellet was washed with cold 70% ethanol, dried, and re-suspended in 50 μl of distilled water.

PCR and sequencing

Two primers were used to amplify and sequence the cyt *b* gene of weather loach: L14724 (5′-GACTTGAAAAA CCACCGTTG-3′) and H15915 (5′-CTCCGATCTCCG GA- TTACAAGAC-3′) (Xiao et al. 2001), synthesized by Shanghai Sangon Biological Engineering Technology and Service. PCR reactions were carried out in a 50 μl final volume containing 2.5 μl DNA, 5 μl 10×buffer, 0.5 μl dNTP, 3 μl of each primer, 1 μl Taq DNA, 4 μl Mg^{2+} and 31 μl H_2O.

The thermal cycling profile started with 94°C for 3 min followed by 35 cycles of denaturation at 94°C for 30 s, annealing at 54°C for 45 s, extension at 72°C for 1 min, with a final extension at 72°C for 10 min. The amplified DNA fragments were checked in 1% agarose gel and then purified for direct sequencing by an automatic DNA Sequencer (ABI PRISM 377, Perkin-Elmer) of Shanghai Sangon Biological Engineering Technology and Service.

Sequence analyses

The sequences were aligned using the CLUSTAL W (Thompson et al. 1994), and then were manually edited using the SEAVIEW (Galtier et al. 1996). Pairwise sequence comparisons to determine the distribution and amount of variation and the degree of saturation among codon position were performed using MEGA 3.0 (Kumar and Gadagkar 2001). Phylogenetic reconstructions were obtained using the neighbor-joining (NJ) and maximum parsimony (MP) trees. The reliability of the tree topology was assessed by 1000 bootstrap replications (Felsenstein 1985). Two diploid species *Paramisgurnus dabryanus* (PD) and *Botia superciliar* (BS) were used as out-groups in the phylogenetic analysis. AMOVA analysis was used to analyze the distribution of the molecular genetic variation among samples and their relationship with geographical or chromosome data (Excoffier et al. 1992).

Results

Ploidy level of loaches samples

After counting the chromosome number, only diploid and tetraploid weather loaches were found in the present study. No natural triploid loaches were detected in the present study. The number of chromosome in diploidy is 2n=50, and that of tetraploidy is 4n=100. The ploidy distribution of weather loach in each sampling site is presented in Table 1. Diploid loaches were found in ES city, YX city, TM city, SY city and ZG city, while tetraploid loaches were detected in the

Table 1 Ploidy distribution of *M. anguillicaudatus* in central China

Sampling location (abbr.)	Specimen no.	Ploidy level	Chromosome no.
Enshi (ES)	6	Diploidy	2n=50
Yangxin (YX)	6	Diploidy	2n=50
Tianmen (TM)	6	Diploidy	2n=50
Shiyan (SY)	6	Diploidy	2n=50
Zigui (ZG)	6	Diploidy	2n=50
Wuhan (WH)	6	Tetraploidy	4n=100
Shashi (SS)	6	Tetraploidy	4n=100
Xiaogan (XG)	6	Tetraploidy	4n=100
Chibi (CB)	6	Tetraploidy	4n=100

rest of sampling sites (WH city, SS city, XG city and CB city).

Base composition and variation

After alignment, 1,140 bp of the complete cyt *b* gene sequences were used for analysis. The average nucleotide composition for all sequences were A=27.3%, T=31.6%, C=26.1%, and G=15.0% (Table 2). The content of $A+T$ (58.9%) was higher than that of $G+C$ (41.1%). In the triplet codons, the first position has even usage in four nucleotides; the second position has high content of T (41.0%) but low content of G (13.2%); the lowest G content (6.4%) was found in the third position (Table 3). There were 45 sites for all sequences, 23 of which were variable among diploid loaches and 14 among tetraploids. Base substitutions happened mostly at the third codon position. The average transition/transversion ratio across all the pairwise sequence comparisons was 0.9. Usually the level of transition bias less than 2.0 is considered that the sequence variation begins to saturation (Zhang and Ryder 1997; Zhou et al. 2001). Most of the variability among sequences was observed in the third codon position. First and second positions show very slight saturation, whereas substitutions at the third position are strongly saturated.

Table 2 Average base composition of mtDNA cyt *b* gene in *M. anguillicaudatus*

Sampling location (abbr.)	A (%)	T (%)	C (%)	G (%)	GeneBank accession no.
Enshi (ES)	27.2	31.7	26.1	15.1	EF424608
Yangxin (YX)	27.4	31.5	26.2	14.9	EF424604
Tianmen (TM)	27.5	31.4	26.3	14.8	EF376188
Shiyan (SY)	27.5	31.4	26.2	14.9	EF424605
Zigui (ZG)	27.5	31.6	26.1	14.8	EF424603
Wuhan (WH)	27.1	31.6	26	15.3	EF088651
Shashi (SS)	27.2	31.6	26.1	15.1	EF424606
Xiaogan (XG)	27.1	31.7	26	15.2	EF424607
Chibi (CB)	26.9	31.5	26.2	15.4	EF424602
AVERAGE	27.3	31.6	26.1	15.0	

Amino acid variation

There were 380 amino acids deduced from the nucleotide sequences of the cyt *b* genes in both diploid and tetraploid loaches. Most of the nucleotide substitutions were located at the third codon position and were silent mutations. In addition, within 18 amino acid substitutions, 10 were in the diploids while 8 were in tetraploid loaches (Table 4).

Genetic distances and genetic variation

Among all loaches sequences, the nucleotide pairwise distance ranged from 0.001 to 0.378. The nucleotide pairwise distance between the tetraploid and diploid ranged from 0.001 to 0.025. The nucleotide pairwise distance between tetraploid and diploid was smaller than that between *M. anguillicaudatus* and the outgroup species. Amino acid pairwise distances among these fish ranged from 0.003 to 0.359 (Table 5). Based on the AMOVA analysis, there were no significant genetic variations within diploid loaches (Fst=0.2529, $P>0.05$) and within tetraploid loaches (Fst=0.0564, $P>0.05$), neither. However, significant genetic differences were found between diploid and tetraploid loaches (Fst=0.7634, $P<0.05$).

Phylogenetic trees

NJ phylogenetic tree (Fig. 2) and MP phylogenetic tree (Fig. 3) of the cyt *b* genes from diploid and tetraploid *M. anguillicaudatus* in cypriniformes was reconstructed. The branches of two kinds of phylogenetic trees had the same topology and similar bootstrap probabilities. Phylogenetic analyses using NJ identified two major lineages supported by moderate to high bootstrap values: diploid and tetraploid loaches (51 and 99% bootstrap, respectively). In the MP tree, tetraploid (XG, CB, SS, WH) loaches were at the derivative clade, while diploid (TM, SY, YX, ES, ZG) loaches

Table 3 Base composition (%) at first, second, and third codon positions of cyt *b* gene in *M. anguillicaudatus*

Sampling location	First position				Second position				Third position			
	T	C	A	G	T	C	A	G	T	C	A	G
CB	26.3	23.9	23.9	25.8	41.1	25.8	20.0	13.2	26.8	28.9	36.8	7.4
SS	26.1	24.2	24.5	25.3	41.1	25.8	20.0	13.2	27.4	28.4	37.1	7.1
XG	26.3	23.9	23.9	25.8	41.1	25.8	20.0	13.2	27.4	28.4	37.4	6.8
WH	26.3	24.2	23.9	25.5	40.8	25.8	20.0	13.4	27.4	28.2	37.4	7.1
ZG	25.5	24.5	24.2	25.8	41.1	26.1	19.7	13.2	27.9	27.9	38.7	5.5
YX	25.8	24.5	24.2	25.5	41.1	25.8	19.7	13.4	27.4	28.4	38.2	6.1
SY	25.8	24.2	24.2	25.8	41.1	25.8	20.0	13.2	27.1	28.7	38.4	5.8
TM	25.8	24.5	23.9	25.8	41.1	25.8	20.0	13.2	27.1	28.4	38.4	5.8
ES	25.8	24.7	23.7	25.8	41.1	25.8	19.7	13.4	27.9	27.9	38.2	6.1

were at the basal clade of MP tree, which was similar to the NJ tree.

Discussion

Ploidy distribution

In the present study, only diploid and tetraploid loaches were detected. However, none of triploid loaches were observed. This result is in agreement with the former research (Yin et al. 2005) that there were probably only two cytotypes of loaches in China, including diploid and tetraploid loaches. The occurrence of natural polyploid individuals was firstly recognized by Ojima and Takai (1979) in Japan. Weather loach includes polyploid and unisexual biotypes in nature (Zhang and Arai 1999; Morishima et al. 2002; Yin et al. 2005) and appears to tolerate genomic changes by means of artificial chromosome manipulation from diploid to polyploid state (Arai 2001). However, in Japan, a relatively high frequency of diploid and triploid (3n=75) individuals has been recorded in several places; out of approximately 40 localities examined for ploidy status by the measurement of relative DNA content with flow cytometry, but no tetraploid has been discovered in wild populations in Japan (Zhang and Arai 1999; Morishima et al. 2002). The actual origin of these polyploids is still unknown. Since the natural tetraploid with 100 chromosomes was reported to occur among the specimens from Yangtze River (Li et al. 1983), the tetraploid specimens are likely to originate from the Asian continent and thus are genetically different from common diploid loach in Japan. Probably due to the geographical differences between China and Japan, triploid loaches do not adapt to establish itself in China.

Sequence variation and gene flow

In the present study, among 1,140 nucleotides of the cyt *b* gene, the content of $A+T$ (58.9%) was higher than that of $C+G$ (41.1%), which fell within the range of GC content typical for vertebrates (Nei and Kumar 2000). Base compositional biases were found which

Table 4 Different amino acids among the deduced amino acid sequences of the mitochondrial cyt *b* gene

Amino acid sites	42	155	180	240	265	267	290	292	309	315	316	332	341	352	356	357	358	369
YX	I	Y	A	T	T	S	R	P	D	P	S	H	N	P	Y	P	I	G
TM	I	Y	A	T	T	S	R	P	H	P	S	H	N	T	C	P	I	G
SY	I	Y	A	T	T	S	R	P	H	P	S	H	N	T	C	H	I	G
ES	I	Y	A	A	T	S	R	S	D	P	G	P	D	P	Y	P	I	G
XG	I	Y	A	T	A	S	G	S	H	P	S	H	D	T	C	P	I	G
SS	I	H	T	T	A	S	G	S	H	P	S	H	D	T	C	H	I	A
CB	I	Y	A	T	A	P	G	S	H	P	S	H	D	T	C	P	I	G
WH	S	Y	A	T	A	S	G	S	H	T	T	H	D	T	C	P	I	G
ZG	I	Y	A	T	T	S	R	S	H	P	S	H	D	T	C	P	N	G

Table 5 The percent divergences of nucleotide sequences (below diagonal) and amino acid sequences (above diagonal) of the mitochondrial cyt *b* gene

Sampling location	YX	TM	SY	ES	PD	BS	XG	SS	CB	WH	ZG
YX		0.008	0.011	0.014	0.307	0.356	0.019	0.030	0.022	0.028	0.017
TM	0.008		0.003	0.022	0.307	0.356	0.011	0.022	0.014	0.019	0.008
SY	0.009	0.001		0.025	0.307	0.356	0.014	0.019	0.017	0.022	0.011
ES	0.011	0.017	0.018		0.309	0.359	0.022	0.033	0.025	0.028	0.019
PD	0.325	0.328	0.328	0.329		0.083	0.304	0.307	0.304	0.304	0.307
BS	0.371	0.372	0.372	0.378	0.200		0.354	0.356	0.354	0.354	0.356
XG	0.021	0.017	0.018	0.019	0.329	0.376		0.011	0.003	0.008	0.008
SS	0.024	0.019	0.018	0.022	0.331	0.378	0.004		0.014	0.019	0.019
CB	0.025	0.020	0.021	0.021	0.332	0.376	0.005	0.008		0.011	0.011
WH	0.025	0.020	0.021	0.021	0.330	0.378	0.005	0.008	0.007		0.017
ZG	0.015	0.011	0.011	0.010	0.328	0.375	0.013	0.016	0.015	0.015	

showed a strong bias against *G* in the cyt *b* gene of weather loaches, particularly at the third position. Our results fit well with Meyer (1993), who investigated the base compositional biases of several species of fishes and found that all species of fishes investigated showed a strong bias against *G* in the cyt *b* gene. The variable sites number (14) in tetraploid loaches is less than that of diploids (23). According to our results, tetraploid loaches were found to cluster in three sites close to WH city, but diploid loaches distributed sporadic in sampling sites. There was significant difference among diploid loaches from geographically distinct populations, while less difference observed among tetraploid loaches. The presence of genetically distinct populations in Chinese loaches has been suggested (Yin et al. 2005), but the results were fragmentary and inconclusive due to the small number of localities examined and specimens used. Khan and Arai (2000) conducted allozyme analyses in total 923 individual collected from 44 populations all over Japan. They found that genetic differentiation among some groups was likely to be of inter-subspecies level, whereas divergence in other groups seemed to be a local-race level. Analyses of the genetic population structure in many fish species revealed that individuals from different areas of a lake are typically genetically substructured into different local populations (Douglas et al. 1999; Garant et al. 2000; Gerlach et al. 2001). Genetic differentiation of local populations is caused, among other reasons, by environmental barriers and limited dispersal (Johnson and Gaines 1990), which restrict migration and therefore limits gene flow (Barluenga and Meyer 2005). However, in the present study, no intra-population variability was found. Many geographically distant populations frequently showed genetically close relationship. Such inconsistencies between genetic and geographical relationships might be

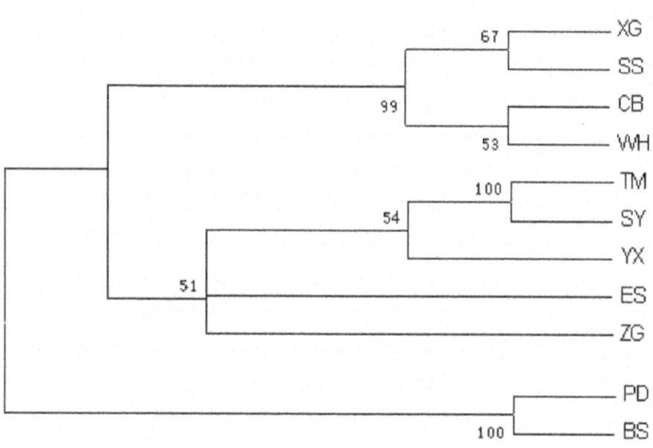

Fig. 2 Neighbor-joining (NJ) tree of diploid and tetraploid *M. anguillicaudatus*: the *number above branches* is 1,000 bootstrap value of NJ tree; *P. dabryanus* (*PD*) and *B. superciliar* (*BS*) are two out-groups

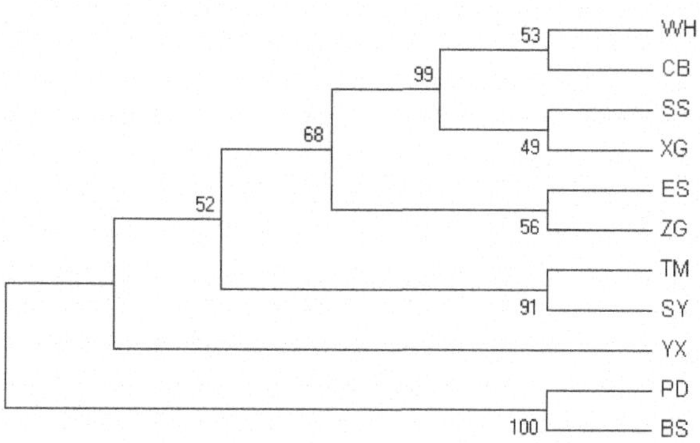

Fig. 3 Maximum parsimony (MP) tree of diploid and tetraploid *M. anguillicaudatus*: the *number above branches* is 1,000 bootstrap value of MP tree; *P. dabryanus* (*PD*) and *B. superciliar* (*BS*) are two out-groups

due to insufficient number of samples or less sensitivities of analytical method (Khan et al. 2005). Based on the AMOVA analysis, there were no significant genetic variations within diploid loaches (Fst=0.2529, $P>0.05$) and within tetraploid loaches (Fst=0.0564, $P>0.05$). However, significant genetic differences were found between diploid and tetraploid loaches (Fst=0.7634, $P<0.05$). Thus, it is concluded that no reproductive isolation was found within the same cytotypes of different localities, but there was reproductive isolation between these two cytotypes.

Phylogenetic relationship

The phylogenetic relationships among branching lineages of weather loaches were resolved by cyt *b* sequences. In Figs. 2 and 3, the phylogenetic trees (NJ and MP) were both divided into two branches including diploid and tetraploid branches, which formed a monophyletic lineage in both analyses. The diploid branch in the basal clade of NJ and MP phylogenetic trees and tetraploid branches at the derivative clade of the trees indicate that diploids are earlier existing cytotype than tetraploids in nature. Among diploid and tetraploid loaches, the cyt *b* gene showed highly homology. High levels of substructuring between adjacent populations might be expected to facilitate speciation (Van and Turner 1997). Particularly in small isolated populations, drift, adaptation to local conditions or sexual selection will have important effects on gene frequencies. Therefore, one would expect that species with poor dispersal capabilities and high levels of substructuring exhibit high levels of intraspecific variation in some traits (Van and Turner 1997). This appears, however, not to be the case for stone loach, where no appreciable variation in morphological, physiological, or behavioral traits has been detected so far. Similar discordances between morphological and genetic differences have been documented in other fish taxa (Shaw et al. 2000; Wang et al. 2000; Taylor and Verheyen 2001). Therefore, high levels of genetic substructuring alone do not seem to be sufficient to promote high levels of speciation. This suggests that other species-specific factors might influence whether groups speciate or not, such as breeding behavior (Mayr 1984). This is the first report to describe the phylogenetic relationships of natural polyploidy weather loach using mtDNA cyt *b* gene, which provide a new insight on the evolution of two cytotypes. Analysis of cyt *b* gene and elucidation of the variation of cyt *b* gene in different fish species prove that cyt *b* is a useful genetic marker to monitor the variations in the progeny of the crosses.

Conclusion

Based on our research, 45 variable sites among cyt *b* gene sequences and 18 amino acid substitution occurred within the tetraploid and diploid loaches which deduced from the nucleotide sequences of the cyt *b* genes. The nucleotide pairwise distance between diploid and tetraploid loach ranged from 0.001 to 0.025. Based on the AMOVA analysis, it is concluded that no reproductive isolation was found within the same cytotypes of different localities. But there was reproductive isolation between these two cytotypes. Phylogenetic analysis revealed an evolutionary relationship between diploid and tetraploid loaches. The diploid loach existed before tetraploid loaches in nature. This is the first report to describe the phylogenetic relationships of natural weather loach using mtDNA cyt *b* gene. To further understand genetic and cyto geographical relationships in weather loaches, the distribution of diploids and diploid–tetraploid mosaics needs to be investigated with more sensitive and reliable DNA marks.

Acknowledgement This research is a component of the Aquaculture Collaborative Research Support Program (ACRSP), supported by the US Agency for International Development Grant, and by contributions from University of Michigan, Asian Institute of Technology, and Huazhong Agricultural University. The accession number is 1329. The opinions expressed herein are those of the authors and do not necessarily reflect the views of the US Agency for International Development. This study is also funded by "National Key Technology R&D Program in Eleventh-Five-Year Plan" titled as "The integrated demonstration of safe fishery production". The accession number is 2006BAK02A22.

References

Alves MJ, Coelho MM, Collares PMJ (2001) Evolution in action through hybridisation and polyploidy in an Iberian freshwater fish: a genetic review. Genetica 111:375–385

Arai K (2001) Genetic improvement of aquaculture finfish species by chromosome manipulation techniques in Japan. Aquaculture 197:205–228

Barluenga M, Meyer A (2005) Old fish in a young lake: stone loach (Pisces: *Barbatula barbatula*) populations in Lake

Constance are genetically isolated by distance. Mol Ecol 14:1229–1239

Briolay J, Galtier N, Brito RM, Bouvet Y (1998) Molecular phylogeny of Cyprinidae inferred from cytochrome *b* DNA sequences. Mol Phylogenet Evol 9:100–108

Crespi BJ, Fulton NJ (2004) Molecular systematics of Salmonidae: combined nuclear data yields a robust phylogeny. Mol Phylogenet Evol 31:658–679

David L, Blum S, Feldman MW, Lavi U, Hillel J (2003) Recent duplication of the common carp (*Cyprinus carpio* L.) genome as revealed by analyses of microsatellite loci. Mol Biol Evol 20:1425–1434

Douglas MR, Brunner PC, Bernatchez L (1999) Do assemblages of Coregonus (Teleostei: Salmoniformes) in the Central Alpine region of Europe represent species flocks. Mol Ecol 8:589–603

Excoffier L, Smouse PE, Quattro JM (1992) Analysis of molecular variance inferred from metric distances among DNA haplotypes applications to human mitochondrial DNA restriction data. Genetics 131:479–491

Felsenstein J (1985) Confidences limits on phylogenies: an approach using the bootstrap. Evolution 39:783–791

Galtier N, Gouy M, Gautier C (1996) SeaView and Phylo-Win: two graphic tools for sequence alignment and molecular phylogeny. Comput Appl Biosci 12:543–548

Garant D, Dodson JJ, Bernatchez L (2000) Ecological determinants and temporal stability of the within-river population structure in Atlantic salmon (*Salmo salar* L.). Mol Ecol 9:615–628

Gerlach G, Schardt U, Eckmann R, Meyer A (2001) Kin-structured subpopulations in Eurasian perch (*Perca fluviatilis* L.). Heredity 86:213–221

He SP, Chen YJ, Zhang YP (1999) Preliminary study on mitochondrial cytochrome b DNA sequences and phylogeny of formalin fixed sisorid fishes. Zool Res 20:81–87 (in Chinese)

Janko K, Kotlik P, Ráb P (2003) Evolutionary history of asexual hybrid loaches (Cobitis: Teleostei) inferred from phylogenetic analysis of mitochondrial DNA variation. J Evol Biol 16:1280–1287

Johnson ML, Gaines MS (1990) Evolution of dispersal: theoretical models and empirical tests using birds and mammals. Ann Rev Ecolog Syst 21:449–480

Khan MR, Arai K (2000) Allozyme variation and genetic differentiation in the loach *Misgurnus anguillicaudatus*. Fish Sci 66:211–222

Khan MMR, Arai K, Kuroda K, Umino T, Nakagawa H (2005) Genetic variation of Japanese loach inferred from restriction fragment length polymorphism analysis of mitochondrial DNA. Afr J Biotechnol 4:318–325

Kocher TD, Thomas WK, Meyer A, Edwards SV, Paabo S, Villablanca FX, Wilson AC (1989) Dynamics of mitochondrial DNA evolution in animals: amplification and sequencing with conserved primers. Proc Natl Acad Sci U S A 86:6196–6200

Kumar S, Gadagkar SR (2001) Disparity index: a simple statistic to measure and test the homogeneity of substitution patterns between molecular sequences. Genetics 158:1321–1327

Le Comber, SC, Smith, C (2004) Polyploidy in fishes: patterns and processes. Biol J Linn Soc 82:431–442

Levan A, Fredga K, Sandberg A (1964) Nomenclature for centrometric position on chromosomes. Hereditas 52:201–220

Li K, Li Y, Zhou D (1983) A comparative study of the karyotypes in two species of mud loaches. Zool Res 4:75–81 (in Chinese with English abstract)

Ludwig A, BelWore NM, Pitra C (2001) Genome duplication events and functional reduction of ploidy levels in sturgeon (*Acipenser*, *Huso* and *Scaphirhynchus*). Genetics 158:1203–1215

Matsuo T, Ogawa Y, Kumamaru A, Ochi K, Adachi Y (2001) Complete Nucleotide Sequence of the cytochrome *b* gene of channel catfish *Ictalurus punctatus* and comparison of sequence homology among channel catfish and other fishes. Mol Biol 63:207–210

Mayr E (1984) Evolution of fish species flocks: a commentary. In: Echelle AA, Kornfield IRV (eds) Evolution of fish species flocks. University of Maine at Orono Press, Orono Maine, pp 3–11

Meyer A (1993) Evolution of mitochondrial DNA in fishes. In: Hochachka PW, Mommsen TP (eds) Biochemistry and molecular biology of fishes, vol 2. Elsevier, Amsterdam, pp 1–38

Morishima K, Horie S, Yamaha E, Arai K (2002) A cryptic clonal line of the loach *Misgurnus anguillicaudatus* (Teleostie: Cobitidae) evidenced by induced gynogenesis, interspecific hybridization, microsatellite genotyping and multi locus DNA fingerprinting. Zool Sci 19:565–575

Nei M, Kumar S (2000) Molecular evolution and phylogenetics. Oxford Univ. Press, New York

Ojima Y, Takai A (1979) The occurrence of spontaneous polyploid in the Japanese common loach, *Misgurnus anguillicaudatus*. Proc Japan Acad Ser B Phys Biol Sci 55:487–491

Peng ZG, He SP, Zhang YG (2004) Phylogenetic relationships of glyptosternoid fishes (Siluriformes:Sisoridae) inferred from mitochondrial cytochrome *b* gene sequences. Mol Phylogenet Evol 31:979–987

Phillips RB, Ráb P (2001) Chromosome evolution in the Salmonidae (Pisces): an update. Biol Rev (Cambridge) 76:1–25

Quan CG, Wang J, Ding SX, Su YQ, Yao JG (2000) The karyotypes of *Pseudosciaena crocea* (Richardson). J Xiamen Univ (Nat Sci) 39: 107–110 (in Chinese with English abstract)

Shao J, Shi GQ, Jin XL, Song MY, Shi JB, Jiang GB (2005) Development and Validation of an enzyme-linked immunosorbent assay for vitellogenin in Chinese loach (*Misgurnus anguillicaudatus*). Environ Int 31:763–770

Shaw PW, Turner GF, Idid MR, Robinson RL, Carvalho GR (2000) Genetic population structure indicates sympatric speciation of Lake Malawi pelagic cichlids. Proc R Soc Lond B Biol Sci 267:2273–2280

Slechtova V, Bohlen J, Freyhof J, Rab P (2006) Molecular phylogeny of the Southeast Asian freshwater fish family Botiidae (Teleostei: Cobitoidea) and the origin of polyploidy in their evolution. Mol Phylogenet Evol 39:529–541

Soltis DE, Soltis PS (1995) The dynamic nature of polyploid genomes. Proc Natl Acad Sci U S A 92:8089–8091

Soltis DE, Soltis PS (1999) Polyploidy: recurrent formation and genome evolution. Trends Ecol Evol 14:348–352

Song CB, Near TJ, Page LM (1998) Phylogenetic relations among percid fishes as inferred from mitochondrial cytochrome b DNA sequence data. Mol Phylogecetic Evol 10:343–35

Tang QY, Liu HZ, Mayden R, Xiong BX (2006) Comparison of evolutionary rates in the mitochondrial DNA cytochrome *b* gene and control region and their implications for phylogeny of the Cobitoidea (Teleostei: Cypriniformes). Mol Phylogenet Evol 39:347–357

Taylor MI, Verheyen E (2001) Microsatellite data reveals weak population substructuring in *Copadichromis* sp. 'virginalis kajose', a demersal cichlid from Lake Malawi, Africa. J Fish Biol 59:593–604

Thompson JD, Higgins DG, Gibson TJ (1994) CLUSTAL W: improving the sensitivity of progressive multiple sequence alignment through sequence weighting, position-specific gap penalties and weight matrix choice. Nucl Acid Res 22:4673–4680

Tsigenopoulous CS, Ráb P, Naran D, Berrebi P (2002) Multiple origins of polyploidy in the phylogeny of southern African barbs (Cyprinidae) as inferred from mtDNA markers. Heredity 88:466–473

Ueno K, Nagase A, Ye YJ (1988) Tetraploid origin of the karyotype of the Asian sucker, *Myxocyprinus asiaticus*. Jpn J Ichthyol 34:512–514

Van OMJ, Turner GF (1997) Unusually fine-scale genetic structuring found in rapidly speciating Malawi cichlid fishes. Proc R Soc Lond B Biol Sci 264:1803–1812

Wang JP, Hsu KC, Chiang TY (2000) Mitochondrial DNA phylogeography of *Crossocheilus paradoxus* (Cyprinidae) in Taiwan. Mol Ecol 9:1483–1494

Xiao WH, Zhang YP, Liu HZ (2001) Molecular systematics of Xenocyprinae (Teleostei: Cyprinidae): taxonomy, biogeography, and coevolution of a special group restricted in East Asia. Mol Phylogene Evol 18(2):163–173

Yin J, Zhao ZS, Chen XQ, Li YQ, Zhu LY (2005).Karyotype comparison of diploid and tetraploid loach. Acta Hydrobiol Sin 29(4):469–472 (in Chinese with English abstract)

Zhang Q, Arai K (1999) Distribution and reproductive capacity of natural triploid individuals and occurrence of unreduced eggs as a cause of polyploidization in the loach, *Misgurnus anguillicaudatus*. Ichthyol Res 46:153–161

Zhang YP, Ryder OA (1997) Molecular phylogeny of the superfamily arctoidea. Acta Genet Sin 24(1):15–22 (in Chinese with English abstract)

Zheng L, Liu CW, Li CL (2005) Studies on the karyotype of 4 groupers. Mar Sci 29(4):51–55 (in Chinese with English abstract)

Zhou JL, Zhang YP, Huang MH (2001) Phylogenetic relationships among crotalinae based on mitochondrial cytochrome b gene sequence variations. Acta Zool Sin 47(4):361–366 (in Chinese with English abstract)

Environ Biol Fish (2009) 86:155–162
DOI 10.1007/s10641-008-9370-6

Age and growth of *Schizopygopsis younghusbandi younghusbandi* in the Yarlung Zangbo River in Tibet, China

Feng Chen · Yifeng Chen · Dekui He

Received: 4 June 2007 / Accepted: 7 March 2008 / Published online: 25 June 2008

Abstract *Schizopygopsis younghusbandi younghusbandi* is an endemic species whose distribution is restricted to the middle reaches of the Yarlung Zangbo River, being one of the most important commercial fishes in this area. Age and growth of 606 specimens captured between October 2002 and April 2005 were studied. The range in standard length (L) was 65.7–387.3 mm and total weight (W) was 3.3–772.0 g. The relationship between L and W was $W=0.000909L^{2.2493}$ for males and $W=0.000259L^{2.4781}$ for females. Age, determined from anal scales and lapillus otoliths, ranged from 3 to 18 years. The parameters of von Bertalanffy growth functions, estimated by back-calculated length, were $L_\infty = 442.7mm\ L$, k=0.0738 year^{-1} and t_0=−1.4 year for males, and $L_\infty = 471.4mm\ L$, k=0.0789 year^{-1} and t_0=0.2 year for females. Males and females exhibited statistically significant differences in growth. χ^2-test indicated that von Bertalanffy growth functions could well describe the growth of *S. y. younghusbandi*. The longevities were 39.2 and 38.2 years for males and females, respectively. Growth inflexion points were 10.2 and 12.0 years for males and females, respectively, but 84.8% of the captures were at the smaller ages. So conservation and management schemes for this population should be considered urgently. In addition, we found that populations from the upstream of the Lhasa River, the downstream of the Lhasa River and the middle reaches of the Yarlung Zangbo River showed statistically significant differences in growth patterns.

Keywords *Schizopygopsis younghusbandi younghusbandi* · Age · Growth · Otolith · Anal scale · Tibet

F. Chen · Y. Chen (✉) · D. He
Institute of Hydrobiology,
Chinese Academy of Sciences,
Wuhan 430072, China
e-mail: chenyf@ihb.ac.cn

F. Chen
e-mail: windchenfeng@sina.com

F. Chen · D. He
Graduate School of Chinese Academy of Sciences,
Beijing 100039, China

Introduction

The Yarlung Zangbo River (28°00′–31°16′ N, 82°00′–97°07′ E) is the largest river in Tibet (2057 km long), and is the river with the highest altitude in the world. It runs from west to east in the south Tibet. The Lhasa River (29°20′–30°15′ N, 90°05′–1°09′ E) is the biggest tributary of the Yarlung Zangbo River (551 km long) lies on the left of the Yarlung Zangbo River (Fig. 1). The drainage area of the Yarlung Zangbo River from Xigaze to Sangri has a large population, and is the most important agricultural and pasturing area. Lhasa the center of politics, culture and economy of Tibet is located here (the compre-

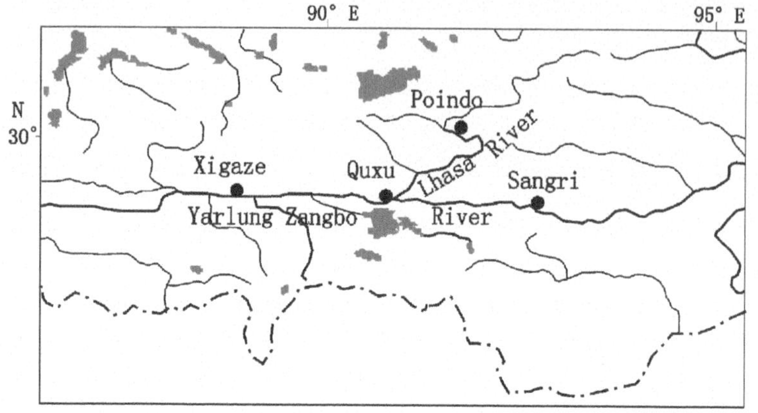

Fig. 1 Map of sampling area of *S. y. younghusbandi*

hensive scientific expedition to the Qinghai-Xizang Plateau of Chinese Academy of Sciences 1984). There is a high demand for fish in this area. Moreover, a large proportion of the fishing from that area is also sent to Chengdu another big city in China. In addition to large-scale exploitation, other threats to the native fishes are present, such as dam construction and exotic fishes. Most of the fishes of the Qinghai-Xizang Plateau are endemic species. They grow slowly and mature late (Wu et al. 1992; Bureau of Aquatic Products 1995; Chen et al. 2002b), making them very vulnerable to these threats and making it difficult to replenish the population if it is depleted. Therefore, understanding the biology of these fishes is focused on their conservation.

Schizopygopsis younghusbandi younghusbandi, belonging to the subfamily Schizothoracinae, family Cyprinidae, is an endemic species found only in the middle reaches of the Yarlung Zangbo River, and is one of the most important commercial fishes in this area (Bureau of Aquatic Products 1995). Studies on the biology and ecology of this fish cannot only provide vital information to achieve a sustainable use of this species, but also better understand its general biology.

Materials and methods

Sampling of fish and processing

A total of 606 individuals of *S. y. younghusbandi* were captured between October 2002 and April 2005 in the Lhasa River and the middle reaches from Xigaze to Sangri of the Yarlung Zangbo River in Tibet (Fig. 1) by floating gill nets and set nets.

Each specimen was measured for standard length (L) to the nearest 0.1 mm and total weight (W) to the nearest 0.1 g. Sex was checked by observing the gonad with the naked eye.

Anal scales and the subcutaneous tissue were removed and kept in 95% alcohol, and otoliths were removed and kept dry in small numbered plastic cuvettes for subsequent age-determination.

Age determination

Since the asterisci otolith of *S. y. younghusbandi* has a concave shape, and the sagittae otolith is very fragile, they are both not suitable for age determination. In this study, we used the lapilli otolith which is larger and can be easily burnished. The processing of otoliths and anal scales was following Chen et al. (2002a). All of the otolith sections and anal scales were photographed using a digital camera (Leica-DC180).

Growth

Otolith sections with regular and clear rings were selected to back-calculate length at different ages. The radius and the distances between adjacent annulus of each otolith section on the photos were measured on a consistent direction and then converted to the factual lengths.

The following function was used to calculate back-calculate length (Chen et al 2002b):

$$L_i = \mathrm{EXP}[\ln L_c(\ln O_i + a/b)(\ln O_c + a/b)]$$

where L_i is the standard length of fish at age i, $-a/b$ the intercept on the X axis of linear function $\ln L_c = a + b \ln O_c$, L_c the standard length of fish at capture and O_i the radius of the otolith at age i, and O_c the radius of the otolith at capture.

The von Bertalanffy growth parameters (L_∞, k, t_0) were estimated from the mean back-calculated standard length at each age using two iterative least squares procedures. These estimated parameters were fit to the back-calculated trajectories using the von Bertalanffy growth function (VBGF):

$$L_t = L_\infty\{1 - \exp[-k(t - t_0)]\}$$

where L_t is the standard length at age t, L_∞ is the asymptotic standard length, k is the growth coefficient, t is the age, and t_0 is the theoretical age at zero length (von Bertalanffy 1938).

The growth rate functions, growth accelerated rate functions and growth inflexion point (t_{IP}) were calculated subsequently following Yin (1995).

The following function was used to calculate longevity to define the longevity of *S. y. younghusbandi* as the time required to attain $95L_\infty$. (Taylor 1958; Natanson et al. 2006):

$$\mathrm{Longevity} = 1/\mathrm{k} \ln[(\mathrm{L}_\infty - \mathrm{L}_0)/\mathrm{L}_\infty/(1 - \mathrm{x})]$$

where L_0 is the standard length when t=0, $x = L_t/L_\infty = 0.95$.

The relationships of W-L of both sexes were compared by likelihood test (Joung et al. 2004). The difference of VBGFs between the sexes was compared by an analysis of the residual sum of squares (Chen et al. 1992). The difference between the standard length calculated from VBGF and the observed standard length was tested using χ^2-test (Yin 1995; Wang et al. 1998).

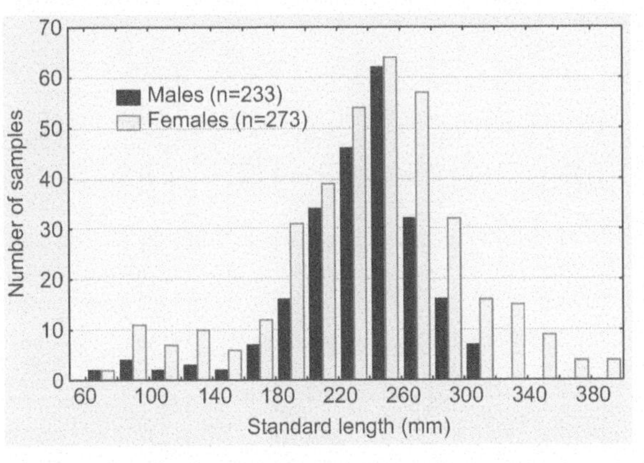

Fig. 2 Distributions of the standard length frequency of *S. y. younghusbandi*

Comparison of growth patterns among populations from different sampling sites

We divided all of the specimens into three populations according to the sampling sites: population A (up to Poindo, n=130), population B (from Poindo to Quxu, n=216) and population C (from Xigaze to Sangri, n= 68; Fig. 1). The difference of growth pattern among these three populations was compared using χ^2-test.

Results

Population structure and sex ratio

Distributions of the standard length frequency for males and females are presented in Fig. 2. The standard lengths of the captures were mainly 180–300 mm (75.2%). More females than males were captured especially in larger group. The sex ratio was 1 (233♂):1.60 (373♀).

Age determination

Five hundred thirty-six otolith sections were prepared, in which 439 (81.9%) were interpretable. Four

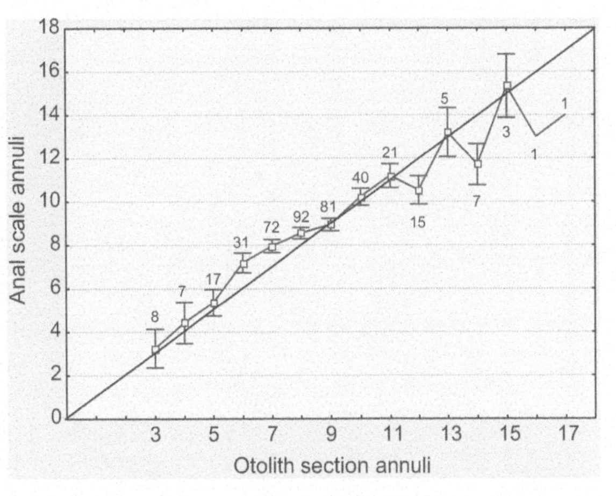

Fig. 3 Mean standard errors calculated using pooled variance between otolith sections and anal scales of *S. y. younghusbandi*. The sample size and 95% confidence limits at each age group and the 1:1 equivalence line were indicated

hundred forty-eight individuals had clear anal scales, of which 248 had two or more clear anal scales. Of these, 106 (42.7%) had anal scales showing different age. The difference was between 1–2 years (93 (87.7%) within 1 year and 13 [12.3%] within 2 years). We chose the larger value as the final result because the anal scale with the smaller age may be the result of abrasion.

When comparing the ages determined from the otolith sections and those from the anal scales, we found that the two aging structures were relatively similar from age 3 to 11 (Fig. 3); 45.8% were the same, 73.1% were within 1 year and 90.3% were within 2 years.

Standard lengths and total weights at age of 414 specimens are shown in Table 1. Ages were determined by comparing and integrating the anal scales and otolith sections. The results of age determination were 3–16 years, with a range of standard lengths ranging from 65.7 mm to 311.1 mm and a range of total weight ranging from 3.3 to 435.0 g for males. For females, the age determination was from 3–18 years with standard lengths ranging from 84.0 to 387.3 mm and total weight ranging from 8.1 to 739.0 g. We could conclude that there was a large variation of the standard length and total weight at the same age from Table 1.

Growth

Relationships between standard length and total weight were calculated as follows:

$W = 0.000909L^{2.2493} (n = 233, r = 0.8242)$ for males,
$W = 0.000259L^{2.4781} (n = 373, r = 0.9123)$ for females and
$W = 0.000161L^{2.5638} (n = 606, r = 0.8923)$ for both sexes.

Table 1 Observed standard length and total weight at ages of *S. y. younghusbandi*

Age (year)	Number	Standard length (mm)				Total weight (g)			
		Min	Max	Mean	±SE	Min	Max	Mean	±SE
Males									
3	4	65.7	101.9	81.7	18.7	3.3	15.4	7.5	2.7
4	1			116.3				17.8	
5	2	141.3	167.6	154.5	20.4	37.0	60.0	48.5	16.2
6	9	167.6	233.0	193.3	21.2	62.0	167.0	105.4	36.8
7	29	169.3	262.0	219.0	26.2	71.0	289.0	148.1	51.2
8	48	186.1	271.1	227.6	25.2	94.0	297.0	168.4	50.5
9	38	202.1	305.6	246.9	19.1	117.0	489.0	221.2	62.6
10	27	201.2	298.3	259.1	26.3	132.0	353.0	264.8	57.6
11	11	250.2	315.6	277.4	22.5	258.0	498.0	343.2	80.9
12	1			289.3				350.0	
14	1			256.6				571.0	
16	1			311.1				435.0	
Females									
3	4	84.0	96.4	92.7	6.4	8.1	14.4	10.8	2.6
4	1			94.8				10.0	
5	8	122.1	188.0	148.9	27.9	25.6	97.0	50.2	26.1
6	14	151.3	222.1	197.3	23.1	46.3	136.0	105.0	28.5
7	28	181.4	265.7	214.4	24.6	75.0	242.0	144.5	45.4
8	45	181.2	298.3	231.5	24.6	82.0	336.0	177.0	54.6
9	51	184.3	337.4	250.0	27.4	92.0	505.0	224.7	65.6
10	33	233.9	353.7	269.0	26.6	172.0	671.0	292.1	97.7
11	21	227.5	371.0	286.5	32.3	164.0	613.0	341.3	100.2
12	21	268.4	364.6	308.0	26.5	298.0	555.0	420.1	67.0
13	10	281.1	352.8	321.1	24.4	358.0	546.0	473.5	54.7
14	2	320.1	328.3	324.2	6.3	526.0	548.0	537.0	1.5
16	3	301.1	387.3	337.7	49.1	429.0	739.0	569.3	157.0
18	1			363.7				693.0	

Table 2 Back-calculated standard length for males of *S. y. younghusbandi*

Age	Number											
3	2	36.3	51.5	67.2								
4	1	44.1	62.2	80.4	107.6							
5	6	40.9	63.9	83.5	97.7	116.7						
6	9	42.6	63.1	81.8	104.9	127.6	149.2					
7	13	46.8	72.0	96.7	124.0	152.1	177.2	207.3				
8	14	45.2	68.0	94.0	120.4	146.7	172.8	191.6	229.1			
9	12	47.3	71.9	102.2	130.3	157.6	184.7	191.0	220.2	249.8		
10	8	45.7	69.6	94.0	121.3	146.7	171.3	176.6	207.2	236.7	272.1	
11	4	40.6	59.1	85.7	110.4	136.7	166.2	152.0	177.7	209.7	237.8	272.2
Average of total		45.0	68.0	93.1	120.4	146.5	173.8	188.1	216.3	238.8	260.7	272.2

There was no significant difference between the sexes (χ^2=0.90, P<0.05).

Anal scale radius was not correlated with standard length because of the abnormity of its size and figure. Otolith sections had more well-regulated sizes than anal scales and the rings on otolith sections were more regular than those on anal scales. Therefore, otolith sections were used to do the back-calculation. Back-calculation functions used in this study were:

$L_i = \mathrm{EXP}[\ln L_c(\ln O_i + 5.5955)(\ln O_c + 5.5955)]$ for males and
$L_i = \mathrm{EXP}[\ln L_c(\ln O_i + 5.1373)(\ln O_c + 5.1373)]$ for females.

Back-calculated standard length for males and females were showed in Table 2 and Table 3, respectively.

The VBGFs fitted to length-at-age data were:

$L_t = 442.7\left[1 - e^{-0.0738(t+1.4)}\right]$ for males,
$L_t = 471.4\left[1 - e^{-0.078928(t-0.2)}\right]$ for females and
$L_t = 524.5\left[1 - e^{-0.06702(t-0.1)}\right]$ for both sexes.

The ARSS indicated that the VBGFs of males and females were significantly different (F=20.59, P< 0.05). χ^2-test indicated that the VBGFs can well describe the growth of *S. y. younghusbandi* (♂: χ^2= 14.18, P<0.05; ♀: χ^2=6.67, P<0.05).

The VBGFs fitted to weight-at-age data were:

$W_t = 840.6\left[1 - e^{-0.0738(t+1.4)}\right]^{2.2493}$ for males and
$W_t = 1137.3\left[1 - e^{-0.0789(t-0.2)}\right]^{2.4781}$ for females.

Table 3 Back-calculated standard length for females of *S. y. younghusbandi*

Age	Number													
3	1	26.7	51.8	84.5										
4	1	24.9	46.3	73.0	94.9									
5	3	30.2	52.6	78.8	114.0	154.4								
6	6	30.6	58.1	88.3	119.4	150.3	184.8							
7	14	33.1	61.8	97.7	130.3	168.8	198.9	228.8						
8	26	32.2	58.2	88.0	120.4	154.0	189.4	216.6	251.8					
9	9	34.0	61.1	93.1	127.6	160.7	195.4	221.1	256.3	288.5				
10	7	31.5	57.2	87.8	114.9	138.5	168.8	186.4	217.9	248.8	276.9			
11	8	30.9	57.2	82.2	110.4	136.2	163.0	181.3	209.5	239.2	268.1	297.3		
12	7	27.2	53.3	82.2	108.6	136.6	161.1	174.7	201.0	227.4	252.7	276.1	298.9	
13	1	30.9	51.0	77.7	99.3	126.0	141.9	167.8	186.3	208.1	238.0	263.5	286.0	305.6
Average of total		31.5	58.0	88.5	120.0	152.1	184.0	207.9	235.3	251.6	264.8	285.9	297.3	305.6

Growth rate functions were:

$$dL/dt = 31.4649e^{-0.0738(t+1.4)} \text{ for males and}$$
$$dL/dt = 36.0222e^{-0.0789(t-0.2)} \text{ for females.}$$

$$dW/dt = 145.6283e^{-0.0738(t+1.4)}\left(1 - e^{-0.0738(t+1.4)}\right)^{1.2493} \text{ for males and}$$
$$dW/dt = 228.5109e^{-0.0789(t-0.2)}\left(1 - e^{-0.0789(t-0.2)}\right)^{1.4781} \text{ for females.}$$

Growth accelerations were:

$$d^2L/dt^2 = -1.1361e^{-0.0738(t+1.4)} \text{ for males and}$$
$$d^2L/dt^2 = -1.6641e^{-0.0789(t-0.2)} \text{ for females.}$$

$$d^2W/dt^2 = 10.7430e^{-0.0738(t+1.4)}\left(1 - e^{-0.0738(t+1.4)}\right)^{0.2493}\left(2.3483e^{-0.0738(t+1.4)} - 1\right) \text{ for males and}$$
$$d^2W/dt^2 = 18.0342e^{-0.0789(t-0.2)}\left(1 - e^{-0.0789(t-0.2)}\right)^{0.4781}\left(-0.0789e^{-0.0789(t-0.2)} - 1\right) \text{ for females.}$$

Growth inflexion points, which were the time when d^2W/dt^2 equaled to zero, were 10.2 years and 12.0 years for males and females, respectively. The standard lengths and total weights were 254.5 mm and 242.1 g for males and 315.5 mm and 328.9 g for females when the ages attained t_{IP}. Standard lengths of 84.8% captures were less than those at t_{IP} (70.8% of males and 91.2% of females were less than those at respective t_{IP}).

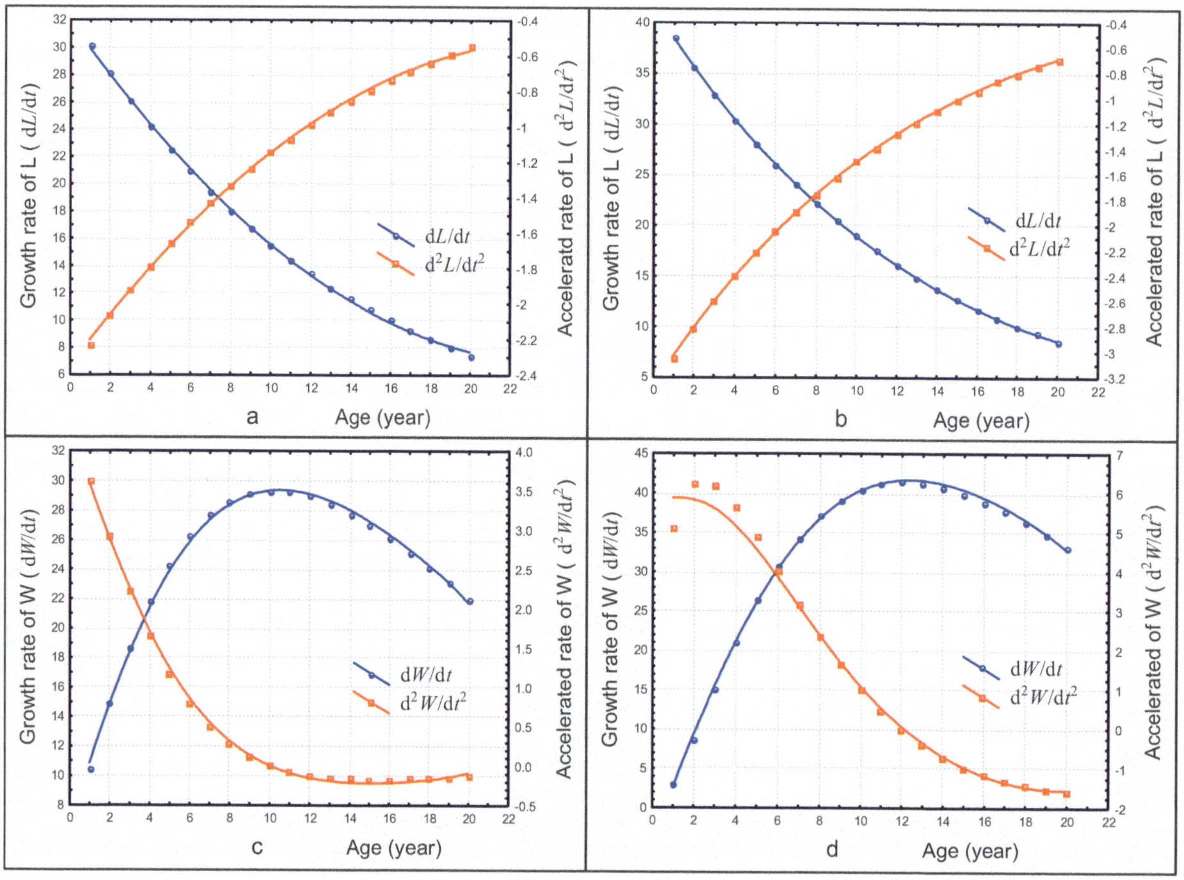

Fig. 4 Curves of growth rate and its acceleration in von Bertalanffy functions of *S. y. younghusbandi*. (**a**: curves of growth rate and accelerate of standard length of males, **b**: curves of growth rate and accelerate of standard length of females, **c**: curves of growth rate and accelerate of total weight of males, **d**: curves of growth rate and accelerate of total weight of females)

Growth rate curves and its acceleration of VBGFs were showed in Fig. 4. The standard length growth rate gradually decreased with age, and the acceleration was below zero at all times and inclined to zero, which indicated that the standard length was approaching L_∞. The total weight growth rate gradually increased before t_{IP} and then decreased, and the acceleration was above zero and decreased before t_{IP}, and then below zero but approaching to zero in the end.

Put the value k, L_∞ and L_0 into the function $\text{Longevity} = 1/k \ln\left[(L_\infty - L_0)/L_\infty/(1-x)\right]$. Longevities were 39.2 and 38.2 years for males and females, respectively.

Comparison of growth patterns among populations from different sampling sites

The difference of growth pattern among these three populations was significant (χ^2=36.24, P>0.05). Growth rates of population C, B and A gradually became rapid through observing the curves in Fig. 5.

Discussion

Environmental factors influence growth rate and ultimate size among fishes (Hodgkiss and Man 1977). In the present study, *S. y. younghusbandi* living in the upstream reaches of the Lhasa River grew slower than all other areas studied, while those captured in the middle reaches of the Yarlung Zangbo River grew most rapidly. The phenomenon may be related to altitude. Water temperature increases as altitude decreases for these waterbodies (the comprehensive scientific expedition to the Qinghai-Xizang Plateau of Chinese Academy of Sciences 1984). Warmer water accelerates fish growth. Additionally, we must consider that the lower reaches of the Lhasa River and the middle reaches of the Yarlung Zangbo River undergo increased fishing pressures, while there is very little fishing pressure on the populations found in the upper reaches of the Lhasa River. Fishing can decrease the population density and thereby decrease the intraspecific competition, which, in turn, increases individual growth.

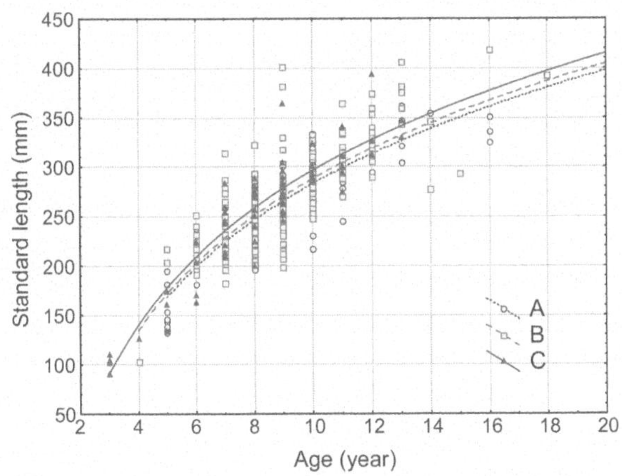

Fig. 5 Relationship between age and standard length of three populations of *S. y. younghusbandi*: upstream of the Lhasa River (*A*), downstream of the Lhasa River (*B*) and the middle reaches of the Yarlung Zangbo River (*C*)

From the age and growth analysis of *S. y. younghusbandi*, we can see that there is a large variation in the standard length and total weight of individuals of the same age (Table 1), which may be attributed to the following factors (1) the hard environment of the plateau may cause fishes to grow slower, which may produce greater variability in age structure; (2) sampling for this study has been done over a very wide geographic area, ranging from headstream (4400 m above sea level) to its connection to the Yarlung Zangbo River (3580 m above sea level) of the Lhasa River, and from Xigaze to Sangri of the Yarlung Zangbo River (Fig. 1). In such a wide area, there is a great deal of variability in terms of the environment such as altitude and water temperature resulting in greater variability in growth rate; (3) according to the study by Li (2005), *S. y. younghusbandi* have a high parasitic infection rate. Parasites can influence the growth of *S. y. younghusbandi*, which may also cause differences in growth.

From a fishery management perspective, determination of age and growth patterns is essential, particularly in species of commercial value (Joung et al. 2004). Most of the fishes of the Qinghai-Xizang Plateau grow slowly and mature late (Bureau of Aquatic Products, Tibet, China 1995; Chen et al. 2002b), making them very vulnerable to overfishing and making it difficult to replenish the population if it is depleted. We currently know very little on the biology and ecology of fishes in Tibet, although they are being exploited at an increasing rate. The growth inflexion point signifies the age the growth rate, as a function of body weight is the greatest (Chung et al. 1999), while considering the growth potential of fish. The growth inflexion point is usually considered as the catching standard for the fishery management (Ma et al. 1999). In this study, the growth inflexion points were 10.2 years for males and 12.0 years for females of *S. y. younghusbandi*, and the corresponding average standard lengths were approximately 286 and 340 mm, respectively. But the standard lengths of the captures are generally 200–280 mm for males and 200–320 mm for females which are below those at t_{IP}. Therefore, it is essential and urgent to establish science-based management of this resource to guarantee its sustainable use.

Acknowledgements We thank Fuhua Hao and Xiuqi Li for the help in collecting the specimens. This study was supported by the Knowledge Innovation Project of the Chinese Academy of Sciences (grant no. KSCX2-SW-125, KSCX1-SW-13-04) and the National Natural Science Foundation of China (grant no. 30471345).

References

Bureau of Aquatic Products (1995) Fishes and fish resources in Xizang, China. China agriculture press, Beijing (In Chinese)

Chen Y, Jackson DA, Harvey HH (1992) A comparison of von Bertalanffy and polynominal functions in modeling fish growth data. Can J Fish Aquat Sci 49:1228–1235

Chen YF, He DK, Chen YY (2002a) Age discrimination of Selincuo schizothoracini (*Gymnocypris selincuoensis*) in Selincuo Lake. Tibet. Acta Zool Sin 48(4):527–533 In Chinese

Chen YF, He DK, Cao WX, Duan ZH (2002b) Growth of Selincuo schizothoracini (*Gymnocypris selincuoensis*) in Selincuo Lake. Tibet Platean. Acta Zool Sin 48(5):667–676 In Chinese

Chung KC, Woo NYS (1999) Age and growth by scale analysis of *Pomacanthus imperator* (Teleostei: Pomacanthidae) from Dongsha Islands, Southern China. Environ Biol Fishes 55:399–412 doi:10.1023/A:1007571532648

Hodgkiss IJ, Man HSH (1977) Age composition, growth and body condition of the introduced *Sarotherodon mossambicus* (Cichlidae) in Plover Cove Reservoir, Hong Kong. Environ Biol Fishes 2(1):35–44 doi:10.1007/BF00001414

Joung S-J, Liao Y-Y, Chen C-T (2004) Age and growth of sandbar shark, Carcharhinus plumbeus, in northeastern Taiwan waters. Fish Res 70:83–96 doi:10.1016/j.fishres.2004.06.018

Li WX (2005) Ecology of endohelminths in the yellowhead catfish, *Pelteobagrus fulvidraco*. Doctoral Dissertation. Chinese Academy of Sciences, Beijing (In Chinese)

Ma TW, Xie CX, Gong SJ (1999) Studies on population dynamics and aximum sustainable yield of Chinese snakehead fish *Channa argus*, in the Liangzi Lake. Acta Hydrobiologica Sin 23(6):591–599 In Chinese

Natanson LJ, Kohler NE, Ardizzone D, Cailliet GM, Wintner SP, Mollet HF (2006) Validated age and growth estimates for the shortfin mako, *Isurus oxyrinchus*, in the North Atlantic Ocean. Environ Biol Fishes 77:367–383 doi: 10.1007/s10641-006-9127-z

Taylor CC (1958) Cod growth and temperature. J Conseil Int Explor Mer 23:366–370

The research team of the Qinghai-Xizang Plateau of Chinses Academy of Sciences (1984) Rivers and lakes in Tibet. Science, Beijing (In Chinese)

von Bertalanffy L (1938) A quantitative theory of organic growth (inquiries on growth laws II). Hum Biol 10:181–213

Wang Y, Wang Y, Wang C, Peng QY, Chen YR, Yang JX (1998) Study on the correlation between the growing characters of *Neosalanx taihuensis* and its resources in Lake Dianchi. Zool Res 19(4):289–295 In Chinese

Wu YF, Wu CZ (1992) The fishes of the Qinghai-Xizang Plateau. Sichuan Publishing House of Science & Technology, Chengdu (In Chinese)

Yin MC (1995) Ecology of fish. China agriculture press, Beijing, pp 11–63 (In Chinese)

Environ Biol Fish (2009) 86:163
DOI 10.1007/s10641-008-9361-7

Threatened fishes of the world: *Sinocyclocheilus anophthalmus* (Chen and Chu, 1988) (Cyprinidae)

Yahui Zhao · Chunguang Zhang

Received: 17 July 2007 / Accepted: 22 April 2008 / Published online: 23 May 2008

Common name: Blind golden-line barbel. *Conservation status*: Rare—China Red Data Book of Endangered Animals, Pisces (Yue and Chen 1998); Vulnerable—China Species Red List (Wang and Xie 2004). *Identification*: D iii, 7; A iii, 5; P i, 15–18; V ii, 8–9. Body almost naked, lateral line scales 46–59. Blind, either eyeless or with small eye dots. Snout blunt, duck-billed. Pectoral fin long, reaching pelvic fin insertion. Living fish generally semi-transparent. Adults reach 113 mm standard length. *Distribution*: The fish is only found in Xiaogou Cave in Jiuxiang Town, Yiliang County, Yunnan Province, China, connected to the Xijiang River system. *Abundance*: Population a few hundred at most (Yue and Chen 1998). *Habitat and ecology*: The Karst limestone cave is located in altitude 1,638 m above sea level. Cave water is shallow, around 1 m deep. Some insects and bats also inhabit the cave. Pieces of insect body parts were detected in stomach contents of fishes. *Reproduction*: Estimated spawning season is late May or early June. Ripe eggs are about 2 mm (Chen et al. 1988). *Threats*: Over-capture was considered to be one of threats (Yue and Chen 1998). Habitat degradation and environmental changes are even more serious, especially from a nearby hydropower station. The very limited distribution puts it at risk for rapid or sudden extinction due to any changes in the cave. *Conservation recommendations*: Capture for *Sinocyclocheilus anophthalmus* should be strictly prohibited except for scientific research. The border of Jiuxiang National Park, currently only a few kilometers away from the cave *S. anophthalmus* inhabits, should be extended to encompass the cave. Effective habitat protection and public awareness programs will be crucial for the survival of this species. *Remarks*: The conservation status of *S. anophthalmus* was previously underestimated, should be "endangered (EN)". A detailed study of the population dynamics is urgently needed.

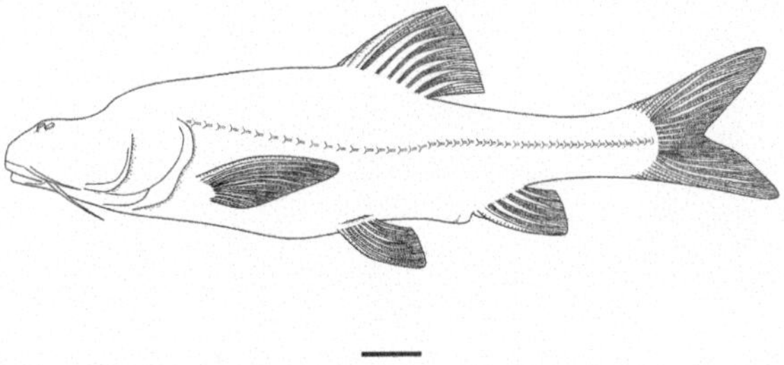

Y. Zhao (✉) · C. Zhang
Institute of Zoology, Chinese Academy of Sciences,
Beijing 100101, People's Republic of China
e-mail: zhaoyh@ioz.ac.cn

References

Chen YR, Chu XL, Luo ZY, Wu JY (1988) A new blind cyprinid fish from Yunnan, China with reference to the evolution of its characters. Acta Zool Sinica 34:64–70

Wang S, Xie Y (2004) China species red list, vol. 1. Red list. Higher Education Press, Beijing, p 160

Yue PQ, Chen YY (1998) China red data book of endangered animals, Pisces. Science Press, Beijing, pp 107–109

Environ Biol Fish (2009) 86:165–183
DOI 10.1007/s10641-009-9499-y

The taxonomy and phylogeny of the cyprinid genus *Opsariichthys* Bleeker (Teleostei: Cyprinidae) from Taiwan, with description of a new species

I-Shiung Chen · Jui-Hsien Wu · Shih-Pin Huang

Received: 13 October 2007 / Accepted: 2 June 2009 / Published online: 28 July 2009

Abstract The morphological and mitochondrial genetic differentiation in the cyprinid genus, *Opsariichthys* Bleeker (Nederlandsch Tijdschrift voor de Dierkunde 1:187–218, 1863) have been surveyed in Taiwan. Among them, there are three valid species can be recognized in Taiwan including *Opsariichthys pachycephalus* Günther (1868) distributed in northern and western Taiwan, *Opsariichthys evolans* (Jordan and Evermann Proc US Nat Mus 25:315–368, 1902) in northern Taiwan and an unnamed species from southern Taiwan described herein as *Opsariichthys kaopingensis* Chen and Wu, new species which can be well distinguished from the related *O. pachycaphalus* by their body proportions, scale counts, and specific coloration patterns. We utilized mitochondrial complete D-loop sequence data to infer phylogenetic relationships within a subset of related genera of opsariichthines, and to examine evidence for genetic differentiation in these two sibling species formerly assigned to "*Zacco*" *pachycephalus* and their genetic relationship with other congeneric species around nearby regions. The clade of *O. pachycephalus* and *O. kaopingensis* in genetically were recovered as more closely related to *Opsariichthys uncirostris* (Temminck and Schlegel 1846) species complex including both *O. uncirostris* and *O. bidens* Günther (1868) from Japan and mainland China than to typical *Zacco* from Japan. This molecular phylogenetic insight strongly supports the assignment for both so-called "*Zacco*" *pachycephalus* and this new species described herein as the typical monophyletic members of *Opsariichthys* and the type species of *Zacco* as *Zacco platypus* (Temminck and Schlegel 1846) from Japan is sister clade for all species groups in *Opsariichthys*. *Opsariichthys pachycephalus* and *O. kaopingensis* were strongly differentiated by large mitogenetic distances and phylogenetic support from distance and discrete method and Bayesian inference based on complete mtDNA D-loop sequences, with an average mitogenetic divergence of 3.3%, which may suggest that the separation of the two species happened much earlier than the last glacial period. *Opsariichthys evolans* seems to share the close genetic relationship with *O. acutipinnis* (Bleeker Nederlandsch Tijdschrift voor de Dierkunde 1:187–218, 1863) from the Yangtsi River basin.

I.-S. Chen (✉) · S.-P. Huang
Institute of Marine Biology and Department of Life Science,
National Taiwan Ocean University,
Keelung 202 Taiwan, Republic of China
e-mail: isc@mail.ntou.edu.tw

J.-H. Wu
Institute of Marine Resources,
National Sun Yst-sen University,
Kaohsiung 804 Taiwan, Republic of China

Present Address:
J.-H. Wu
Dongsha Marine National Park Headquarters,
No. 24, Demin Rd.,
Kaohsiung City 811 Taiwan, Republic of China

Keywords *Opsariichthys* · New species · Cyprinidae · Mitochondrial DNA · Control region · Molecular phylogeny · Taiwan

Introduction

The freshwater and estuarine fish fauna of Taiwan was recently surveyed by Chen and Fang 1999, who documented at least 224 fish species, including 209 native and 15 alien species. Among them, fishes of the family Cyprinidae are the most speciose group in all components of primary freshwater fishes (Tzeng 1986; Chen and Fang 1999). High endemicity can be seen in Taiwanese Cypriniformes, including several species of cyprinids (Cyprinidae) and loaches (Balitoridae) are endemic (Chen and Fang 1999, 2000; Chen et al. 2002a; Chen and Chang 2007). More recently, Chen and Chang (2005) provided more comprehensive fauna list of freshwater fishes for Cypriniformes including several recently described and unnamed species.

Opsariichthys pachycephalus was first described by Günther (1868) from Taiwan, but was transferred to *Zacco* by Jordan and Evermann (1902). *Zacco pachycephalus* has been regarded as a widely-distributed species in western Taiwan (Oshima 1919, 1923; Tzeng 1986), where it is native to at least from the river drainages of Taipei to Pingtung County along the west slope of Central Mountain Ridge as well as Ilan County (Chen and Fang 1999). However, the morphological differences between these Taiwanese populations of so-called "*Zacco*" *pachycephalus* had previously not been investigated in detail except the genetic studies by allozymes (Wang et al. 1999). More recently, Chen and Chang (2005) have considered the so-called "*Zacco*" *evolans* Jordan and Evermann (1902) should be a member of *Opsariichthys* which have been concurred by molecular phylogenetic support of Chen et al. (2008a, b) although some authors (Ma et al. 2006) still tentatively treated as "*Zacco*" *evolans*.

Howes (1980) suggested that *Opsariichthys* and *Zacco* might not be monophyletic as currently recognized based on the conclusion of his osteological research. In recent 10 years, Ashiwa and Hosoya (1998) examined the osteology of so-called "*Zacco*" *pachycephalus* in detail and suggested that it may be a member of the genus *Opsariichthys* which reconfirmed the previous views of Howes (1980, 1983).

In recent years, several papers have concern the phylogenies of *Zacco* and *Opsariichthys* in China by different genetic markers (Perdices et al. 2004, 2005; Berrebi et al. 2005, 2006; Perdices and Coelho 2006; Wang et al. 2007), but none of them combined any of morphological studies. Chen et al. (2008a, b) employed both morphological and mitogenetic studies for taxonomy and phylogeny of Hainanese *Opsariichthys* group and Taiwanese *Candidia* group respectively. Therefore, the aim of this paper is to verify the taxonomy of *Opsariichthys* in Taiwan and reconstruct their molecular phylogeny employing mtDNA D-loop sequences with other members of opsariichthines in nearby geographical regions. Herein we provide morphological and molecular evidence for recognizing three valid species as *Opsariichthys* and describe a distinct species from southern Taiwan with comments for their interspecific relationships among congeners.

Materials and methods

Sample collection and morphological survey

The specimens of cyprinids examined were collected by cast-net, gill-net or electric fishing during recent years from most of the different river basins of Taiwan and other comparative species of cyprinids were collected from China and Japan. Counts and measurements mostly follow the methods of Hosoya et al. (2002) with minor modifications of including more morphometric characters as body depth in dorsal origin, body depth in anal origin, head width, head depth in midline of orbit, as well as upper jaw length. The head illustrations of two related *Opsariichthys* species were made by camera lucida on the dissecting microscope as Leica MZ 75. Type specimens and comparative materials examined in the study are deposited in the Museum, Institute of Zoology, Academia Sinica, Taipei (ASIZP); British Museum of Natural History, London (BMNH); and the Pisces collection of National Museum of Marine Biology & Aquarium, Pingtung (NMMBP).

mtDNA amplification and sequence analysis

Samples of cyprinid fishes for molecular studies were preserved in 95% ethanol at the time of field collection and then transferred to a refrigerator. Seventy-five individuals of *Opsariichthys pachycephalus* complex from across their entire range and two individuals of *Opsariichthys evolans* (Jordan and Evermann 1902) in Taiwan collected for DNA sequences analysis were employed for mitochondrial sequence analysis (Fig. 1). Other related cyprinids including *Zacco platypus* (Temminck and Schlegel 1846), *Opsariichthys uncirostris* (Temminck and Schlegel 1846) from Japan served as the comparative OTUs of type species for each above genus; other congeneric species of *Opsariichthys* from China including *Opsariichthys acutipinnis* Bleeker (1871); *Opsariichthys bidens* Günther (1868); *Opsariichthys evolans* (Jordan and Evermann 1902); and *Opsariichthys hainanensis* (Nichols and Pope 1927); two related genera of *Opsariichthys* genus group as *Candidia barbatus* (Regan 1908) and *Parazacco spilurus* (Günther 1868); two assigned outgroups of *Aphyocypris kikuchii* (Oshima 1919) and *Pararasbora normalis* (Nichols and Pope 1927); all of them are used and included herein for molecular biological analysis. The sampling localities for those other species of *Opsariichthys* genus group and two related outgroups from China, Taiwan and Japan are shown in Fig. 2. All available mtDNA sequences of current studies herein were either deposited in the Genbank listed either as APPENDIX or cited from Chen et al. (2008a, b).

Fig. 1 Sampling localities of *Opsariichthys pachycephalus* (black triangle) and *O. kaopingensis* (black square) from Taiwan for molecular sequence studies. Numbers for names of river basins indicate in Table 3

Crude DNA extraction followed the methods of Sambrook et al. (1989) and Chen et al. (1998, 2002b). The mtDNA fragment that included the control region sequence was amplified by the primer set as P-CPTHRA (5′-AAAGCATCGGTCTTGTAATCC GAAG-3′) and P-CP12SB (5′-AGCTAAGTGGGG TATCTAATCCCAGTTTG-3′) which are located in the THR-tRNA and 12 S ribosomal RNA genes respectively. PCRs were performed in 50 uL reaction volumes. PCR amplification was carried out in a MODEL 2700 or 9700 thermal cycler (Perkin-Elmer) using parameters similar to those of Chen et al. (2002b) and Chen and Chang (2007). A negative control without template was carried out for each run of PCR. The PCR products were run on a 1.0% L 03 agarose gel (Takara) and stained with ethidium bromide for band characterization under ultraviolet trans-illumination. The double stranded DNA product was purified by the High Pure Purification PCR kit (Roche). DNA sequencing was performed by the dideoxy nucleotide chain termination method (Sanger et al. 1977) using a Big Dye cycle sequencing kit (Applied Biosystems). Sequencing primers were employed by either the same original primer set as above or the following sequencing primers: CP-DLMF1 (5′- CATGCGGAGTTTCTTGTC -3′) and CP-DLMR2 (5′- GCTCGGCATGTTGGGTAA -3′). The DNA sequencing reactions were performed on a ABI Model 377-36, 377-64 and 3100 automated DNA sequencer (Applied Biosystems).

Mitochondrial D-loop sequences were verified and aligned with sequences of other cyprinids from Genbank. Nucleotide sequence alignment was veri-

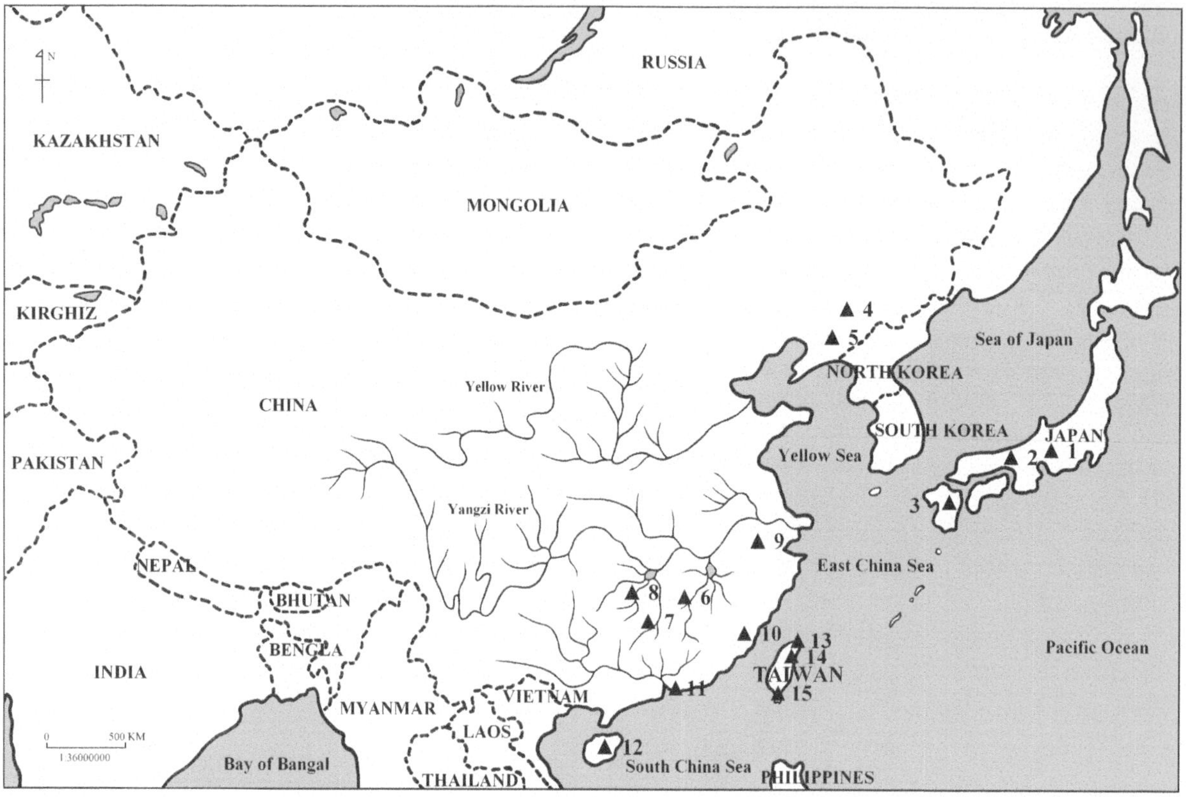

Fig. 2 Sampling localities of other species of *Opsariichthys* and related genera (black triangle) from China, Taiwan and Japan for molecular sequence studies. Numbers for names of localities indicate in Table 4

fied manually after running through CLUSTAL W (Thompson et al. 1994) and BIOEDIT version 5.9 (Hall 2001). The analysis of aligned mutation sites were conducted using Molecular Evolutionary Genetics Analysis (MEGA) version 3 (Kumar et al. 2004) for aligned mutation sites analysis.

The neighbor joining (NJ) analysis (distance method) was also conducted by MEGA version 3 based on Kimura 2-parameter distance matrix. The maximum parsimony (MP) analysis (discrete method) was carried out using PAUP* version 4.0 B10 (Swofford 2003) using heuristic search. Branch support was established via bootstrap analysis (2000 replications). For the Bayesian (BI) analysis, the best-fitting model for sequence evolution was determined for mtDNA D-loop sequences using MrMODELTEST version 2.2 (Nylander 2005). The analyses reconstructed by Bayesian Inference (BI) were performed using MrBayes 3.0 (Ronquist and Huelsenbeck 2003). The posterior probabilities of each node were computed from remaining 75% of all sampled trees.

Systematics

Opsariichthys evolans Jordan and Evermann (1902) (Fig. 3a–c)

Zacco evolans Jordan and Evermann 1902: 322 (Taihoku, Taiwan).

Zacco evolans, Jordan and Hubbs 1925: 185.

Zacco platypus, (not Temminck and Schlegel 1846) Oshima 1919: 235; Tzeng 1986: 66; Shen and Tzeng in Shen 1993: 141; Chen and Fang 1999: 95.

Opsariichthys evolans, Chen and Chang 2005: 20.

Material examined

NTOU P-2006-3-457, 3 specimens, 61.1-89.9 mm SL, Yu-Rui Stream, Keelung River, Tanshuei River basin, Keelung City, Taiwan, Coll. I-S. Chen et al., 17 March, 2006; NTOU P-2008-11-459, 2 specimens, 84.8-89.6 mm SL, Rui-Fang, Keelung River, Tanshuei River basin, Taipei County, Taiwan, Coll. S.P. Huang, 10 July, 2008; NTOU P-2008-11-460, 70.2 mm SL, Kong-Liao,

Fig. 3 *Opsariichthys evolans,* A, male, NTOUP 2008-11-459, holotype, Keelung River, Taipei County, Taiwan, 89.6 mm SL; B, female, NTOUP 2006-3-457, paratype, Keelung River, Keelung City, Taiwan, 69.7 mm SL; and C, alive male, Holong River, Miaoli County, Taiwan, about 70 mm SL; *Opsariichthys kaopingensis* new species, D, male, NMMBP 6965, holotype, Kaoping River, Pingtung County, Taiwan, 76.2 mm SL; E, female, NMMBP 2911, paratype, Kaoping River, Pingtung County, Taiwan, 72.1 mm SL; and F, alive male, Kaoping River, Pingtung County, Taiwan, about 70 mm SL. The photos of alive fishes by Y.C. Chang

Shi-Ting Stream, Shungshi River basin, Taipei County, Coll. S.P. Huang, 12 Aug. 2008; NTOU P-2008-11-461, 79.3 mm SL, Keelung River, Tanshuei River basin, Taipei County, Taiwan, Coll. S.H. Su, 19 June, 2008; NTOU P-2008-11-462, 3 specimens, 52.4-59.7 mm SL, Pau-Chong Bridge, Fongshan River, Shinchu County, Taiwan, Coll. S.H. Su, 26 Dec., 2007.

Diagnosis

Opsariichthys evolans can be distinguished from the congeneric species by the following combination of morphological features: (1) no maxillary barbels; (2) no anterior notch on tip of upper lip; (3) lateral-line scales usually 43–45 (always 43–44); (4) predorsal scales 15–17 (modally 16); (5) very narrow body width; (6) maxillary very short, never extending to vertical of anterior margin of orbit; (7) a series of 6–7 (modally 6) rounded tubercles under lower jaw and a series of tubercles on cheek well separate in adult male; (8) pectoral fin extending far beyond origin of ventral fin in adult male; and (9) snout with blackish-purple background in adult male in life.

Distribution

The species is distributed around the main drainages of northern and northwestern Taiwan (including Taipei, Taoyuan, Shinchu, and Miaoli Counties; Taipei and Keelung Cities). The species (including its sibling species) can be also found in the main drainages of both Fujian Province and southern region of Zhejiang Province, PR China.

Remarks

Although this species, *Opsariichthys evolans*, has been confused with the typical Japanese pale chub, *Zacco platypus*, for a very long time after Oshima (1919) and the suggested result have been followed by recent decades of studies (Tzeng 1986; Shen 1993; Chen and Fang 1999) which was apparently ignored the original description of Jordan and Evermann 1902. After Oshima's studies, *Opsariichthys evolans* had ever been pointed out the errors of identification by Oshima and confirmed the specific validity by Jordan and Hubb (1925) under "*Zacco*" *evolans*. Jordan and Hubb (1925) also mentioned that the holotype is a typical male as original description but the paratype apparently is referable rather to so-called "*Zacco*" *pachycephalus*. However, Chen and Chang (2005) clearly addressed it as a native species in Taiwan and stated its diagnostic features which have been reconfirmed independently by the evidence of molecular data as Ma et al. (2006) and with the several distinct features of morphological differences between the both *Opsariichthys evolans* and *Zacco platypus*. "*Zacco*" *evolans* should been treated as a member of *Opsariichthys* followed Chen and Chang (2005).

Opsariichthys pachycephalus Günther (1868) (Figs. 4 and 5)

Opsariichthys pachycephalus Günther 1868: 296 (Taiwan).

Fig. 4 *Opsariichthys pachycephalus*, A, alive male, Tanshuei River, Taipei County, Taiwan, about 120 mm SL; B, alive female, Tanshuei River, Taipei County, Taiwan, about 110 mm SL; C, syntype, female, BMNH 1865.5.2.31-34, 76.4 mm SL, Taiwan. The photos of alive fish by Y.C. Chang; the photo of syntype by H.C. Ho

Opsariichthys pachycephalus, Ashiwa and Hosoya 1998:163; Chen and Chang 2005: 26.

Zacco pachycephalus, Jordan and Evermann 1902: 322; Jordan and Richardson 1909: 170; Oshima 1919: 240; Banarescu 1968: 306; Tzeng 1986: 66; Chen and Yu 1986: 293; Shen 1993:141; Chen and Chu in Chen 1998: 45; Chen and Fang 1999: 94.

Zacco taiwanensis Chen 1982: 296 (Joshuei River, Nantou County, Taiwan); Chen and Chu in Chen 1998: 44.

Material examined

Syntypes.—BMNH 1865.5.2.31-34, 4 specimens, 76.4–87.4 mm SL (Standard length), Taiwan.

Nontypes.—NMMBP 2048, 6 specimens, 52.9–99.6 mm SL, Holong River, Miaoli County, Taiwan, Coll. I-S. Chen, May, 1995; NMMBP 3980, 2 specimens, 59.1–64.3 mm SL, Tzengwen River, Tainan County, Taiwan, Coll. Y.M. Ju, 19 Aug. 2002; NMMBP 3989, 1 specimen, 110.6 mm SL, Tzengwen River, Tainan County, Taiwan, Coll. Y.M. Ju, 18 Aug. 2002; NMMBP 6027, 2 specimens, 55.1–95.3 mm SL, Suau River, Ilan County, Taiwan, Coll. J.H. Wu, 29 Oct. 2002; NMMBP 6028, 3 specimens, 75.9–95.2 mm SL, Lanyang River, Ilan County, Taiwan, Coll. C.W. Wang, 28 Dec. 2002; NMMBP 6029, 2 specimens, 61.7–62.5 mm SL, Lanyang River, Ilan County, Taiwan, Coll. J.H. Wu, 28 Mar. 2003; NMMBP 6030, 23 specimens, 59.0–90.2 mm SL, Kanfang River, Ilan County, Taiwan, Coll. J.H. Wu, 31 Mar. 2003; NMMBP 6031, 35 specimens, 49.7–116.7 mm SL, Shuanshi River, Taipei County, Taiwan, Coll. J.H. Wu, 31 Mar. 2003; NMMBP 6032, 27 specimens, 67.2–114.5 mm SL, Parlian River, Taipei County, Taiwan, Coll. J.H. Wu, 31 Mar. 2003; NMMBP 6036, 20 specimens, 56.7–116.8 mm SL, Tanshuei River, Taipei County, Taiwan, Coll. J.H. Wu, 28 Mar. 2003; NMMBP 6038, 22 specimens, 56.4–122.9 mm SL, Tochien River, Hsinchu County, Taiwan, Coll. J.H. Wu, 29 Mar. 2003; NMMBP 6039, 23 specimens, 60.8–95.3 mm SL, Chonkang River, Miaoli County, Taiwan, Coll. J.H. Wu, 29 Mar. 2003; NMMBP 6040, 4 specimens, 49.9–70.8 mm SL, Holong River, Miaoli County, Taiwan, Coll. I-S. Chen, 31 Mar. 1999; NMMBP 6041, 11 specimens, 52.1–79.1 mm SL, Holong River, Miaoli County, Taiwan, Coll. J.H. Wu, 30 Mar. 2003; NMMBP 2049, 3 specimens, 57.3–75.5 mm SL, Taan River, Miaoli

Fig. 5 Illustration of the heads of two *Opsariichthys* species from Taiwan. *Opsariichthys pachycephalus* (upper half): male (left), NMMBP 6036, Tanshuei River, Taipei County, Taiwan, 116.0 mm SL; female (right), NMMBP 6036, Tanshuei River, Taipei County, Taipei, 106.4 mm SL. *Opsariichthys kaopingensis* (lower half): male (left), NMMBP 2050, paratype, Fongkong River, Pingtung County, Taiwan, 77.1 mm SL; female (right), NMMBP 2911, paratype, Kaoping River, Pingtung County, Taiwan, 72.1 mm SL. (Bar=1 mm)

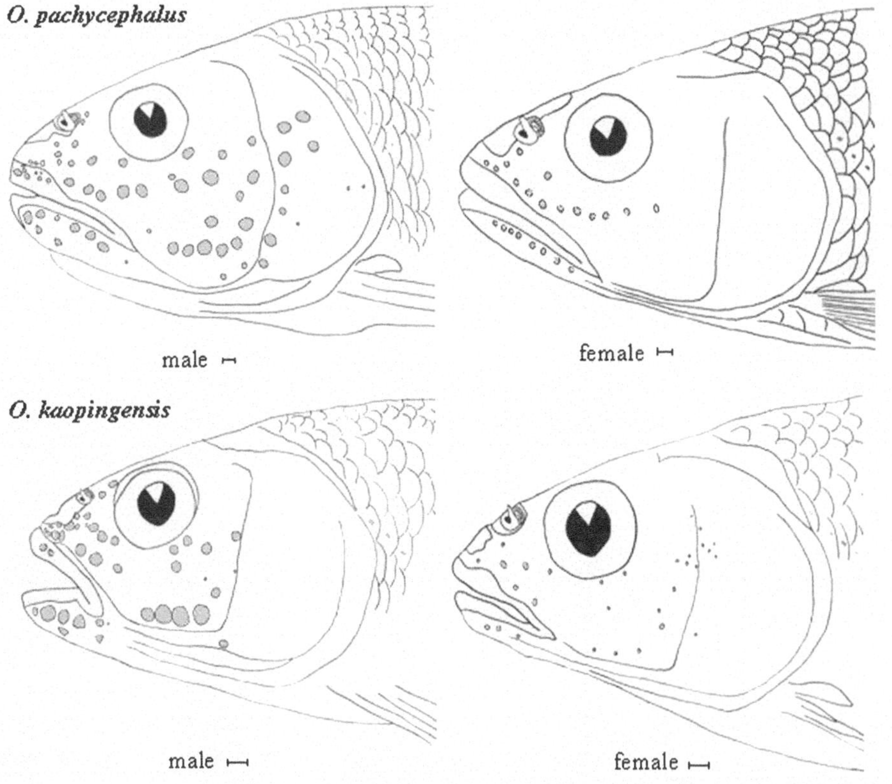

County, Taiwan, Coll. I-S. Chen, July, 1999; NMMBP 6042, 14 specimens, 57.1–85.7 mm SL, Taan River, Miaoli County, Taiwan, Coll. J.H. Wu, 30 Mar. 2003; NMMBP 6043, 18 specimens, 62.8–120.4 mm SL, Taan River, Miaoli County, Taiwan, Coll. J.H. Wu, 16 Mar. 2003; NMMBP 6044, 11 specimens, 63.5–88.7 mm SL, Tajar River, Taichung County, Taiwan, Coll. J.H. Wu, 15 Mar. 2003; NMMBP 6045, 21 specimens, 61.0–85.2 mm SL, Tajar River, Taichung County, Taiwan, Coll. J.H. Wu, 30 Mar. 2003; NMMBP 3933, 1 specimen, 82.8 mm SL, Wu River, Taichung County, Taiwan, Coll. I-S. Chen, July, 1999; NMMBP 6046, 11 specimens, 72.3–88.9 mm SL, Wu River, Taichung County, Taiwan, Coll. C.W. Wang, 13 Feb. 2003; NMMBP 6047, 9 specimens, 76.0–100.7 mm SL, Wu River, Taichung County, Taiwan, Coll. J.H. Wu, 13 Feb. 2003; NMMBP 6048, 12 specimens, 73.0–107.0 mm SL, Wu River, Taichung County, Taiwan, Coll. J.H. Wu, 13 Feb. 2003; NMMBP 6049, 25 specimens, 63.3–108.5 mm SL, Joshui River, Nantou County, Taiwan, Coll. I-S. Chen, 14 Feb. 2003; NMMBP 6050, 2 specimens, 49.5–52.1 mm SL, Joshui River, Nantou County, Taiwan, Coll. J.H. Wu, 15 Mar. 2003; NMMBP 6055, 10 specimens, 61.8–72.3 mm SL, Parchan River, Chiayi County, Taiwan, Coll. C.W. Wang, 15 Feb. 2003; NMMBP 6056, 33 specimens, 65.1–102.1 mm SL, Parchan River, Chiayi County, Taiwan, Coll. J.H. Wu, 27 Mar. 2003; NMMBP 6057, 3 specimens, 72.8–96.5 mm SL, Parchan River, Tainan County, Taiwan, Coll. Y.M. Ju, 15 Sep. 2002; NMMBP 6058, 2 specimens, 84.5 mm SL, Tzengwen River, Tainan County, Taiwan, Coll. C.W. Wang, 15 Feb. 2003; NMMBP 6060, 33 specimens, 63.3–111.6 mm SL, Tzengwen River, Chiayi County, Taiwan, Coll. J.H. Wu, 10 Apr. 2003.

Diagnosis

Opsariichthys pachycephalus can be distinguished from the congeneric species by the following combination of morphological features: (1) no maxillary barbels; (2) no anterior notch on tip of upper lip; (3) lateral-line scales usually 50–54 (modally 52); (4) predorsal scales more than 20–23 (modally 21); (5) maxillary long, always extending to or beyond middle vertical of orbit in adult female; (6) a series of 6–7 (modally 6) rounded tubercles under lower jaw in adult male; and (7) opercle and ventral side of head orange red to pinkish red in adult male in life.

Distribution

The species is widely distributed across northern and western Taiwan (including Ilan, Taipei, Taoyuan, Shinchu, Miaoli, Taichung, Chanhwa, Yunlin, Charyi, and Tainan Counties), except river basins south of the Tzengwen River. This species did not occur in eastern Taiwan (Hualien and Taitung Counties) until fifteen years ago, when it was artificially introduced. It has subsequently expanded its range in these basins and caused greatly substantial declines of the locally endemic cyprinid (*Aphyocypris kikuchii*) and gobiid fishes (two species of *Rhinogobius* spp.).

Opsariichthys kaopingensis Chen and Wu, new species (Figs. 3d–f and 5)

Opsariichthys sp. Chen and Chang 2005: 32. (Kaohsiung and Pingtung Counties, Taiwan)

Material examined

Holotype—NMMBP 6965, 76.2 mm SL, Kaoping River, Pingtung County, Taiwan, Coll. J.H. Wu, 24 Feb. 2002.

Paratypes—ASIZP 062619, 5 specimens, 52.9–64.0 mm SL, Linbien River, Pingtung County, Taiwan, Coll. J.H. Wu, 11 Apr. 2003; ASIZP 062620, 5 specimens, 47.0–74.0 mm SL, Fansan River, Pingtung County, Taiwan, Coll. J.H. Wu, 11 Apr. 2003; NMMBP 2050, 13 specimens, 61.6–86.0 mm SL, Fongkan River, Pingtung County, Taiwan, Coll. I-S. Chen, May 2000; NMMBP 2051, 2 specimens, 72.5–84.8 mm SL, Fongkan River, Pingtung County, Taiwan, Coll. I-S. Chen, Aug. 2000; NMMBP 2911, 8 specimens, 58.0–72.0 mm SL, Kaoping River, Pingtung County, Taiwan, Coll. J. H. Wu, 24 Feb. 2002; NMMBP 3000, 16 specimens, 56.8–77.1 mm SL, Fongkan River, Pingtung County, Taiwan, Coll. C.W. Wang, 5 Apr. 2003; NMMBP 3952, 2 specimens, 59.2–65.8 mm SL, Linbien River, Pingtung County, Taiwan, Coll. I-S. Chen, 1 Aug. 2002; NMMBP 6061, 16 specimens, 50.1–64.0 mm SL, Kaoping River, Pingtung County, Taiwan, Coll. J. H. Wu, 27 Mar. 2003; NMMBP 6062, 1 specimen, 42.6 mm SL, Tongkong River, Pingtung County, Taiwan, Coll. J.H. Wu, 11 Apr. 2003; NMMBP 6063, 16 specimens, 46.3–77.7 mm SL, Linbien River, Pingtung County, Taiwan, Coll. J.H. Wu, 11 Apr. 2003; NMMBP 6064, 2 specimens, 63.8–64.3 mm SL, Fansan River, Pingtung County, Taiwan, Coll. J. H. Wu, 11 Apr. 2003; NMMBP 6066, 15 specimens, 50.0–72.1 mm SL, Szuchung River, Pingtung County, Taiwan, Coll. C.W. Wang, 5 Apr. 2003; NMMBP 6067, 16 specimens, 55.3–113.7 mm SL, Kankou River, Pingtung County, Taiwan, Coll. C.W. Wang, 5 Apr. 2003.

Diagnosis

The new species, *Opsariichthys kaopingensis* can be distinguished from the congeneric species by the following combination of morphological features: (1) no maxillary barbels; (2) no notch on anterior tip of upper lip; (3) lateral-line scales 44–48 (always 45–47); (4) predorsal scales less than 20 (always 18–19; modally 19); (5) maxillary extending to or slightly beyond the vertical of anterior margin of orbit; (6) a series of 4–5 (modally 5) rounded tubercles under lower jaw in adult male; and (7) opercle and ventral side of head shiny yellow in adult male in life.

Description

Body proportions listed in Table 1. Body moderately elongated and compressed. Body depth slightly shorter than head length. Body width rather thick. No maxillary barbels. Gill rakers 3 + 9–10 (in 10 specimens). Mouth oblique, maxillary always extending to or slightly beyond the vertical of anterior margin of orbit, but never reaching the vertical of anterior margin of pupil. The size of mouth in both sexes is about equal. Eye rather large, upper-lateral. Upper lip slightly concave but not notched. Breeding tubercles distinct on head and anal fin rays of adult male, the lateral region beneath lower jaw with a row of 4–5 large tubercles; cheek with two mainly longitudinal rows of tubercles but lower row separate and interrupted below eye; opercle only with few small tubercles, a series of 4–5 (modally 5) rounded tubercles under lower jaw. Other details shown in Fig. 3. Pharyngeal teeth arranged into two to three rows, 4–5 (modally 4), 3–4 (3), 0–1 (1); 0–2 (1), 3–4 (4), 4–5 (4) (in 15 specimens). Vertebral count 4 + 34–36 (modally 4 + 35; in 118 specimens).

Table 1 Morphometric measurements of three *Opsariichthys* species from Taiwan

	Opsariichthys pachycephalus						*Opsariichthys evolans*						*Opsariichthys kaopingensis* new species						
	Male			Female			Male			Female			Holotype	Male			Female		
N	39			303			5			5			1	14			100		
Standard length(mm)	66.9~120.4			49.5~122.9			70.3~89.9			52.4~69.7			72.1	55.0~113.7			42.6~86.0		
% in SL	min	max	mean	min	max	mean	min	max	mean	min	max	mean		min	max	mean	min	max	mean
Head length	26.2	31.5	28.1	24.4	34.8	29.3	21.2	24.1	22.8	21.7	23.1	22.4	27.0	25.9	30.1	27.8	25.2	31.5	27.5
Body depth	24.7	30.8	27.3	20.8	30.8	25.7	25.4	28.8	26.8	23.0	25.4	24.5	31.2	24.3	29.6	26.3	21.7	31.2	26.1
Body width in dorsal origin	13.1	20.9	15.5	10.7	19.2	14.8	10.0	13.5	12.1	11.6	12.6	12.0	16.8	11.7	18.2	15.0	12.2	17.7	14.7
Body width in anal origin	9.3	15.1	11.8	8.0	14.5	10.7	8.4	11.1	9.7	8.8	9.8	9.2	11.2	9.4	13.9	11.7	8.5	12.9	10.4
Depth of caudal peduncle	9.9	12.5	10.8	9.5	12.5	11.0	9.1	10.9	9.9	9.3	10.3	9.7	10.9	9.8	12.2	10.8	9.3	12.0	10.6
Length of caudal peduncle	18.1	21.1	19.4	16.2	22.5	19.3	18.6	20.8	19.8	19.3	21.0	20.1	20.3	17.8	19.7	18.8	16.6	21.2	18.7
Predorsal length	47.4	53.5	50.1	45.7	55.2	51.6	47.8	51.4	49.6	47.2	50.4	49.5	53.3	48.2	52.1	50.0	48.2	53.5	51.4
Preanal fin length	63.8	71.1	67.7	64.5	75.6	70.2	65.3	69.0	67.0	68.4	69.8	69.1	68.9	65.5	71.0	67.9	66.2	75.2	69.6
Distance of snout to anus	61.6	69.6	65.5	61.0	73.0	67.5	62.9	66.5	64.9	65.3	67.8	66.8	67.1	62.5	69.4	65.7	64.5	70.9	67.3
Prepelvic length	47.6	53.4	50.5	46.8	57.1	51.8	48.4	51.6	49.7	48.6	51.3	49.9	52.3	48.3	53.2	50.7	48.6	57.9	51.2
Pelvic fin ray length	15.6	19.5	17.4	12.5	19.6	15.2	17.3	19.9	18.6	14.1	15.7	14.9	15.4	14.7	18.3	17.1	12.9	18.6	14.9
Pectoral fin ray length	20.3	26.6	23.4	17.9	25.7	21.0	25.2	28.8	27.0	19.6	21.6	20.2	21.9	21.9	27.1	24.7	20.1	25.5	22.0
Anal fin ray length	28.2	42.1	34.7	17.6	37.8	25.6	34.3	38.2	36.5	19.1	29.7	23.8	25.9	23.4	41.5	34.4	18.5	34.7	25.2
Dorsal fin base length	10.6	15.2	12.9	7.8	13.8	11.0	12.2	14.3	13.6	10.5	11.8	11.2	12.2	11.4	15.9	14.1	10.3	14.4	11.7
Anal fin base length	12.3	18.1	15.8	9.2	18.0	12.3	17.2	20.1	18.9	12.0	15.9	13.7	13.7	12.4	20.0	16.7	11.1	16.4	13.5
% in HL																			
Snout length	26.1	35.7	30.9	24.0	36.3	30.1	34.9	38.8	36.2	33.2	36.5	35.5	31.8	28.2	34.2	30.4	25.4	34.6	29.9
Interorbital width	17.9	28.2	23.0	17.5	30.3	24.9	36.6	40.6	38.6	34.4	37.3	36.1	24.4	19.6	30.4	25.0	23.7	32.3	27.4
Eye diameter	17.9	28.2	23.0	17.5	30.3	24.9	29.2	32.8	31.0	32.1	35.4	33.5	24.4	19.6	30.4	25.0	23.7	32.3	27.4
Head width	44.4	64.5	53.0	39.8	60.7	50.0	56.3	58.0	57.1	50.8	56.4	53.7	50.8	47.2	58.3	52.3	46.0	60.2	52.0
Head depth in midline of orbit	48.7	63.4	56.3	45.1	60.4	53.1	53.5	61.0	58.9	53.7	60.3	57.2	58.5	55.1	61.2	57.2	49.0	63.8	54.6
Upper jaw length	40.7	48.8	45.1	38.5	50.2	44.3	40.2	43.7	41.9	43.0	46.2	44.8	39.9	39.5	46.6	42.0	36.5	46.5	41.7

Table 2 Frequency distribution of meristic features of *Opsariichthys pachycephalus* and *O. kaopingensis*

Species	D			A				P1				P2				
	iii, 7	8	M	iii, 8	9	10	M	i, 13	14	15	M	i, 8	9	M		
Opsariichthys pachycephalus	421	2	7.0	17	398	12	9.0	251	251	38	13.6	251	123	8.3		
Opsariichthys kaopingensis	118	–	7.0	3	115	–	9.0	5	147	72	14.3	210	14	8.1		
Species	LL												TRa			
	44	45	46	47	48	49	50	51	52	53	54	M	9	10	11	M
Opsariichthys pachycephalus	–	–	–	–	2	59	121	129	137	120	108	51.7	–	251	38	10.1
Opsariichthys kaopingensis	13	84	66	57	2	–	–	–	–	–	–	45.8	95	17	–	9.2
Species	TRb			PredS								VC				
	3	4	M	17	18	19	20	21	22	23	M	38	39	40	41	M
Opsariichthys pachycephalus	251	20	3.1	–	–	–	90	139	89	25	21.1	–	57	260	108	40.1
Opsariichthys kaopingensis	104	7	3.1	7	47	60	–	–	–	–	18.6	8	77	33	–	39.2

M: mean of all listed values

Fins

Dorsal fin rays iii, 7–8 (modally 7); anal fin rays iii, 8–9 (modally 9); pectoral fin rays i, 13–15 (modally 14); and pelvic fin rays i, 8–9 (modally 8) (frequency distribution listed in Table 2). Pelvic fin origin vertical of the dorsal fin origin. Pectoral fin almost reaching pelvic fin origin when depressed in male, shorter in female. Pelvic fin margin rounded, extending to anal fin origin when depressed in male. Caudal fin forked, lower lobe slightly larger than upper.

Scales

Body with moderately small cycloid scales. Lateral-line complete, depressed downward above pectoral fin and extending along lower half of body to mid-lateral on caudal peduncle. Lateral-line scales 44–48 (always 45–47; modally 45); scale series above lateral-line 9–10; scale series below lateral-line 3–4; and predorsal scale rows 17–19 (always 18–19; modally 19). Belly with tiny cycloid scales.

Coloration in life

Body silver white to silver grey on sides. In adult male, body with 10–12 greenish blue bars, with two wider bars on caudal peduncle. In female, lateral body lacking distinct bands, a few indistinct bars on caudal peduncle. Some of bars extend downward and bifurcate below the middle axis of body. A wider greenish blue mark crossing anterior 5–6 scales just behind gill-opening on anterior trunk. Head yellow or shiny yellow on ventral side of head and lower margin of cheek and opercle in male. Upper region of iris yellow in male. Dorsal fin with light yellow coloration, each fin membrane between rays with a long, oblong black mark in male. Outer 2–3 rays of pectoral fin shiny orange-yellow, caudal fin and anal fin light yellow in male. All fins whitish or somewhat yellow and unmarked in female.

Etymology

The specific name, *kaopingensis*, is referred to the type locality of holotype: "the Kaoping River" in southern Taiwan.

Distribution

River basins of both Kaohsiung and Pingtung Counties, southern Taiwan: including the Kaoping, Tongkong, Linbien, Fansan, Fongkong, Szuchung and Kankou rivers. All basins drain into the Formosan Strait, except the Kankou River which drains into the western Pacific Ocean.

Remarks

This new species, *O. kaopingensis*, can be well distinguished from following congeneric species from *O. uncirostris* from Japan, Korea and northeastern

China; *O. bidens* and *O. hainanensis* from eastern and southern China; *O. bea* Nguyen (1987), *O. heini* Nguyen (1987), *O. dienbienensis* Nguyen and Nguyen (2000) and *O. songmaensis* Nguyen and Nguyen (2000) from Vietnam by the absence of anterior notch on upper lip vs. the presence of distinctly anterior notch on that. This new species shared the morphological similarity of lacking the anterior notch on upper lip as *O. pachycephalus* and *O. evolans* in Taiwan. It can be well distinguished from *O. evolans* by the following features: (1) predorsal scales always 19 (vs. 15–17); (2) scale series below lateral-line 3–4 (vs. 2); (3) maxillary extending to or slightly beyond vertical margin of orbit in female (vs. maxillary never extending to vertical of anterior margin of orbit in female); and (4) opercle and ventral side of head bright yellow in adult male (vs. snout with blackish-purple background in adult male). This new species seems to be more similar to *O. pachycephalus* than any other congeners by overall morphological and mitogenetic evidence, and it can be well distinguished from *O. pachycephlaus* by the following features: (1) lateral-line scales always 45–47 (vs. 50–54); (2) predorsal scales always 19 (vs. 20–21); (3) vertebral count always 39 (vs. 40–41); and (4) maxillary extending to or slightly beyond vertical margin of orbit in female (vs. maxillary always extending to or beyond vertical midline of orbit in female).

A diagnostic key to three species of *Opsariichthys* from Taiwan:

1a. Lateral-line scales always 43–44; predorsal scales 15–17 (modally 16); rather narrow body width; maxillary never extending to vertical of anterior margin of orbit; pectoral fin extending far beyond origin of ventral fin; snout with blackish-purple background in adult male ——————————(northern Taiwan) *O. evolans*

1b. Lateral-line scales always more than 45; predorsal scales 18–23; rather thick body width; maxillary extending to or far beyond the vertical of anterior margin of orbit; pectoral fin not extending beyond origin of ventral fin; snout in lacking blackish-purple background in adult male ——————2

2a. Lateral-line scales always not less than 50; predorsal scales always 20–22; vertebral count always 4 + 36–37 (= 40–41); maxillary always extending to or beyond vertical midline of orbit in female; opercle and ventral side of head orange-red to pink-red in adult male ———————— (northern, middle and western Taiwan) *O. pachycephalus*

2b. Lateral-line scales always 45–47; predorsal scales always 18–19; vertebral count always 4 + 35 (= 39); maxillary extending to or slightly beyond vertical margin of orbit in female; opercle and ventral side of head bright yellow in adult male —————————— (southern Taiwan) *O. kaopingensis* new species

Molecular phylogenetic analysis

The 75 samples of Taiwanese *Opsariichcthys pachycephalus* complex yielded 21 different haplotypes (Table 3) from 21 discrete river basins across Taiwan (Fig. 1). Other related taxa of *Opsariichthys* genus group used for molecular phylogenetic analysis from China, Taiwan and Japan (Fig. 2) are listed in Table 4. The length of mtDNA D-loop sequences of *Opsariichthys* species from China, Taiwan and Japan ranged from 923 to 926 bp. The haplotypes (n=21) formed two groups, A (n=15) and B (n=6), which corresponded completely with the species allocations in Taiwan discussed below. The uncorrected genetic distance between these groups ranged from 2.6 to 3.8% (average 3.3%). Within the A group (*O. pachycephalus*), the genetic divergence among 15 haplotypes was 0.1–1.0% (average 0.42%). In the B group (*O. kaopingensis*), genetic divergence among the 6 detected haplotypes ranged from 0.4–1.6% (average 1.0%).

The molecular phylogenetic tree of *Opsariichthys* genus group constructed by distance method (NJ) based on the Kimura 2-parameter distance model shown on Fig. 6, and almost all nodes above interspecific level with very high bootstrap support from mostly 93 to 100 except one with 67. The consistency index being 0.678 and homoplasy index 0.323; and retention index is 0.831. The heuristic research for discrete method (MP) has yielded 5941 trees with same minimal tree length 586. The majority-rule consensus tree of MP trees with equal minimal tree length provided the very similar topology to NJ tree above interspecific level with rather

Table 3 Distribution of different haplotypes from mitochondrial D-loop sequence of two *Opsariichthys* species in Taiwan

River basin	NMMBA Cat. No.	Simple No.	Haplotype No.	A1	A2	A3	A4	A5	A6	A7	A8	A9	A10	A11	A12	A13	A14	A15	B1	B2	B3	B4	B5	B6
1. Tanshuei River	6036	4	2	3	1																			
2. Shuanshi River	6031	2	1			2																		
3. Tochien River	6038	5	4		1		1	2	1															
4. Sihu River	UC	2	2							1	1													
5. Chonkang River	6039	4	1									4												
6. Taan River	2049	2	1					2																
7. Tajar River	6045	5	1					5																
8. Wu River	6046	4	3					1					2	1										
9. Joshui River	6049	1	1												1									
10. Putsi River	UC	5	2					4							1									
11. Parchan River	6055	6	3					2	3							1								
12. Tzengwen River	3989	1	1														1							
13. Lanyang River	6028	1	1					1																
14. Shiukuluan River*	UC	1	1		1																			
15. Peinan River*	UC	1	1						1															
16. Taju River*	UC	2	2		1													1						
17. Kaoping River	2911	15	2																14	1				
18. Fansan River	6064	4	2																3	1				
19. Fongkan River	3000	3	1																		3			
20. Szuchung River	6066	5	1																			5		
21. Kankou River	6067	2	2																				1	1
Total number		75	21	3	4	2	1	17	5	1	1	4	2	1	2	1	1	1	17	2	3	5	1	1

PS. *: the mark indicates the introduced population; UC: uncatalogued

Table 4 The OTU codes of cyprinid species of *Opsariichthys* genus group from China, Taiwan and Japan used for mitogenetic analysis

Code	Species name	Locality	Locality code	Source
APKITH1	*Aphyocypris kikuchii*	Hualian River, Hualian, Taiwan	14	Chen et al. 2008b
CABATT1	*Candidia barbatus*	Tanshuei River, Taipei, Taiwan	13	Chen et al. 2008b
CAPITF1	*Candidia pingtungensis*	Fongkang River, Pingtung, Taiwan	15	Chen et al. 2008b
OPACHS1	*Opsariichthys acutipinnis*	Shiangjiang, Yangtsi River, Hunan Province, China	7	this study
OPACHZ1	*Opsariichthys acutipinnis*	Lishuei, Yangtsi River, Hunan Province, China	8	this study
OPACJG1	*Opsariichthys acutipinnis*	Jian City, Gang Jiang, Yangtsi River, Jiangxi Province, China	6	this study
OPBIZC1-2	*Opsariichthys bidens*	Changtanjiang, Zhejiang Province, China	9	Chen et al. 2008b
OPEVFM1	*Opsariichthys evolans*	Mulan River, Fujian Province, China	10	Chen et al. 2008b
OPEVTT1	*Opsariichthys evolans*	Tanshuei River, Taipei, Taiwan	13	Chen et al. 2008b
OPHAHN1-2	*Opsariichthys hainanensis*	Nandujiang, Hainan Island, China	12	Chen et al. 2008a
OPUNJB1-2	*Opsariichthys uncirostris*	Biwa Lake, Otsu, Honshu, Japan	1	Chen et al. 2008b
PASPHK1	*Parazacco spilurus*	Hongkong, China	11	Chen et al. 2008b
PRNOHK1	*Pararasbora normalis*	Hongkong, China	11	Chen et al. 2008b
ZAPLLD1	*Zacco* cf. *platypus*	Dandong City, Liaoning Province, China	5	this study
ZAPLLK1	*Zacco* cf. *platypus*	Kuandian County, Liaoning Province, China	4	this study
ZAPLJK1	*Zacco platypus*	Kyoto, Honshu, Japan	2	Chen et al. 2008b
ZAPLJK2	*Zacco platypus*	Kagoshima, Japan	3	Chen et al. 2008b

high bootstrap support (73–100) except two nodes with lower support, 67. Overall, the well genetic differentiation of *O. pachycephalus* complex can be seen the support for two currently recognized allopatric species: *O. pachycephalus* and *O. kaopingensis* in any of clustering methods. Two distinctly main clades can be observed as both *Opsariichthys-Zacco* clade and *Candida-Parazacco* clade with good bootstrap support (93–99 in NJ; 88–97 in MP). In the group of *Opsariichthys uncirostris* complex, the *O. uncirostris* and *O. bidens* is well differentiated genetically with high bootstrap support 93–99 which is congruent to the view of Okazaki et al. (2002) for recognizing the distinct endemicity of Japanese *O. uncirostris* from the continental group from China and Korea. *O hainanensis* is supported as a sister OTU of *O. acutipinnis*—*O. evolans* clade with bootstrap values 67 and 89.

The molecular phylogenetic analyses by BI method employed the "GTR + I + G model" suggested by ModelTest with rather high posterior probability mostly 99–100 except one as 94 above interspecific differentiation (Fig. 7). In contrast, the *O. pachycephalus* complex shares as the sister group of the *O. hainanensis*—*O. evolans* clade with 94 support value unlike the result of NJ and MP trees it shared with *O. uncirostris* complex with bootstrap value 67.

On the other hand, the molecular phylogenetic analyses from above three methods are congruent and suggest that the Taiwanese *O. pachycephalus* complex should be recognized as the member of *Opsariichthys*, rather than *Zacco* (Figs. 6 and 7). This phylogenetic result is congruent with the osteological view of Ashiwa and Hosoya (1998) grouping *Opsariichthys pachycephalus* complex (*O. pachycephalus* and *O. kaopingensis*) into the clade of *Opsariichthys* (*O. uncirostris*), not that of *Zacco* (*Zacco platypus*). Furthermore, the molecular phylogenetic result of *O. acutipinnis*—*O. evolans* clade is also supported the assignment into *Opsariichthys* as Chen and Chang (2005) and Chen et al. (2008b).

Therefore, the traditional morphological character of "the zigzag shape of mouth with distinct anterior notch on upper lip" defining species *Opsariichthys* seem to be not represented a good diagnostic character for the whole genus while including such three Taiwanese cyprinids: *O. pachycephalus*, *O. kaopingensis* and *O. evolans*, and other Chinese *Opsariicthys* species at least for *O. acutipinnis*

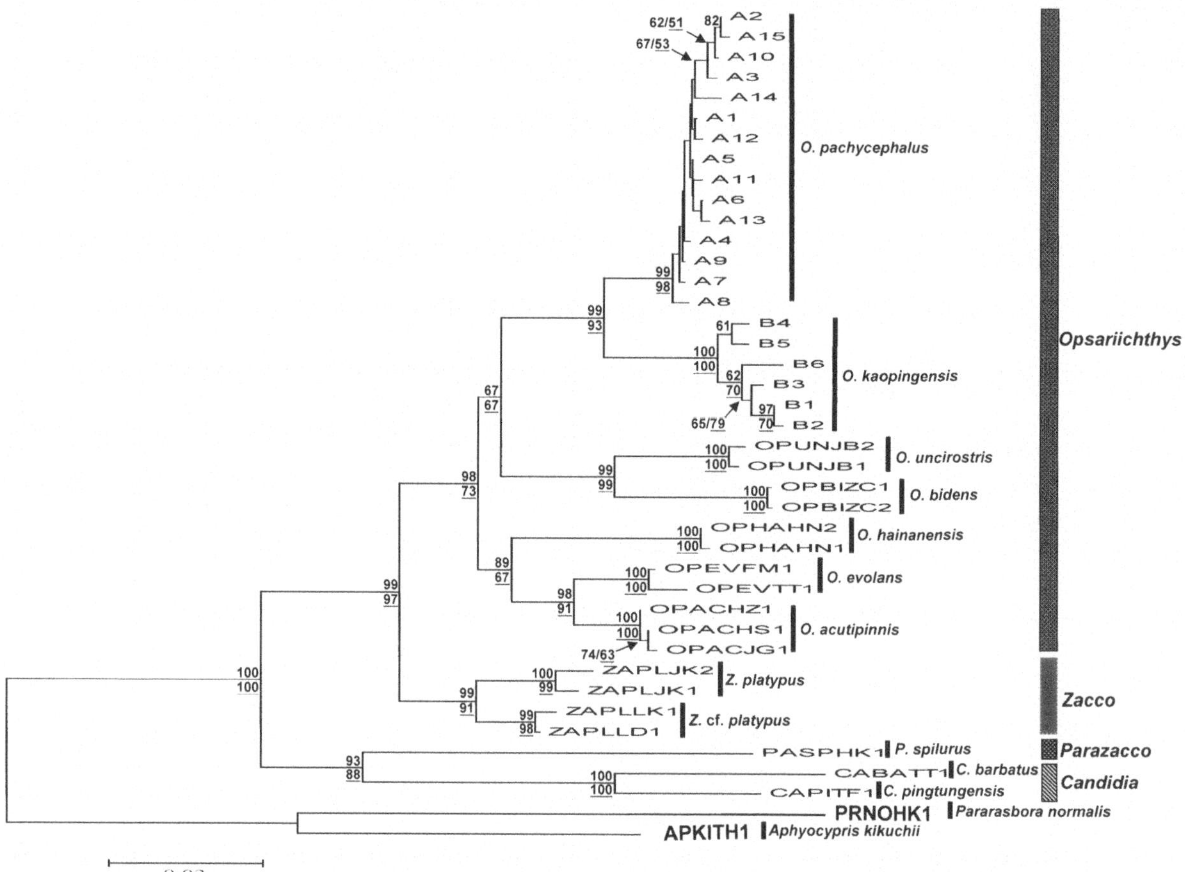

Fig. 6 Molecular phylogenetic tree of cyprinid fishes in *Opsariichthys* genus group from China, Taiwan and Japan inferred from complete mtDNA D-loop sequences. The topology indicated here is the reconstruction by the distance method (NJ). The values above branches are bootstrap proportion with 1000 replications by the NJ method if above 50%; the values below branch with underline represent as bootstrap proportion with 1000 replications by the MP if above 50%

complex. The paraphyletic clade beside *Opsariicthys* clade is merely as the sister clade with two taxa: *Zacco platypus* from Japan and *Zacco* cf. *platypus* from Northeast China. The male body coloration pattern with separated transverse bands and its common head tubercle alignment may be the main characters to redefine the whole member for this genus if we are convincing the molecular phylogenetic perspectives of this mitogenetic marker.

Within *O. pachycephalus*, the haplotype A5 is most common, represented in samples from the Tochien, Taan, Tajar, Wu, Putsi and Lanyan rivers. Within *O. kaopingensis*, the haplotype B1 is most common and represented in samples from Kaoping and Fansan Rivers. Across the range of *Opsariichthys pachycephalus*, there is only minor differentiation between different haplotypes. They have shared some common haplotypes although minor mutations still exist. In contrast, *Opsariichthys kaopingensis*, considerable geographic structure was observed, with populations in the southern tip of Henchuen Peninsula (Fongkang to Kankou rivers) rather differentiated from those in the Kaoping to Fangsan Rivers with higher divergences compared than that of *O. pachycephalus*. This suggests that the southern populations of *Opsariichthys kaopingensis* have been isolated from the northern population for a long time. The congruent, biogeographical patterns of *Opsariichthys pachycephalus* complex by the main separation between the Tzengwen and Kaoping River basins was also reported by part of the results from allozyme data of Wang et al. (1999) which share the great distinction from the region south to the Kaoping River basin.

Introduced populations of *O. pachycephalus* in the Shiukuluan and Taju Rivers of eastern Taiwan Chen and Chang (2005) shared mitogenetic haplotypes with

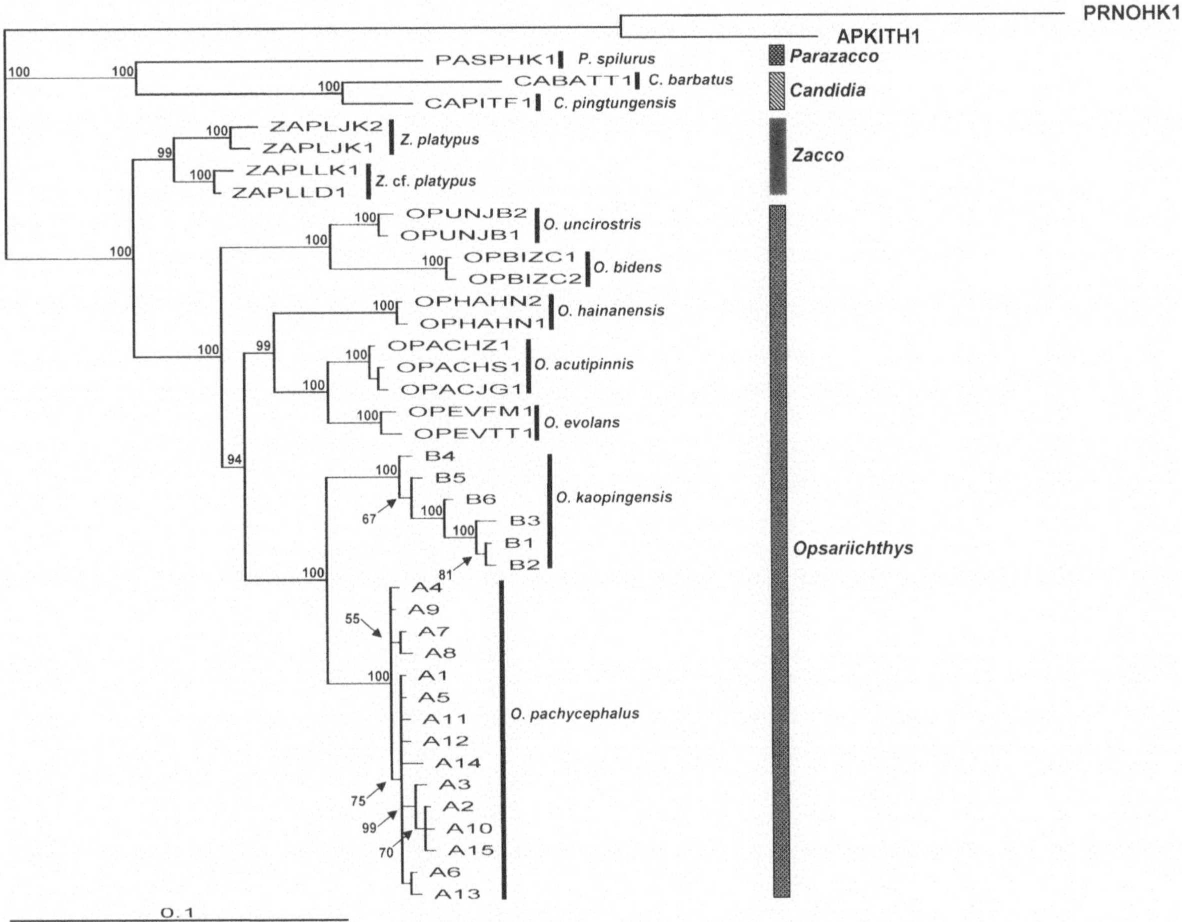

Fig. 7 Molecular phylogenetic tree reconstructed by BI method based on complete mtDNA D-loop sequences of *Opsariichthys* related genus groups from China, Taiwan and Japan. The posterior probability listed above the branch if above 50%

populations from the Tanshui and Tochien Rivers, in northern Taiwan. This suggests that the likely origin of the introduced populations is northern Taiwan.

On the generic assignment of *O. pachycephalus*, another independent work by Wang et al. (2007) used mtDNA makrer: Cytochrome *b* gene data to suggest that the *pachycephalus* complex is also actually close related to typical *Opsariithchys*, not *Zacco* which has been also proved by our team employed D-loop marker as Chen et al. (2008a, b) more recently.

Discussions

Generic limitation of *Zacco*

If we follow and accept the monophyletic assemblages for *Opsariichthys* based on current molecular phylogenetic respective, we can not find any of native Taiwanese opsariichthines should be grouped into the clade of *Z. platypus*, but all of them represent their great genetic affinity and would be grouped to either the clade with *O. uncirostris* complex or *O. acutipinnis* complex. Based on the priority of genetic nomenclature, *Opsariichthys* Bleeker (1863) is published much earlier than *Zacco* Jordan and Evermann (1902). However, the possibility for assigning the single *Zacco* type species into one earlier genus, *Opsariichthys* is still remained. In Japanese waters, both so-called "*Z*". *temminckii* and "*Z*" *sieboldii* had recently been proved by molecular phylogenetic analysis that they are very closely related to genus *Candidia*, but not typical *Zacco* by morphological and mitogenetic comparisons (Wang et al. 2007; Chen et al. 2008b) which have been assigned as a newly described genus: *Nipponocypris* (Chen et al. 2008b).

Based on current research, we strongly suggest the generic redefinition of *Zacco* should be restricted only to the type species, *Zacco platypus* complex which may contain about 2–3 very closely related species in Chinese waters, if we considered the validity for generic level of "*Zacco*" originally assigned from "*Zacco*" *platypus*. Therefore, the diagnostic features of *Zacco* would be considered to restrict into *Zacco platypus* complex from the feature as: (1) "the nuptial tubercles on the cheeks are united basally to from a plate in male" commented by Jordan and Hubbs (1925); and (2) the fused light green lateral cross-bars into fewer large patches which can be well separate from other members of *Opsariichthys* (Chen and Chang 2005). However, the taxonomy of *Zacco platypus* complex in mainland China, Korea and Japan is still needed for further revision work for such widely geographical areas. Therefore, the real range of typical *Zacco platypus* complex is not like that suggested by previous research from Japan to southern China (Perdices et al. 2004; Perdices and Coelho 2006; Wang et al. 2007); this species complex may be merely found from Japan to northern region of Zhejiang Province, China. The previous recognized group: most of so-called "*Zacco platypus*" complex as "*Zacco* spp. A to D" which is not fit above diagnostic generic definition of *Zacco* and revealed by current molecular evidence of our samples from similar localities from Yangtsi river basin in southern China (Perdices et al. 2004; Perdices and Coelho 2006) should be regarded as the members for *O. acutipinnis*—*O. evolans* complex which collected from the Yangtsi River to northern Taiwan (Chen et al. 2008b), not belonging to the true Japanese *Zacco platypus*. Therefore, we would consider that "*Zacco* spp. A to D" should be regarded as *Opsariichthys* spp. A to D with the feature of lacking anterior notch of upper jaw. However, the more detail survey of several large River basins in mainland China for true *Zacco* species is very essential for resolving such taxonomic problems.

The divergent-time estimation of *Opsariichthys pachycephalus* complex

Both morphological and mitochondrial sequence data in present studies provide clear evidence of a deep biogeographical disjunction between north-western and southern regions of Taiwan. Divergence rates of the mitochondrial control region (= D-loop sequences) has been estimated in the fishes range from 1.4 to 3.6% my^{-1} (walleyes in Stepien and Faber 1998; snooks in Donaldson and Wilson 1999; sticklebacks in Takahashi and Goto 2001). Based on the above estimation, our result for average of interspecific genetic divergences for both *O. pachycephalus* and *O. kaopingensis* is 3.3%, and it may be employed to estimate the divergent time as 0.9–2.3 Mya. It may strongly suggest that the separation of both *O. pachycephalus* and *O. kaopingensis* evolved much earlier speciation event around late Pliocene Epoch, rather than the last glacial period about 15 000–18 000 years ago (Boggs et al. 1979; Wang et al. 1999).

Comparison of distribution patterns for primary freshwater fish fauna

The two allopatric species of *Opsariichthys* discussed above are separated by the geographical boundary between the Tzengwen and Kaoping Rivers. Although Wang et al. (1999) shown the similar results for great disjunction from between the Tzengwen and Kaoping Rivers of *Opsariichthys pachycehpalus* complex which is congruent with our current mtDNA data, their data shown the distinct separation between both northern subgroup from the Tanshuei River (including Nanshi and Peishi Rivers) to Shuanshi basins and middle subgroup from the Tochien (Touchien) to Tzengwen River basins observed from allozyme data which can not be detected and separate by our mtDNA phylogenetic studies. However, our molecular results seem to be very similar on our morphological studies which can not be found any diagnotic feature for defining any morphological subgroup for current recognized *Opsariichthys pachycephalus* which is common seen in most of drainages which are northern to Kaoping River basin.

The geographic break in southern Taiwan can be also observed with congruent pattern in several other genera of the cyprinid and loach's fauna (Chen and Fang 1999; Chen and Chang 2005). For examples of middle-reaches of cyprinids and loaches, *Gobiobotia cheni* Banarescu and Nalbant (1966) is endemic and distributed in middle Taiwan while *Gobiobotia intermedia* Banarescu and Nalbant

(1968) is endemic and only found in the Kaoping River basin of southern Taiwan; and *Cobitis cf. sinensis* Sauvage and Dabry de Thiersant (1874) is widely distributed in northern and western Taiwan while *Cobitis* sp. occurs in the region from both the Kaoping to Linbien Rivers of southern Taiwan (Chen and Fang 1999; Chen and Chang 2005). However, recent molecular analyses of the Taiwanese endemic and high mountain/hill-stream cyprinid, *Candidia barbatus* complex, have revealed a different pattern of species boundaries between the Kaoping and Tongkong Rivers (Wu et al. 2007; Chen et al. 2008b). Such results suggest that both high mountain/hill-stream and middle-reaches species may utilize slightly different colonization pathways and mechanisms, by both glacial unification of river basins and river piracy.

The congruent distribution patterns suggest that similar mechanisms have occurred in the common dispersal history of primary freshwater fish in western drainages of Taiwan. Differentiation in fishes of the southern region of Taiwan may suggest that long-term isolation of freshwater fishes began well before the last glacial period.

Geographical isolation of these two *Opsariichthys* species appears to date much earlier than the last glacial period, during which the Kaoping basin (including the Henchuen Peninsula) was well isolated from the other geographical regions of Taiwan. Boggs et al. (1979) pointed out that river basins south of the Kaoping River could not have joined other river basins in the western postulated paleo-drainage system on the Taiwan continental shelf during subaerial exposure approximately 15000 years ago. Sea level is assumed to be 140 meters below the present level. The very deep Kaoping submarine canyon is rather far away and well isolated from the major river drainages (Boggs et al. 1979). Rather, the isolated population of *O. kaopingensis* may have entered this region earlier, as suggested by estimation of divergence time using molecular clock analysis as above. This geographical feature may be responsible for speciation of several species/population of freshwater fishes occurring in southern Taiwan independently from those in other areas. Detailed morphological and molecular surveys of other groups of freshwater fish fauna such as cyprinids, loaches and even catfishes may provide more comprehensive information for speciation and dispersal history of freshwater fishes in Taiwan.

Comparative materials

Opsariichthys bidens (Günther)

NMMBP 6222, 109.6 mm SL, Joulongjiang River, Fuchien Prov., China, Coll. C.W. Wang, 1 Sep. 2002; NMMBP 6223, 2 specimens, 56.7–94.8 mm SL, Joulongjiang River, Fuchien Prov., China, Coll. C.W. Wang, 1 Sep. 2002; NMMBP 6224, 98.5 mm SL, Chujiang River, Kuangsi Prov., China, Coll. I-S. Chen, 16 Jul. 1996; NMMBP 6225, 3 specimens, 73.0–78.9 mm SL, Nanliujiang River, Kuangsi Prov., China, Coll. I-S. Chen, 20 Jul. 1996; NMMBP 6227, 146.0 mm SL, Minjiang River, Fuchien Prov., China, Coll. C.W. Wang, 1 Aug. 2002; NMMBP 6228, 62.7 mm SL, Minjiang River, Fuchien Prov., China, Coll. C.S. Chuang, 29 Nov. 2002.

Opsariichthys hainanensis (Nichols and Pope)

NTOU P-2008-6-382, 10 specimens, 39.7–71.7 mm SL, Tengchiaotung River, Pouting County, Hainan Province, China, Coll. I-S. Chen and S.P. Huang, 9 Sep. 2005; NTOU P-2008-6-383, 2 specimens, 46.8–54.3 mm SL, Partsun, Linshui River basin, Pouting County, Hainan Province, China, Coll. I-S. Chen and S.P. Huang, 10 Sep. 2005; NTOU P-2008-6-384, 2 specimens, 54.4–71.7 mm SL, Linshui River basin, Linshui County, Hainan Province, China, Coll. I-S. Chen and S.P. Huang, 10 Sep. 2005; NTOU P-2008-6-385, 2 specimens, 54.1–65.5 mm SL, Leunan, Nandujiang basin, Chungjung County, Hainan Province, China, Coll. I-S. Chen and S.P. Huang, 11 Sep 2005; NTOU P-2008-6-386, 57.2 mm SL, Nankai, Nandujiang basin, Paishar County, Hainan Province, China, Coll. I-S. Chen and S.P. Huang, 12 Sep. 2005; NTOU P-2008-6-387, 8 specimens, 54.8–63.2 mm SL, Wangchuang River basin, Chungjung County, Hainan Province, China, Coll. I-S. Chen and S.P. Huang, 11 Sep. 2005.

Opsariichthys uncirostris (Temminck and Schlegel)

NMMBP 6226, 3 specimens, 213.0–248.0 mm SL, Biwa lake, Japan, Coll. J.H. Wu, 6 Oct. 2002.

Zacco platypus (Temminck and Schlegel)

NMMBP 6229, 131.5 mm SL, Coll. I-S. Chen, Biwa Lake, Japan, 17 Jun. 2002; NMMBP 6230, 2 specimens, 94.9–108.9 mm SL, Coll. I-S. Chen, Biwa Lake, Japan, 17 Jun. 2002.

Aphyocypris kikuchii (Oshima)

NMMBP 2680, 44.0 mm SL, Coll. I-S. Chen, Pei-nan River, Taitung County, Taiwan, 1 Oct. 1995; NMMBP 3905, 47.9 mm SL, Coll. HH Tan and YM Ju, Ta-chu River, Taitung County, Taiwan, 2 Dec. 2001.

Acknowledgements The first author is very grateful for the grant support from the National Science Council, Taipei, Taiwan in 2003–2005, and from the Agriculture Council, Taiwan in 2001 to 2003. He also extends his thanks to both K.T. Shao and H.C. Ho for kindly providing the very important information and photo for the type specimens of *Opsariichthys pachycephalus* from BMNH; M.J. Grygier for his kindly help for collecting Japanese specimens and L.S. Fang for his kindly support for of this research; N.H. Jang-Liaw for providing some Chinese cyprinids; Y.C. Chang for his kindly providing his excellent, alive photos of cyprinid fishes; and P.H. Liu for her assistance on part of lab. sequencing work for these cyprinid fishes.

Appendix

NCBI accession numbers of complete mitochondrial D-loop sequences used in this study (given in parenthesis). All used alcoholic specimens and tissues for newly sequencing here were deposited in either the Pisces collection of National Museum of Marine Biology & Aquarium, Pingtung (NMMBA) or National Taiwan Ocean University, Keelung (NTOUP). *Opsariichthys acutipinnis* (OPACHS1: FJ648641; OPACHZ1: FJ649642; OPACJG1: FJ648643). *Opsariichthys kaopingensis* (haplotypes B1 to B6: AY3131195 to AY3131200). *Opsariichthys pachycephalus* (haplotypes A1 to A15: AY313180 to AY313195). *Opsariichthys uncirostris* (AY3131201). *Zacco platypus* from Japan (AY313202). *Zacco* cf. *platypus* from China (ZAPLLD1: FJ64939; ZAPLLK1: FJ64940). *Aphyocypris kikuchii* (AY313203). Other mtDNA sequences are cited from Chen et al. (2008a, b).

References

Ashiwa H, Hosoya K (1998) Osteology of *Zacco pachycephalus*, sensu Jordan & Evermann (1903), with special reference to its systematic position. Environ Biol Fish 52:163–171

Banarescu P (1968) Revision of the genera *Zacco* and *Opsariichthys* (Pisces, Cyprinidae). Vest Cs Zool Spol 32:305–311

Banarescu P, Nalbant TT (1966) Notes on the genus *Gobiobotia* (Pisces, Cyprinidae) with description of three new species. Annot Zool Bot Bratislava 27:1–16

Banarescu P, Nalbant TT (1968) Some new Chinese minnows (Pisces, Cypriniformes). Proc Biol Soc Wash 81:335–346

Berrebi P, Boissin E, Fang F, Cattaneo-Berrebi G (2005) Intron polymorphism (EPIC-PCR) reveals phylogeographic structure of *Zacco platypus* in China: a possible target for aquaculture development. Heredity 94:589–598

Berrebi P, Retif X, Fang F, Zhang CG (2006) Population structure and systematics of *Opsariichthys bidens* (Osteichthyes: Cyprinidae) in southeast China using a new molecular marker: the introns (EPIC-PCR). Biol J Linn Soc 87:155–166

Bleeker P (1863) Systema Cyprinoideorum revisum. Nederlandsch Tijdschrift voor de Dierkunde 1:187–218

Bleeker P (1871) Mémoire sur les cyprinoïdes de Chine. Verh K Akad Wet (Amsterdam) 12:1–91

Boggs SJ, Wang WC, Lewis FS, Chen JC (1979) Sediment properties and water characteristics of the Taiwan shelf and slope. Acta Oceanograph Taiwan 10:10–49

Chen Y (1982) A revision of opsariichthine cyprinid fishes. Oceangr Limnol Sin 13:293–299 (in Chinese with English summary)

Chen Y (ed) (1998) Fauna Sinica. Osteichthys. Cypriniformes II. Science Press, Beijing (in Chinese)

Chen JTF, Yu MJ (1986) A synopisis of the vertebrates of Taiwan, 3rd edn. Commercial Books Press, Taipei (in Chinese)

Chen I-S, Fang LS (1999) The freshwater and estuarine fishes of Taiwan. Nat Mus Mar Biol Aquar, Pingtung (in Chinese)

Chen I-S, Fang LS (2000) Redescription of the types of *Ischikauia macrolepis*, Regan, 1908, an extinct cyprinid from Taiwan and replacement in *Rasborinus* Oshima, 1920. Zool Stud 39:13–17

Chen I-S, Chang YC (2005) A photographic guide to the island-water fishes of Taiwan. The Sueichan Press, Keelung

Chen I-S, Chang YC (2007) Taxonomic revision and mitochondrial sequence evolution of the cyprinid genus *Squalidus* (Teleostei: Cyprinidae) in Taiwan with description of a new species. Raffl Bull Zool S14:69–76

Chen I-S, Hsu CH, Hui CF, Shao KT, Miller PJ, Fang LS (1998) Sequence length and variation in the mitochondrial control region of two freshwater fishes belonging to *Rhinogobius* (Teleostei: Gobiidae). J Fish Biol 53:179–191

Chen I-S, Han CC, Fang LS (2002a) A new balitorid fish, *Sinogastromyzon nantaiensis*, from southern Taiwan. Ichthyol Explor Freshw 13:297–200

Chen I-S, Miller PJ, Wu HL, Fang LS (2002b) Taxonomy and mitochondrial evolution in non-diadromous species of *Rhinogobius* (Teleostei: Gobiidae) of Hainan island, southern China. Mar Freshwater Res 53:259–273

Chen I-S, Huang SP, Jang-Liaw NH, Shen CN, Wu JH (2008a) Molecular evidence for genetic differentiation of the

Opsariichthys bidens complex (Teleostei: Cyprinidae) in Southern China around South China Sea and the validity of *Opsariicthys hainanensis*. Raffl Bull Zool Suppl 19:215–223

Chen I-S, Wu JH, Hsu CH (2008b) The taxonomy and phylogeny of the cyprinid genus, *Candidia* (Teleostei: Cyprinidae) from Taiwan, with description of a new species and comments on a new genus. Raffl Bull Zool Suppl 19:203–214

Donaldson KA, Wilson RR (1999) Amphi-panamic geminates of snook (Percoidei: Cetropomidae) provide a calibration of the divergence rate in the mitochondrial DNA control region of fishes. Mol Phylog Evol 13: 208–213

Günther A (1868) Catalogue of the fishes in British Museum, Vol 7. British Museum, London

Hall TA (2001) Bioedit: a user-friendly biological sequence alignment editor and analysis, version 5.09. Dept Microbiol, N Carolina State Univ, Raleigh

Hosoya K, Ashiwa H, Wayanabe M, Mizunguchi K, Okazaki T (2002) *Zacco sieboldii*, a species distinct from *Zacco temminckii*. Ichthyol Res 50:1–8

Howes GJ (1980) The anatomy, phylogeny and classification of the bariline cyprinid fishes. Bull Br Mus Nat Hist (Zool) 37:129–198

Howes GJ (1983) Additional notes on bariline cyprinid fishes. Bull Br Mus Nat Hist (Zool) 45:95–101

Jordan DS, Evermann BW (1902) Note on a collection of fishes from the island of Formosa. Proc US Nat Mus 25:315–368

Jordan DS, Richardson RE (1909) A catalog of the fishes of the island of Formosa, or Taiwan, based on the collections of Dr Hans Sauter. Mem Carneg Mus 4:159–204

Jordan DS, Hubb CL (1925) Record of fishes obtained by David Starr Jordan in Japan, 1922. Mem Carn Mus 10:93–346

Kumar S, Tamura K, Nei M (2004) MEGA 3: Intergrated software for molecular evolutionary genetics analysis and sequence alignment. Pennsyl State Univ, Philadelphia

Ma GC, Watanabe K, Tsao HS, Yu HT (2006) Mitochondrial phylogeny reveals the artificial introduction of the pale chub *Zacco platypus* (Cyprinidae) in Taiwan. Ichthyol Res 53:323–329

Nichols JT, Pope CH (1927) The fishes of Hainan. Bull Amer Mus Nat Hist 54:321–394

Nguyen TT (1987) Genus *Opsariichthys* Bleeker, 1863 Leuciscini-Cyprinidae of the Lam River Basin (Prov. Nghe-Tinh). Tap Chi Sinh Hoc (J Biol) 9:32–36 In Vietnamese, English summary

Nguyen VH, Nguyen HD (2000) Two new species of the fish genus of *Opsariichthys* from Vietnam. Tap Chi Sinh Hoc (J Biol) 22:12–16 In Vietnamese, English summary

Nylander JAA (2005) MrModeltest V 2.2. Evol biol centre. Uppsala Univ, Uppsala

Okazaki T, Jeon SR, Kitagawa T (2002) Genetic differentiation of piscivorous chub (genus *Opsariichthys*) in Japan, Korea and Russia. Zool Sci 19:601–610

Oshima M (1919) Contribution to the study of the freshwater fishes of the island of Formosa. Ann Carneg Mus 13:169–328

Oshima M (1923) Studies on the distribution of the freshwater fishes of Taiwan and the geographical relationship of the Taiwan and adjacent area. Zool Mag 35:1–49

Perdices A, Cunha C, Coelho MM (2004) Phylogenetic structure of *Zacco platypus* (Teleostei: Cyprinidae) populations on the upper and middle Chang Jiang (=Yangtze) drainage inferred from cytochrome b sequences. Mol Phylogenet Evol 31:192–203

Perdices A, Sayada D, Coelho MM (2005) Mitochondrial diversity of *Opsariichthys bidens* (Teleostei: Cyprinidae) in three Chinese drainages. Mol Phylogenet Evol 37:920–927

Perdices A, Coelho MM (2006) Comparative phylogeography of *Zacco platypus* and *Opsariichthys bidens* (Teleostei, Cyprinidae) in China based on cytochrome b sequences. J Zool System Evol Res 44:330–338

Regan CT (1908) Description of new species from Lake Candidia, Formosa, collected by Dr. A. Moltrecht. Annal Mag Nat Hist (Ser. 8) 2:358–360

Ronquist F, Huelsenbeck JP (2003) MrBayes 3: Bayesian phylogenetic inference under mixed models. Bioinform 19:1572–1574

Sambrook J, Fritsch EF, Manitatais T (1989) Molecular cloning. A laboratory manual, 2nd edn. Cold Spring Harbor Laboratory, Cold Spring Harbor

Sanger F, Nicklen S, Coulson AR (1977) DNA sequencing with chain-terminating inhibitors. Proc Nat Acad Sci USA 74:5463–5467

Sauvage HE (1874) Notes sur les poissons des eaux douces de la Chine. Annal Sci Nat Paris (Zool.) (Sér. 6) 1(5):1–18

Shen SC (ed) (1993) Fishes of Taiwan. Natl Taiwan Univ Press, Taipei (in Chinese)

Stepien CA, Faber JE (1998) Population genetic structure, phylogeography and spawning philopatry in walleye (*Stizostedion vitreum*) from mitochondrial DNA control region sequences. Mol Ecol 7:1757–1769

Swofford DL (2003) PAUP*: phylogenetic analysis using parsimony (* and other methods). Ver 4. Sinauer Associates Press, Sunderland

Takahashi H, Goto A (2001) Evolution of east Asian ninespine sticklebacks as shown by mitochondrial DNA control region sequences. Mol Phylog Evol 21: 135–155

Temminck GJ, Schlegel H (1846) Pisces in Siebold's Fauna Japonica. Lugduni Batavorum, Leiden

Thompson JD, Higgins DG, Gibson TJ (1994) CLUSTAL X: improving the sensitivity of progressive multiple sequence alignment through sequence weighting, position specific gap penalties and weight matrix choice. Nucleic Acids Res 22:4673–4680

Tzeng CS (1986) The freshwater fishes of Taiwan. Taiwan Educ Inst Press, Taichung (in Chinese)

Wang HY, Tsai MP, Yu MJ, Lee SC (1999) Influence of glaciation patterns of the endemic minnow, *Zacco pachycephalus*, in Taiwan. Mol Ecol 8:1879–1888

Wang HY, Wang CF, Du SY, Lee SC (2007) New insights on molecular systematics of opsariichthines based on cytochrome b sequencing. J Fish Biol 71:18–32

Wu JH, Hsu CH, Fang LS, Chen I-S (2007) The molecular phylogeography of *Candidia barbatus* species complex (Teleostei: Cyprinidae) from Taiwan. Raffl Bull Zool Suppl 14:61–67

Environ Biol Fish (2009) 86:185–192
DOI 10.1007/s10641-009-9452-0

Hemimyzon sheni, a new species of balitorid fish (Teleostei: Balitoridae) from Taiwan

I-Shiung Chen · Lee-Shing Fang

Received: 3 October 2007 / Accepted: 27 January 2009 / Published online: 26 February 2009

Abstract A new balitord fish, *Hemimyzon sheni* n. sp. was collected from the Tar-Ju River basin, Taitung County, southeast Taiwan. The new endemic species can be distinguished from other congeneric species by a combination of the following features: (1) dorsal fin rays 3+8, pectoral fin rays 13+11 (total 24); (2) lateral-line scales 78–80 and predorsal scales 31–32; (3) somewhat pointed head shape; (3) small pelvic fin which does not extend to rear vertical of dorsal fin; (4) position of anus close to anal fin base; and (5) predorsal region with many very small light brown marks and unmarked pectoral fin. A diagnostic key to three nominal, endemic species of *Hemimyzon* from Taiwan is also provided in this paper.

Keywords *Hemimyzon* · New fish · Endemic species · Balitoridae · Fish fauna · Fish taxonomy · Taiwan

I-S. Chen (✉)
Institute of Marine Biology
and Department of Life Science,
National Taiwan Ocean University,
Keelung 202,
Taiwan, Republic of China
e-mail: isc@mail.ntou.edu.tw

L.-S. Fang (✉)
Cheng Shiu University,
No. 840, Cheng-Ching Rd., Kaohsiung County 833,
Taiwan, Republic of China
e-mail: lsfang@csu.edu.tw

Introduction

The freshwater and estuarine fish fauna of Taiwan has recently reviewed by Chen and Fang (1999). There are at least 224 freshwater and brackish fish species recorded in Taiwanese waters. Among them, there are three genera with five valid, endemic species of balitorid loaches: *Formosania lacustre* (Steindachner 1908); *Hemimyzon formosanus* (Boulenger 1894); *Hemimyzon taitungensis* Tzeng and Shen (1982); *Sinogastromyzon puliensis* Liang (1974); and *Sinogastromyzon nantaiensis* Chen et al. (2002) (Shen 1993; Chen and Fang 1999; Chen and Chang 2005).

More recently, three very unusual specimens of *Hemimyzon* send by the zoologist, Dr. S. H. Chen, National Taiwan Normal University to first author for request of fish identification which collected from rural, very remote high-mountain stream of south-eastern Taiwan without any motorway to go in. Therefore, authors have recognised as a *Hemimyzon* species which would be new to science after the examination of them. This aim of paper is to describe the new species herein as well as provide a diagnostic key to three nominal, endemic species of *Hemimyzon* from Taiwan.

Materials and methods

Type specimens were collected by hand-net. Other comparative, congeneric specimens were collected by casting-net or electro-fishing. The counts and measurements followed those of Kottelat and Chu (1988) and Doi and Kottelat (1998), except the anal fin base length followed Chen et al. (2002). Abbreviation used: NTOU P, the Pisces collection of National Taiwan Ocean University, Keelung; and NTUM, National Taiwan University, Taipei.

Systematics

Hemimyzon (Regan 1911)
(Type species, *Homaloptera formosana* Boulenger 1894)

Hemimyzon sheni new species (Figs. 1, 2, 4)

Materials examined

Holotype.- NTOU P-2007-07-077, 50.1 mm SL, small tributary (3 km south to Yi-Ting mountain) in upper reaches of the Tar-Ju River, Tar-Ren Village, Taitung County, Taiwan, ROC, coll. by S.H. Chen, July 11, 1993.

Paratypes.- NTOU P-2007-07-078, 2 specimens, 11.2–30.6 mm SL, other data same as holotype.

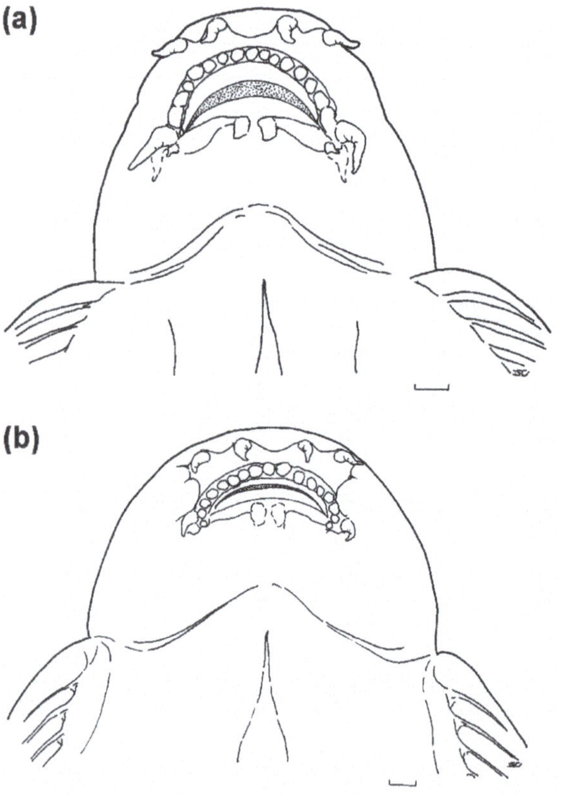

Fig. 1 Ventral view of head, **a** *Hemimyzon sheni*, NTOU P-2007-07-077, holotype, 50.1 mm SL, dotted line indicated the groove under barbel; **b** *Hemimyzon formosanus*, NTUM-05276, 53.5 mm SL, dotted line below lower lip as the inner projection. Bar=1 mm

Fig. 2 *Hemimyzon sheni* new species, NTOU P-2007-07-077, holotype, 50.1 mm SL, Tar-Ju River basin, Taitung County, Taiwan

Diagnosis

The new species can be well distinguished from other congeneric species by following combination of features: (1) dorsal fin rays 3+8, pectoral fin rays 13+11 (total 24); (2) lateral-line scales 78–80, predorsal scales 31–32; (3) head shape somewhat pointed; (3) small pelvic fin which not extending to rear vertical of dorsal fin; (4) position of anus with larger distance of pelvic rear tip to anus about three times to that of anus to anal fin origin; and (5) specific coloration: predorsal region in having many very small, irregularly light brown marks and pectoral fin unmarked.

Description

Morphometrics of this new species as percentages of standard length are listed in Table 1 and meristic features are listed in Table 2. Head and body strongly depressed with flat ventral side anteriorly. Posterior trunk from anus to caudal peduncle rather compressed. See Table 1 for morphometric characters. Head strongly depressed and rounded in dorsal view. Head with a few tiny tubercles. Upper lip with 13–15 papillae; no distinct papillae on lower lip except a pair of round projections on inner side (Fig. 1). Four rostral barbels and two at both corners of mouth which anterior one much larger than posterior one. The length of anterior barbels about equal to eye

Table 1 Morphometry of two *Hemimyzon* species in Taiwan

Species	*H. sheni*			*H. formosanus*			
	H	P					
No.	1	1		28			
Standard length (mm)	50.1	30.6		32.7–61.6			
% in SL			Mean	Min	Max	Mean	STD
Lateral head length	24.5%	28.2%	26.3%	22.9%	26.6%	24.7%	0.9%
Dorsal head length	24.2%	24.6%	24.4%	22.6%	26.7%	24.2%	1.0%
Ventral head length	10.9%	10.6%	10.7%	8.9%	12.1%	10.0%	0.7%
Head width	20.9%	23.6%	22.3%	20.5%	25.0%	22.7%	1.2%
Snout length	13.7%	15.1%	14.4%	13.2%	15.2%	14.3%	0.6%
Eye diameter	4.2%	4.0%	4.1%	3.5%	4.4%	3.9%	0.2%
Interorbital width	10.0%	11.8%	10.9%	9.0%	11.2%	10.2%	0.5%
Mouth width	10.0%	10.5%	10.3%	8.3%	10.5%	9.4%	0.6%
Body width at pectoral fin origin	19.0%	22.2%	20.6%	18.0%	22.5%	20.1%	1.3%
Body width at anal fin origin	6.9%	6.6%	6.8%	7.8%	10.3%	9.2%	0.8%
Body depth at dorsal fin origin	14.7%	15.0%	14.8%	13.6%	18.2%	16.2%	1.2%
Caudal peduncle length	13.4%	13.1%	13.3%	12.4%	16.0%	14.4%	0.9%
Caudal peduncle depth	10.0%	9.7%	9.9%	8.7%	11.4%	9.8%	0.6%
Length of last simple pectoral fin rays	15.5%	16.9%	16.2%	12.3%	17.6%	14.7%	1.2%
Pectoral fin length	31.6%	29.1%	30.3%	31.8%	35.4%	33.2%	0.9%
Pelvic fin length	25.0%	22.4%	23.7%	25.4%	30.2%	27.0%	1.5%
Length of upper caudal fin lobe	25.9%	26.5%	26.2%	23.8%	28.4%	26.2%	1.2%
Length of lower caudal fin lobe	28.2%	27.7%	27.9%	26.4%	31.4%	28.8%	1.5%
Distance between pelvic fin bases	20.7%	17.1%	18.9%	21.3%	26.6%	23.7%	1.4%
Predorsal-fin length	51.5%	51.0%	51.3%	47.6%	54.0%	51.0%	1.6%
Preanal-fin length	81.6%	83.6%	82.6%	77.0%	82.8%	80.3%	1.5%
Prepectoral-fin length	16.4%	18.0%	17.2%	12.8%	15.9%	14.2%	0.9%
Distance between pelvic and pectoral fins	29.9%	30.2%	30.1%	29.0%	33.0%	31.5%	1.1%
Anal fin base length	8.1%	10.8%	9.5%	5.7%	7.6%	6.6%	0.5%

H holotype, *P* paratype

diameter. Inner, posterior rostral barbels short but sharply pointed. Interorbital region rather wide. Gill-opening small and very restricted, merely extending above anterior origin of pectoral fin. The location of anus is rather close to anal fin origin, with large distance of rear tip of pelvic fin to anus about three times to that of anus to anal fin origin.

Dorsal fin 3+8; anal fin 2+5; pectoral fin 13+11 (total 24); pelvic fin 4+10 (total 14). Origin of dorsal fin slightly behind origin of pelvic fin. Pectoral fin rather large, the rear margin extending beyond origin of pelvic fin. Pelvic fins well separated, the gap between the attachments of their innermost rays about 2.0–2.5 times for eye diameter. The rear tip of pelvic fin not extending to the rear vertical of dorsal fin when depressed and also far away from anus. Caudal fin forked, lower lobe longer than upper one.

Dorsal part of body with very small cycloid scales, ventral region between the bases of pectoral and pelvic fins naked. Body scales slightly larger posteriorly. Lateral-line scales 78–80 and predorsal scales 31–32.

Coloration in alcohol

(Figures 2, 4) Dorsum of body olive brown. Head uniform olive brown. Ventrum yellowish gray to pale white. Dorsum with many small, irregularly light brown marks in predorsal region. Lateral body uniform brown. Dorsal fin 3–5 rows of blackish brown spots. Pectoral and pelvic fins pale brown and its membrane unmarked. Anal fin pale white and unmarked. Caudal fin with deep, blackish brown margin on both upper and lower region, and

Table 2 Comparison of distribution frequency of meristic features from three endemic species of *Hemimyzon* in Taiwan

Species	Pectoral fin																		
	Simple rays						Branched rays						total rays							
	10	11	12	13	14	M	9	10	11	12	13	M	20	21	22	23	24	25	26	M
H. sheni n. sp.	–	–	–	3	–	13.0	–	–	3	–	–	11.0	–	–	–	–	3	–	–	24.0
H. formosanus	1	19	12	–	–	11.3	7	18	7	–	–	10.0	1	19	12	–	–	–	–	21.3
H. taitungensis	–	–	1	7	4	13.3	–	–	1	8	3	12.2	–	–	–	–	–	7	5	25.4

	Dorsal fin			Pelvic fin														
	Branched rays			Simple rays					Branched rays				Total rays					
	7	8	M	4	5	6	7	M	9	10	11	M	13	14	15	16	17	M
H. sheni n. sp.	–	3	8.0	3	–	–	–	4.0	–	3	–	10.0	–	3	–	–	–	14.0
H. formosanus	16	–	7.0	16	16	–	–	4.5	20	12	–	9.4	4	27	1	–	–	13.9
H. taitungensis	–	6	8.0	–	–	6	6	6.5	–	6	6	10.5	–	–	–	–	12	17.0

	Lateral-line scales																			Predorsal scales								
	68	69	70	71	72	73	74	–	78	79	80	–	83	84	85	86	87	88	M	0	–	31	32	–	36	37	38	M
H. sheni n. sp.	–	–	–	–	–	–	–	–	1	2	1	–	–	–	–	–	–	–	79.0	–	–	1	1	–	–	–	–	31.5
H. formosanus	1	8	9	6	5	2	1	–	–	–	–	–	–	–	–	–	–	–	70.5	–	–	–	–	–	6	7	3	36.8
H. taitungensis	–	–	–	–	–	–	–	–	–	–	–	–	1	2	4	2	2	1	85.4	6	–	–	–	–	–	–	–	0

M mean of all distributed values

Fig. 3 *Hemimyzon formosanus*, NTUM05278, 58.0 mm SL, Wu River basin, Nan-Tou County, Taiwan

pale brown in middle region. Caudal fin membrane with 4–6 very thin vertical white streaks which somewhat indistinct in middle region based on pale background.

Distribution

This new species is only found from the Tar-Ju River basin, Taitung County, Taiwan (Fig. 5). It may be highly possible to exist in the nearby basins running into the Pacific Ocean, Taitung County although no formal motorway can go through the small tributary of high mountain region for these nearby basins.

Etymology

Named after Prof. Shieh-Chieh Shen, the senior ichthyologist for his great contribution of ichthyology in Taiwan.

Discussion

In the taxonomy of *Hemimyzon*, four endemic nominal species were documented in mainland China (Yue 2000) while two endemic species were reported from Taiwan (Tzeng and Shen 1982; Tzeng in Shen 1993). The new species, *Hemimyzon sheni*, can be well separated from the four Chinese species. *H. sheni* can be easily distinguished from both *H. yaotanensis* (Fang 1931) and *H. megalopseos* (Li and Chen 1985) by the anal fin position: anal fin origin far in front the vertical of dorsal fin origin in *H. sheni* vs. anal fin origin inserted just below dorsal fin origin in those two Chinese species. *H. sheni* shares similarity in the position of anal fin with both *H. macroptera* (Zheng in Zheng et al. 1982) and *H. pumilicorporora* (Zheng and Zhang 1987). However, it can be easily distinguished from those two species by (1) pelvic fin unbranched rays 4 vs. 6–9; (2) shallow forked caudal fin vs. very deep forked one; and (3) caudal peduncle, rather short and stout with a length/depth ratio of 1.3–1.4 vs. very long and slender with a ratio of 2.9–3.8.

Overall, the new species, *H. sheni*, is more similar to *H. formosanus* (Fig. 3) than any other congeneric nominal species in mainland China and Taiwan. While compared to the two nominal endemic species of Taiwan, *H. sheni* shares morphological similarity with *H. formosanus* by following features: pelvic fin simple rays 4–5, well separation of pelvic fin with the large gap between innermost rays of pelvic fins about 2–3 times of eye diameter and the innermost rays of that with very low connection to ventral body on basal 1/4–1/5 region; unlike pelvic fin simple rays 6–7, pelvic fins very close with the very small inner gap less than eye diameter and the innermost rays of that with membrane connecting to ventral body on basal 1/2 region in *H. taitungensis*. *H. sheni* can be well separated from *H. formosanus* by the following

Fig. 4 **a** Dorsal view and **b** ventral view for two endemic *Hemimyzon* species, the upper one: *Hemimyzon sheni* (NTOU P-2007-07-077, holotype, 50.1 mm SL) and the lower one: *Hemimyzon formosanus* (NTUM-05276, 53.9 mm SL)

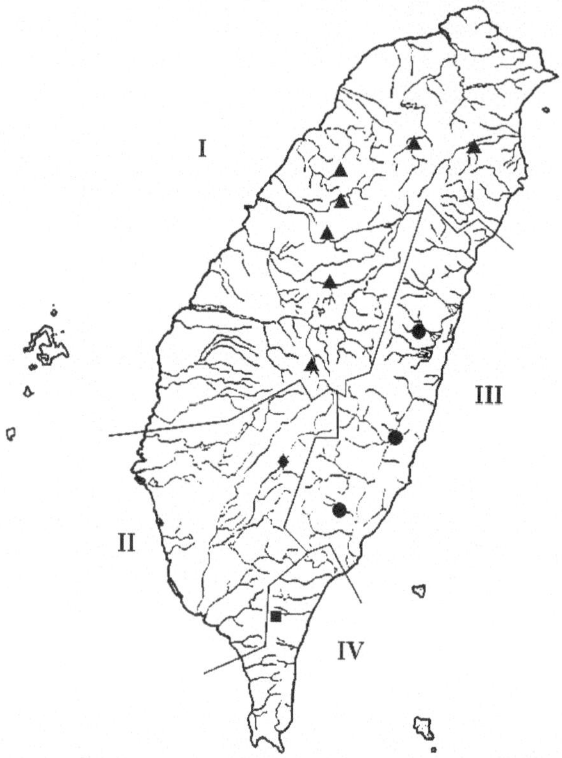

Fig. 5 Distribution pattern of *Hemimyzon* species in Taiwan, **a** ▲ in Region I: *Hemimyzon formosanus*; **b** ♦ in region II: *Hemimyzon* sp. 1; **c** ● in region III: *Hemimyzon taitungensis*; **d** ■ in region IV: *Hemimyzon sheni*

features: (1) fin ray counts: higher counts as dorsal fin 3+8; pectoral fin 13+11 (total 24) vs. dorsal fin 3+7; pectoral fin modally 10–12 (modally 11)+9–11 (modally 10) (total 20–22; modally 21); (2) head shape: somewhat pointed vs. obtuse and rounded (Fig. 1); (3) pelvic fin size: smaller pelvic fin which does not extend to rear vertical of dorsal fin as 22.4–25.0 (average 23.7) % in SL vs. larger pelvic fin always extends beyond that of dorsal fin as 25.4–30.2 (27.0) % in SL; (4) narrower body width between pelvic fin bases 17.1–20.7 (average 18.9) % in SL vs. wider body width between pelvic fin bases 21.3–26.6 (23.7) % in SL; (5) position of anus: larger distance of pelvic rear tip to anus about three times to that of anus to anal fin origin vs. the shorter distance of pelvic rear tip to anus about equal to that of anus to anal fin origin; and (6) specific coloration: predorsal region in having many very small, irregularly light brown marks vs. that always representing several large light brown arcs and bars; and pectoral fin always unmarked vs. always with 2–3 longitudinal light stripes under dark background (Fig. 4). This new species, *H. sheni*, seems to represent narrower and restricted geographical distribution range than *H. formosanus* which generally found in most river basins in northern and western Taiwan. Another new and unnamed specie, *H.* sp. 1, originally mentioned and progressed in Chen and Chang (2005) will be formally described elsewhere by their research teams (Chen et al. in preparation). However, more detailed surveys in southeastern Taiwan may are required for mapping the true species range of *H. sheni* at nearby river basins other than the type locality in southeastern Taiwan. For the comparison of mouth structure among these four species in Taiwan, *H. formosanus* seems to represent shorter, limited barbels in adult than the other two nominal species, *H. sheni* and *H. taitungensis*, and the undecribed one, *H.* sp. 1, in Taiwan.

Although several field trips had been conducted in the middle reaches and small tributaries of Tar-Ju River basins during the last few years, no further specimens had been found, other than three types of *H. sheni* collected from high mountain brook of Tar-Ju River basin in 1993. *H. sheni* can be considered as a cold, head-water species of endemic loach in Taiwan but absent from middle reaches and any small tributaries around middle and lower altitude. In contrast, *H. formosanus* is a rather common species that it can tolerate and survive in middle reaches of river basins of western Taiwan and often colonizing wider range within the basins. Therefore, the ecologically restricted distribution pattern of *H. sheni* seems similar to that of *H. taitungensis* which always inhabits lower water-temperature region around head water and upper reaches of drainages in eastern Taiwan.

From the geographical distribution pattern of these three nominal and one undescribed *Hemimyzon* species in Taiwan (Fig. 5), it is likely that their primitive ancestor may resemble the western Taiwan form because this is close to the origin of continental freshwater loach species from south mainland China. Both the widely-distributed species, *H. formosanus* (region I), and the undescribed species, *H.* sp. 1 (region II) could closely relate to the origin of primitive ancestor group and then, speciation events had separated them from the other two endemic species of eastern Taiwan, *H. taitungensis* (region III) and *H. sheni* (region IV). The speciation process may result from events of river piracy in different geographical region that helped the fishes colonized from the west to east of Taiwan. Further detail survey using molecular phylogenetic approach would provide more comprehensive view to explain the evolutionary history of these four morphological species, especially the two species from extant river basins recently discovered in Taiwan.

An artificial key to 3 endemic, nominal species of *Hemimyzon* in Taiwan:

1a The gap between basal insertion of innermost rays of pelvic fins rather small and less than eye diameter; pelvic fin simple rays 6–7 ———— (San-Jarn, Hua-Lian, and Pei-Nan River basins, eastern Taiwan) *H. taitungensis*

1b The gap between basal insertion of innermost rays of pelvic fins large, more than two times of eye diameter; pelvic fin simple rays 4–5 —————————————— 2

2a Dorsal fin rays 3+8; pectoral fin rays 13+11 (total 24); inner, posterior rostral barbels with pointed tip; the distance of pelvic rear tip to anus about three times to that of anus to anal fin origin ————————————— (Tar-Ju River basin, southeastern Taiwan) *H. sheni* new species

2b Dorsal fin rays 3+7; pectoral fin rays 10–12 (modally 11)+9–11 (modally 10) (total 20–22; modally 21); inner, posterior barbels merely as low rounded projection; the distance of pelvic rear tip to anus about equal to that of anus to anal fin origin ———————— (the most basins of northern and western Taiwan) *H. formosanus*

Comparative materials

Hemimyzon formosanus (Boulenger) NTOU P-2005-09-233, six specimens, 36.8–55.0 mm SL, Fan-Fan Brook, Lany-Yang River basin, I-Lan County, Taiwan, coll. M. Chang, September 5, 2005. NTOU P-2007-09-057, 59.2 mm SL, Hong-Long River basin, Miao-Li County, Taiwan, coll. by I-S. Chen, July 12, 1999. NTUM-05276, ten specimens, 36.4–55.1 mm SL, Her-Sher, Jo-Shuei River basin, Nan-Tou County, Taiwan, coll. by C. S. Tzeng, July 27, 1980. NTUM05278, four specimens, 50.2–58.0 mm SL, Pu-Li, Wu River basin, Nan-Tou County, Taiwan, coll. by C. S. Tzeng, July 20, 1980. NTUM-05283, 14 specimens, 32.7–61.6 mm SL, Tung-Shih, Ta-Chiar River basin, coll. by C. S. Tzeng, January 30, 1981. NTUM-05307, 45.8 mm SL, Chow-Lan, Ta-An River basin, Miao-Li County, Taiwan, coll. by C. S. Tzeng, January 30, 1981.

Hemimyzon taitungensis Tzeng and Shen NTOU P-2007-05-273, 63.0 mm SL, Shin-Wu-Liu River, Pei-Nan River basin, Taitung County, Taiwan, coll. by I-S. Chen, May 16, 2005. NTOU P-2007-09-051, five specimens, 46.0–59.2 mm SL, Shin-Wu-Liu River, Pei-Nan River basin, Taitung County, Taiwan, by I-S. Chen, October 25, 1994.

Hemimyzon **sp. 1** NTOU P-2007-12-175, five specimens, 37.6–55.0 mm SL, Jo-Kou River, Kao-Ping River basin, coll. by I-S. Chen, October 15, 1995.

Acknowledgments The authors are very grateful to S.H. Chen, National Taiwan Normal University, Taipei for his very kindly providing type specimens of this new loach found in high mountain streams of Taiwan, and to S.C. Shen, National Taiwan University, Taipei for providing the opportunity to examine the freshwater fish collection in NTUM. Authors extend the thanks to J.H. Huang, National Taiwan Normal University for making the photographs and also S. H. Su, National Taiwan Ocean University, Keelung for assistance on data assessment. The research is partly supported by the grant of NSC, Taipei in 2007 for first author.

References

Boulenger GA (1894) Descriptions of a new lizard and a new fish obtained in Formosa by Holst. Ann Mag Nat Hist (Ser. 6) 14:462–463

Chen I-S, Fang LS (1999) The freshwater and estuarine fishes of Taiwan. National Museum of Marine Biology and Aquarium, Pingtung (in Chinese)

Chen I-S, Chang YC (2005) Photographic guide of the inland-water fishes of Taiwan. Vol. 1. Cyprinformes. Sueichan, Keelung

Chen I-S, Han CC, Fang LS (2002) *Sinogastromyzon nantaiensis*, a new balitorid fish from southern Taiwan (Teleostei: Balitoridae). Ichthyol Explor Freshwaters 13:239–242

Doi A, Kottelat M (1998) *Hemimyzon nannensis*, a new balitorid fish from the Chao Phraya basin, Thailand. Ichthyol Res 45:7–11

Fang PW (1931) Notes on new species of homalopterin loaches referring to *Sinohomaloptera* from Szechuan, China. Sinensia 1:137–145

Kottelat M, Chu XL (1988) A synopsis of Chinese balitorid loaches (Pisces: Homalopteridae) with comments on their phylogeny and description of a new genus. Rev Suisse Zool 95:181–201

Li ZY, Chen YR (1985) On two new species of Homalopteridae fishes from Yunnan. Zool Res 6:169–173 (In Chinese)

Liang YS (1974) The adaptation and distribution of the small freshwater homalopterid fishes, with description of a new species from Taiwan. Symp Biol Environ Sinica 141–156

Regan CT (1911) The classification of the teleostean fishes of the order Ostariophysi. -I. Cyprinoidea. Ann Mag Nat Hist (Ser 8) 8:13–32

Shen SC (1993) Fishes of Taiwan. Taiwan University Press, Taipei (in Chinese)

Steindachner F (1908) Ueber eine noch unbekannte Art der Gattung *Bergiella* Eig. aus dem La Plata. Anz Akad Wiss Wien 45:110–113

Tzeng CS, Shen SC (1982) Studies on homalopterid fishes of Taiwan, with description of a new species. Bull Inst Zool Acad Sinica 21:161–169

Yue P (2000) (ed) Fauna Sinica. Osteichthyes Cypriniformes Vol. III. Science, Beijing (in Chinese)

Zheng CY, Zhang W (1987) The homalopterid fishes from Guizhou Province, China. J Sci Med Jinan Univ 1987:79–86 (In Chinese)

Zheng CY, Chen YR, Huang S (1982) The homalopterid fishes from Yunnan Province, China. Zool Res 3:393–402 (In Chinese)

Environ Biol Fish (2009) 86:193–201
DOI 10.1007/s10641-008-9388-9

Subspecies validity for genus *Culter* (Teleostei: Cyprinidae) based on morphometrics analysis

Xiao-Gu Zhang · Hui Zhang · Bang-Xi Xiong

Received: 19 January 2007 / Accepted: 24 July 2008 / Published online: 8 October 2008

Abstract The genus *Culter*, found only in East Asia (China, Vietnam, Korea and Russia), is a primary group of freshwater fishes in the subfamily Cultrinae (Cyprinidae). Of a total of 517 specimens studied, 235 were *Culter alburnus* from different localities and were used for morphometric analysis in this study. 17 landmarks were used in geometric morphometrics (GM) analysis of shapes. 13 traditional morphometrics (TM) and 21 truss-based morphometrics (TBM) characteristics were measured. The results of discriminant analysis and cluster analysis showed that four species-groups could be divided into separate groups. They were (1) *C. alburnus*+*C. recurviceps*, (2) *C. dabryi dabryi*+*C. oxycephalus*+*C. dabryi shinkainensis*, (3) *C. oxycephaloides*, (4) *C. mongolicus mongolicus*+*C. mongolicus elongates*+*C. mongolicus qionghaiensis*. The six populations of *C. alburnus* (Basilewsky, *Nouv Mém Soc Imp Nat Moscou*, 10:215–263, 1855), from Lake Poyanghu, River Songhuajiang, River Liaohe, River Jialingjiang, River Lijiang and Lake Taihu, should be classified as different geographical populations, but not subspecies. *C. mongolicus elongates* (He and Liu, *Zool Res*, 1(4):483–485, 1980) from Lake Chenghaihu and *C. mongolicus qionghaiensis* (Ding, *Acta Zootaxon Sin*, 15(2):246–250, 1990) from Lake Qionghaihu should be classified as subspecies. *C. dabryi shinkainensis* differed greatly from *C. dabryi dabryi* in body forms, and we suggest that *C. dabryi shinkainensis* should be changed to species, instead of subspecies, namely *Culter shinkainensis* (Yih and Chu, *Acta Hydrobiol Sin*, 2:170–199, 1959).

Keywords *Culter* · Geometric morphometrics · Discriminant analysis · Cluster analysis · Subspecies validation

X.-G. Zhang · B.-X. Xiong (✉)
College of Fisheries, Huazhong Agriculture University, Wuhan 430070, China
e-mail: bangxix8@mail.hzau.edu.cn

X.-G. Zhang · H. Zhang
College of Life Sciences, Jiujiang University, Jiujiang 332000, China

Introduction

The genus *Culter* is only found in East Asia (China, Vietnam, Korea and Russia). It is a primary group of freshwater fishes in the subfamily Cultrinae (Cyprinidae) (Chen 1998). Some species of the genus are found throughout China except for the Qinghai–Tibetan Plateau and Xinjiang. Ten species or subspecies have been described since the genus *Culter* was first described by Basilewsky (1855). Yih and Chu (1959) moved *C. alburnus* to the genus *Erythroculter*, and described *E. ilishaeformis* Bleeker 1871 (=*Culter alburnus*), *E. pseudobrevicauda* Nichols and Pope 1927, *E. mongolicus* Basilewsky 1855, *E. oxycephalus* Bleeker 1871, *E. oxycephaloides*

Kreyeuberg and Pappenheim 1908, *E. dabryi* Bleeker 1871, *E. ilishaeformis sungarinensis* Yih and Chu 1959, and *E. dabryi shinkainensis* Yih and Chu 1959. They considered that the proportion between body depth and standard length for *E. ilishaeformis sungarinensis* from a section of the river Songhuajiang (i.e., from Zhaoyuan to Yilan) is more than 24%, but the proportion for *E. ilishaeformis* is less than 23%, and that body depth behind head of *E. dabryi shinkainensis* from Lake Xingkaihu is deeper than for *E. dabryi* Bleeker and the body depth and head length are nearly equal. *E. mongolicus elongates* (He and Liu 1980) from Lake Chenghai differs from *E. mongolicus* in having more gill rakers (22–23 versus 17–20), a slender body (standard length 4.5–5.5 versus 3.5–4.3 times the body depth) and shorter caudal peduncle (standard length 6.2–7.4 versus 5.7–6.4 times the caudal peduncle). *E. mongolicus qionghaiensis* (Ding 1990) from Lake Qionghai has most gill rakers (25–28), medium body (standard length 4.3–4.9 times the body depth), shorter caudal peduncle (standard length 6.6–8.5 times the caudal peduncle) and bigger eye (5.13–6.21 times orbital diameter in head length). Luo (1994) re-assigned the genus *Erythroculter* into *Culter*, and eliminated *E. ilishaeformis sungarinensis* as a taxon. In recent years, the systematics of the genus *Culter* has remained problematic. The points of contention are mainly regarding the species or subspecies status, namely validity of *E. ilishaeformis sungarinensis* as a subspecies and whether *C. mongolicus elongates*, *C. mongolicus qionghaiensis* or *C. dabryi shinkainensis* should be elevated to the rank of species.

The landmark-based morphometric analysis has been commonly used in studying shapes in many areas of biological research since geometric morphometrics was proposed. The geometric morphometric methods are increasingly used in fish classification (Hard et al. 2000; Kassam et al. 2002; Parsons et al. 2003; Trapani 2003; Johansson et al. 2006; Costa and Cataudella 2007). Furthermore, multivariate methods are more effective in capturing information about the differences of shape. More recent advances are in the field of image processing techniques, more comprehensive and precise data collection, and new analytical tools, leading to better interpretation of morphometric variables (Cadrin 2000; Adams et al. 2004).

In this paper, the genus *Culter* was studied using geometric morphometrics with discriminant analysis and cluster analysis to investigate morphometric differences among species, subspecies or populations. The validity of the subspecies is also discussed.

Materials and methods

The shape data was collected from a total of 517 specimens, of which 352 were from specimens collected in commercial fisheries from July 2005 to July 2006 in Poyanghu Lake, Jiangxi and 165 preserved in 5% formalin solution were obtained from the Museum of Aquatic Organisms, Institute of Hydrobiology, Chinese Academy of Sciences, and the Collection of Fishes, College of Life Sciences, Nanchang University. The specimens were grouped into nine species or subspecies based on phenotype characters (Chen 1998). The sampling locations and relevant biological data are shown in Table 1 and Fig. 1.

The specimens were placed on their right side on an enamel plate, the body posture and fins were forced into a natural position, and photographs were taken using a digital camera. Seventeen landmarks were used in geometric morphometrics (GM) analysis of shape. Landmarks were: (A) anterior tip of snout on upper jaw; (B) most dorsal point of head depth at center of eye; (C) most posterior point of neurocranium; (D) most dorsal point of head depth at the most posterior point of the operculum; (E) origin of dorsal fin; (F) insertion of dorsal fin; (G) anterior attachment of dorsal membrane from caudal fin; (H) anterior attachment of ventral membrane from caudal fin; (I) insertion of anal fin; (J) origin of anal fin; (K) origin of ventral fin; (L) most ventral point of head depth at the most posterior point of operculum; (M) end of operculum membrane ventrally; (N) anterior tip of body; (O) center of the eye; (P) most posterior aspect of operculum membrane; and (Q) origin of pectoral fin (Fig. 2).

We generated a Thin-plate Spline (TPS) file, from pictures of all specimens using Rohlf's Tps Utility Program Version 1.38. The x and y coordinates of each landmark were collected from every specimen by TpsDig Program Version 2.05 and if sample forms were reasonably comparable they were tested using TpsSmall Program Version 1.20. Landmark coordinates were adjusted using a generalized least squares Procrustes superimposition (GLS). Procrustes superimpositions were computed to average configuration

Table 1 Localities, numbers and total length of specimens studied

Sampling serial	Species or subspecies	Sampling locality	Number of specimens	Total length (mm)	
				Min.–Max.	Mean±SD
1	*Culter alburnus*	Lake Poyanghu	202	135.05–427.50	229.097±74.995
2		River Songhuajiang	12	127.53–419.44	254.859±89.026
3		River Liaohe	11	80.48–232.03	140.082±42.089
4		River Jialingjiang	5	91.44–231.66	178.710±56.376
5		River Lijiang	3	251.00–309.00	279.863±29.001
6		Lake Taihu	2	189.99–234.12	212.055±31.205
7	*C. pseudobrevicauda*	Hainan, Guangxi, Guangdong, Yuannan	14	119.83–369.16	261.516±89.634
8	*C. mongolicus*	Lake Poyanghu	101	106.42–374.53	245.563±65.315
9	*C. mongolicus elongates*	Lake Chenghaihu	10	124.34–322.29	210.214±66.416
10	*C. mongolicus qionghaiensis*	Lake Qionghaihu	20	97.49–486.13	182.945±99.949
11	*C. dabryi*	Jiangxi, Jiangsu, Hubei	77	130.23–359.63	246.147±61.086
12	*C. dabryi shinkainensis*	Lake Xingkaihu	5	248.47–331.71	282.500±33.651
13	*C. oxycephalus*	Hubei, Hunan, Jiangxi, Anhui, Heilongjiang	19	91.82–391.89	278.574±78.334
14	*C. oxycephaloides*	Sichuan, Henan, Hubei, Jiangxi, Zhejiang	36	88.06–403.89	248.441±90.730

of landmarks and relied on translation, scaling, and rotation to remove all information unrelated to shape (Rohlf and Slich 1990; Bookstein 1991; Zelditch et al. 2004). Then we estimated a consensus form composed of the mean coordinates for each landmark averaged across all specimens, deformed and warped the form to the consensus configuration to each specimen and assessed their partial warps through Relative Warps Program Version 1.44. The scores of the partial warp containing more shape information than those of the linear trusses were transformed into relative warp scores saved as a data file.

The data obtained through TPS method were analyzed using multivariate discriminant function analysis (DFA; STATISTICA Version 6.0). The differences among the species or subspecies were tested by multivariate analysis of covariance (MANCOVA) according to standard length of the fish as a covariate. Morphological canonical scores for individual specimens were calculated on each of these axes as roots of scatter plots of scores that summarize shape variation associated with that axis. Using the group means of canonical variable coefficients, we could infer the relationships among these groups from different samples by the nearest neighbor rule of tree clustering analysis (CA). Furthermore, using Thin-plate Spline Program Version 1.20, we deformed grids depicting shape differences. TPS programs used in the analysis were downloaded from the State University of New York at Stony Brook morphometrics website (http://life.bio.sunysb.edu/morph/inedx.html).

The traditional morphometrics (TM) characteristics measured included standard length (SL), body depth (BD), head length (HL), head depth at most posterior point of eye (HD1), head depth at most posterior point of the neurocranium (HD2), head depth at most posterior the of gill cover bones (HD3), snout length (from top of body to most posterior of gill cover bones) (SnL), orbital diameter (OD), interorbital width (IoD), caudal length (CaL), caudal depth (CaD), dorsal fin length (DL), pectoral fin length (PL), and ventral length (VL). The truss-based morphometrics (TBM) measurements based on a truss network protocol were anchored at 10 homologous anatomical landmarks. Landmarks were referred to as: A, C, E, F, G, H, I, J, K, and Q (Fig. 2). Except for interorbital width measured with Vernier calipers, morphometric measurements were made using Motic Images Advanced Version 3.2, with all data taken to the nearest 0.01 mm.

The individual size effect was removed from the data. The transformations used were based on the formula $M_{adj}=M/SL$. Then, multivariate analyses of variance were carried out to test the significance of

Fig. 1 Localities of samples used in the present study

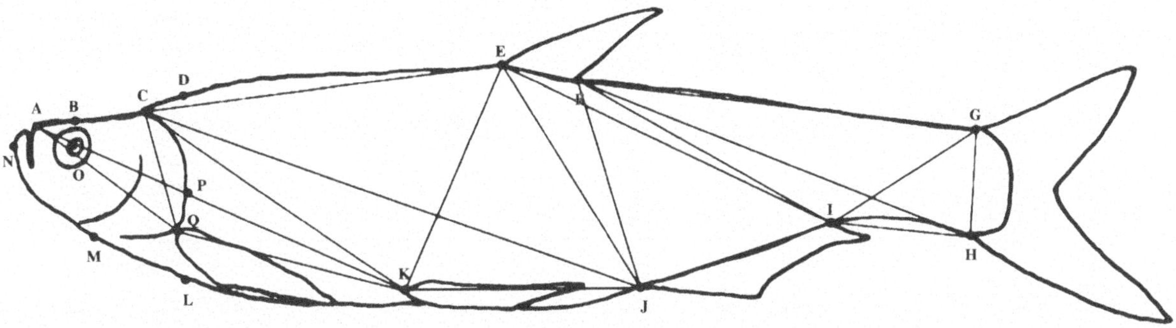

Fig. 2 Outline drawing of *Culter alburnus*, showing the locations of the 17 anatomical landmarks and 21 truss-based characteristics for each specimen

morphological differences, and the adjusted data were submitted to discriminant function analysis (DFA) and cluster analysis (CA) using STATISTICA Version 6.0.

Results

The discriminant function analysis for morphometric characters of the genus *Culter* revealed that the overall percentages of correct classification for all samples using GM, TM and TBM methods were 96.91%, 90.52% and 91.10%, respectively (Table 2).

By using GM, the means of 14 species, subspecies or populations was found to differ significantly (MANCOVA, Wilk's Lambda=0.314, $F[30, 390]=34.382$, $p<0.001$). The first four discriminant functions (DFs) accounted for 47.3%, 30.6%, 7.9%, 5.4%, respectively, of variances inherent in the original data, with a cumulating proportion to 91.2%. UPGMA tree-clustering graph (Fig. 3) by Euclidean distances showed that six geographical groups of *C. alburnus* from the rivers Songhuajiang, Liaohe, and Jialingjiang, and Lake Taihu, River Lijiang, and Lake Poyanghu were clustered successively together. Euclidean distances among them did not exceed 3.0 except for the distance between the two groups from Lake Poyanghu and River Lijiang.

The scatterplot graph of canonical scores (Fig. 4) showed that four species-groups could be clustered. The same outcome could be inferred from the tree-clustering graph (Fig. 3). They were:

1. *C. alburnus*+*C. recurviceps*
2. *C. dabryi dabryi*+*C. oxycephalus*+*C. dabryi shinkainensis*
3. *C. oxycephaloides*
4. *C. mongolicus mongolicus*+*C. mongolicus elongates*+*C. mongolicus qionghaiensis*

In the first species-group, two species had significant upward mouth and head tips (Fig. 5a). In the posterior–ventral region of head, *C. alburnus*, had a slighter longer length, while *C. recurviceps* had significantly shallow body behind head with a deep and long head. In the second group, *C. dabryi dabryi* had slight downward mouth and head tips with a bending upward in the region near to the ventral fin. *C. oxycephalus* had a smaller amount of upward extension than *C. dabryi dabryi* in the occipital

Table 2 Correct percent for all samples of the discriminant functions analysis for the genus *Culter* using GM, TM and TBM

Method	1	2	3	4	5	6	7	8	9	10	11	12	13	14	Total
GM	100.00	83.33	100.00	100.00	100.00	100.00	92.86	99.01	90.00	95.00	98.70	100.00	52.63	100.00	96.91
TM	99.01	75.00	54.55	80.00	0.00	50.00	57.14	93.07	90.00	80.00	90.91	100.00	63.16	94.44	90.52
TBM	100.00	66.67	72.73	40.00	100.00	100.00	85.71	94.06	70.00	95.00	87.01	100.00	42.11	91.67	91.10

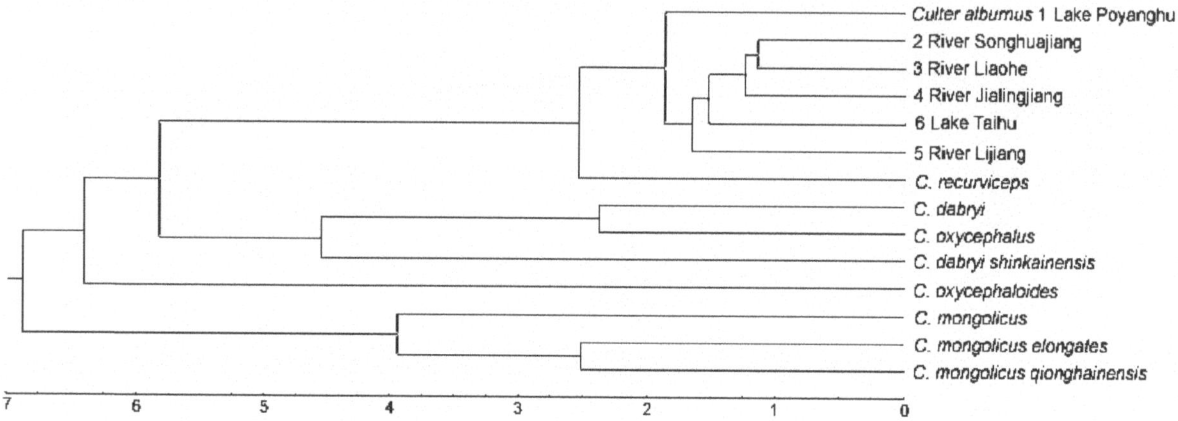

Fig. 3 UPGMA dendrogram constructed from Euclidean distances for the genus *Culter*

region. In the dorsal area behind the head, *C. dabryi shinkainensis* had distinct upward shape (Fig. 5b). *C. oxycephaloides* and the second species-group are similar in form, but the head in *C. oxycephaloides* is flat and slopes downward in the dorsal region (Fig. 5c). *C. mongolicus elongates* and *C. mongolicus qionghaiensis* are more slender than *C. mongolicus mongolicus*, and *C. mongolicus qionghaiensis* has a longer and deeper head than *C. mongolicus elongates* (Fig. 5d).

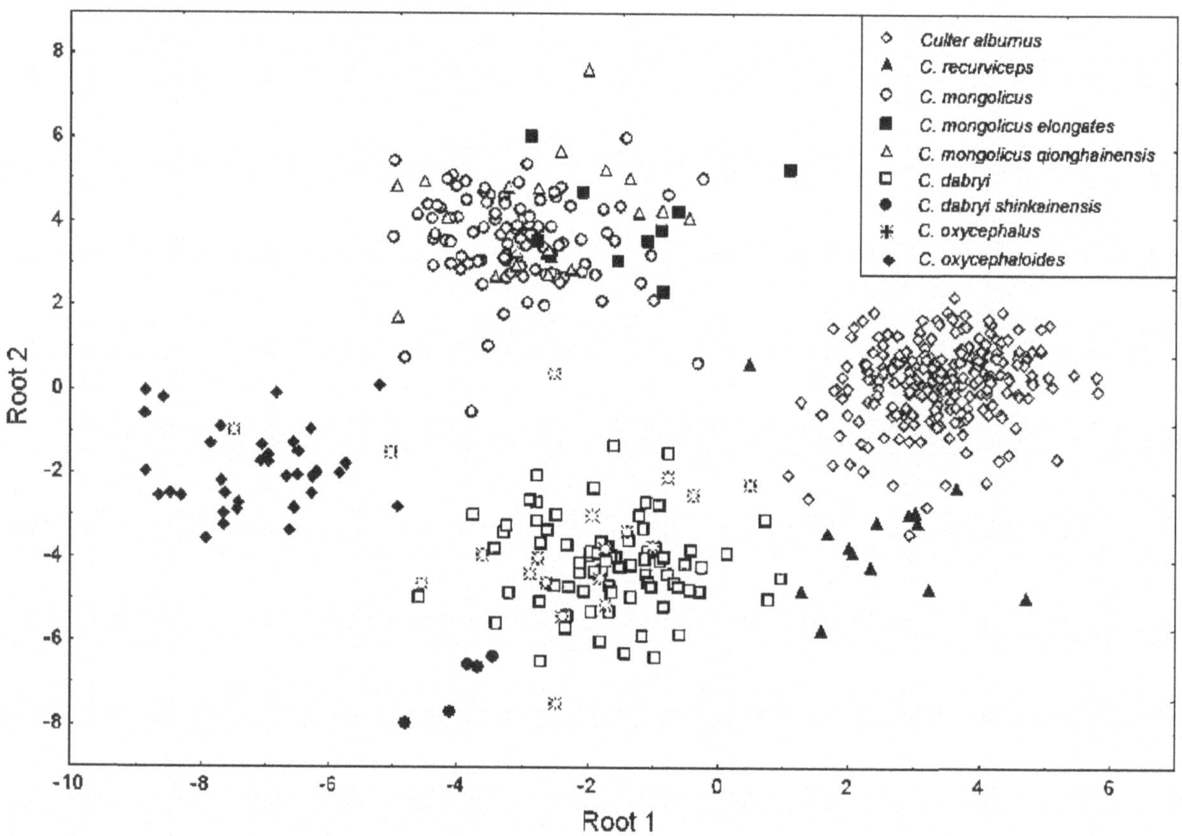

Fig. 4 Scatterplots of canonical scores for the genus *Culter*

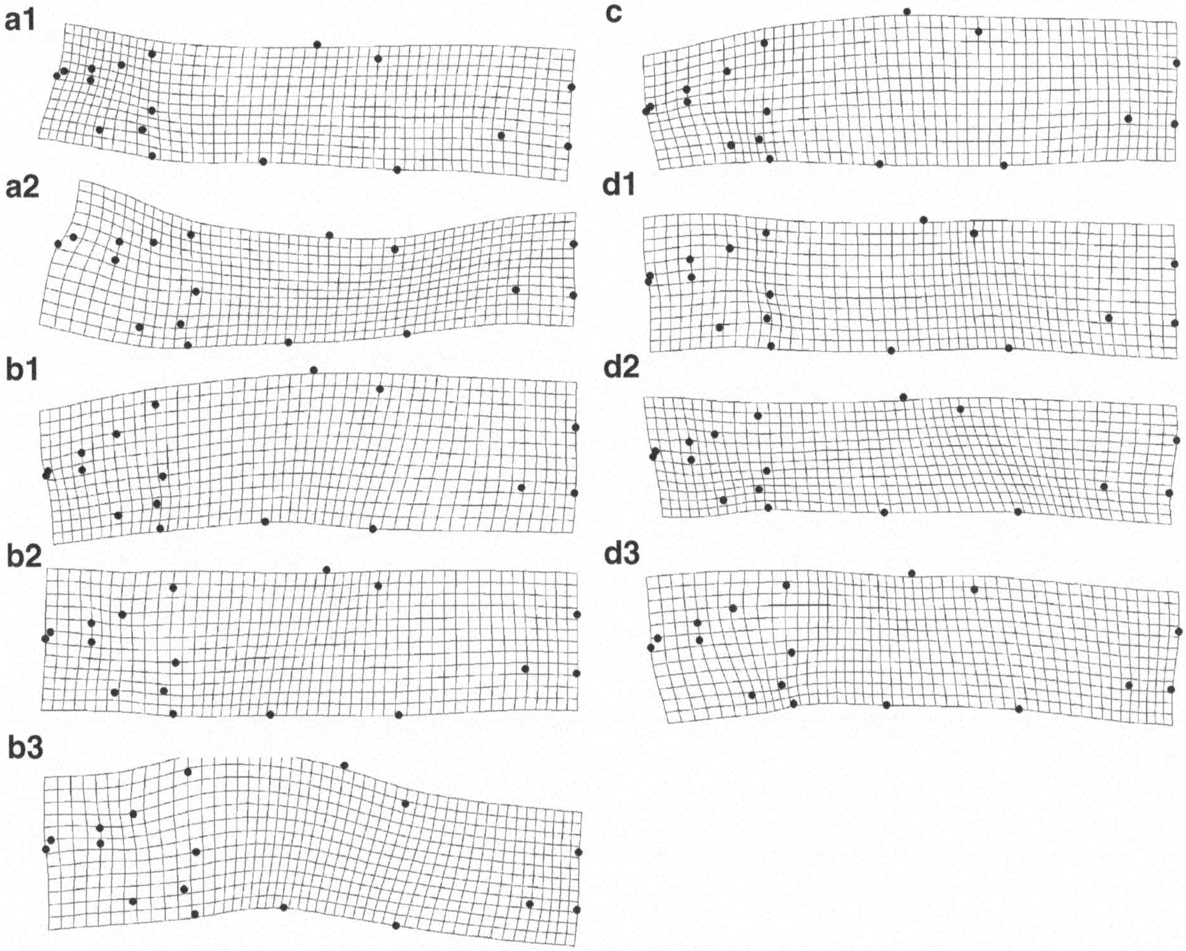

Fig. 5 Deformation grids depicting shape differences using thin-plate splines. **a**–**d** indicate from first to fourth species-group: **a** *1 Culter alburnus*; *2 C. recurviceps*; **b** *1 C. dabryi*; *2 C. oxycephalus*; *3 C. dabryi shinkainensis*; **c** *C. oxycephaloides*; **d** *1 C. mongolicus*; *2 C. mongolicus elongates*; *3 C. mongolicus qionghainensis*

Discussion

This study shows the potential power of GM for the identification of species, subspecies or populations. The means of 14 species, subspecies or populations of the genus *Culter* differ significantly and total percentage when assigning correctly each specimen to the groups based on the discriminating function analysis of body shape is 96.91% (16 out of 517 specimens were misclassified). However, using the TM or TBM methods, the percentages tended to be lower, especially for the samples of *C. alburnus* from the rivers Liaohe, Jialingjiang, and Lijiang and Lake Taihu for *C. pseudobrevicauda* and *C. oxycephalus*. The correct percentage was elevated from 87.23% to 90.52% by adding measurements for HD1 and HD3. Using geometric morphometrics we can determine the shape variables that can be best discriminated, and then translate them into terms of traditional morphometric variables that can be measured with calipers or rulers (Zelditch et al. 2004).

Euclidean distances among fourteen species, subspecies or populations were computed from means of canonical variables. This method, which is commonly applied to type of distance, has certain advantages (e.g., the distance between any two objects is not affected by the addition of new objects to the analysis). Scatterplots of canonical scores (Fig. 4) and UPGMA dendrogram (Fig. 3) showed there are four species-groups, namely the group of *C. alburnus* +*C. recurviceps*, the group of *C. mongolicus* and the group of *C. dabryi*+*C. oxycephalus*, and the group of *C. oxycephaloides*. Furthermore, the group of *C. alburnus* and *C. recurviceps* is close to the group *C. dabryi*+*C. oxycephalus*, while the group of *C. mongolicus* is relatively distinct. The results of the present study, combined with the analyses of body coloration and meristic characters (notably the coloration of the fins, number of scales on the lateral line and number of gill rakes) suggest that *C. alburnus*, *C. recurviceps*, *C. mongolicus*, *C. dabryi*, *C. oxycephalus* and *C. oxycephaloides* are an independent species, and *C. mongolicus elongates* and *C. mongolicus qionghaiensis* should be ranked as subspecies. Though as many as 7 out of 19 specimens of *C. oxycephalus* were misclassified as *C. dabryi*, their characters, such as nacarat caudal fin with black edge on *C. oxycephalus* vs. black caudal fin on *C. dabryi*, were enough to warrant judging them to be different species (Chen 1998). However, based on the fact that *C. dabryi shinkainensis* in the second species-group is different from *C. dabryi dabryi* in body shape, we suggest that *C. dabryi shinkainensis* should be ranked as species, namely *Culter shinkainensis* Yih and Chu 1959.

This study showed evidence of significant morphometric identity among the six populations of *C. alburnus*. The finding that they should be different allopatric populations but not subspecies is consistent with the conclusion derived in a previous study (Luo 1994), but does not agree with Yih's view (in Yih and Chu 1959) that classified the population of *C. alburnus* from River Songhuajiang, from Zhaoyuan to Yilan, as *E. ilishaeformis sungarinensis* (*E. ilishaeformis*=*C. alburnus*). For this reason, we think that differences in the body shapes of *C. alburnus* from different river systems, such as from the rivers Changjiang, Zhujiang, and Heilongjiang, reflect the differences among the allopatric populations. These differences might be related to phenotypic plasticity in *C. alburnus*. This issue should be further explored using molecular genetics. The morphometic comparison carried out here revealed that differences were larger among three populations of *C. mongolicus* from lakes Chenghaihu, Qinghaihu and Poyanghu. These geographical regions may have local ecological characteristics that lead *C. mongolicus* to develop phenotypic specialized characteristics.

The results presented here as morphometric differences among species or subspecies of the genus *Culter* in the head and body, such as the position of the upper jaw, the tip of snout, the most dorsal point of head depth at center of eye or at most posterior of operculum (Fig. 5), are consistent with the characteristics described in earlier research (Yih and Chu 1959; Luo 1994). For instance, if the head is flat or not, if the profile behind head rises or not, if head length is less than body depth or not, etc.

In conclusion, there are nine species or subspecies in the genus *Culter* based upon this morphological analysis. *C. dabryi shinkainensis* from Lake Xingkaihu should be changed to species from subspecies rank, and the population of *C. alburnus* from River Songhuajiang could not be classified as subspecies.

Acknowledgements We would like to thank Y. Chen of School of Marine Science, University of Maine in USA, H.Z. Liu of Institute of Hydrobiology, Chinese Academy of Sciences and two anonymous reviewers for providing valuable comments on our manuscript. We thank SP He and EH Feng of the Institute of Hydrobiology, Chinese Academy of Sciences and ZQ Wu of the College of Life Sciences, Nanchang University in China for assistance in taking photographs of specimens. This research was supported by grants from the Department of Education of Jiangxi Province of China (JYTKJ0201).

References

Adams DC, Rohlf FJ, Slice DE (2004) Geometric morphometrics: ten years of progress following the 'Revolution'. Ital J Zool (Modena) 71:5–16

Basilewsky S (1855) Ichthyographia chinae borelis. Nouv Mém Soc Imp Nat Moscou 10:215–263

Bookstein FL (1991) Morphometric tools for landmark data: geometry and biology. Cambridge University Press, Cambridge

Cadrin SX (2000) Advances in morphometric identification of fishery stocks. Rev Fish Biol Fish 10:91–112 doi:10.1023/A:1008939104413

Chen YY (1998) Fauna sinica (Osteichthyes: Cypriniformes II). Science, Beijing, pp 182–197 (in Chinese)

Costa C, Cataudella S (2007) Relationship between shape and trophic ecology of selected species of Sparids of the Caprolace coastal lagoon (Central Tyrrhenian sea). Environ Biol Fishes 78:115–123 doi:10.1007/s10641-006-9081-9

Ding RH (1990) On a new subspecies of genus *Erythroculter* from Qionghai Lake, Sichuan, China (Cypiniformes: Cyprinidae). Acta Zootaxon Sin 15(2):246–250 (in Chinese with English abstract)

Hard JJ, Berejikian BA, Tezak EP, Schroder SL, Knudsen CM, Parker LT (2000) Evidence for morphometric differentiation of wild and captively reared adult coho salmon: a geometric analysis. Environ Biol Fishes 58:61–73 doi:10.1023/A:1007646332666

He JC, Liu ZH (1980) Description of a new subspecies of *Erythroculter* from Yunnan, China. Zool Res 1(4):483–485 (in Chinese with English abstract)

Johansson F, Radman P, Andersson J (2006) The relationship between ontogeny, morphology, and diet in the Chinese hook snout carp (*Opsariichthys bidens*). Ichthyol Res 53:63–69 doi:10.1007/s10228-005-0316-0

Kassam DD, Sato T, Yamaoka K (2002) Landmark-based morphometric analysis of the body shape of two sympatric species, *Ctenopharynx pictus* and *Otopharynx* sp. "heterodon nankhumba" (Teleostei: Cichlidae), from Lake Malawi. Ichthyol Res 49:340–345 doi:10.1007/s102280200050

Luo YL (1994) Some clarifications on the Cultrinae fishes of China. Acta Hydrobiologica Sin 18(1):45–49 (in Chinese with English abstract)

Parsons KJ, Robinson BW, Hrbek T (2003) Getting into shape: an empirical comparison of traditional truss-based morphometric methods with a newer geometric method applied to New World cichlids. Environ Biol Fishes 67:417–431 doi:10.1023/A:1025895317253

Rohlf FJ, Slich D (1990) Extensions of the procrustes method for the optimal superimposition of landmarks. Syst Zool 39:40–59 doi:10.2307/2992207

Trapani J (2003) Geometric morphometric analysis of body-form variability in *Cichlasoma minckleyi*, the Cuatro Cienegas cichlid. Environ Biol Fishes 68:357–369 doi:10.1023/B:EBFI.0000005763.96260.2a

Yih BL, Chu ZR (1959) К ИЗУЧЕНИЮ РОДОВ *CULTER* И *ERYTHROCULTER* В КИТАЕ. Acta Hydrobiol Sin 2:170–199 (in Chinese with Russian abstract)

Zelditch ML, Swiderski DL, Sheets HD, Fink WL (2004) Geometric morphometrics for biologists: a primer. Elsevier, San Diego, CA, pp 113–119

Environ Biol Fish (2009) 86:203–209
DOI 10.1007/s10641-008-9338-6

A new cavefish species, *Sinocyclocheilus brevibarbatus* (Teleostei: Cypriniformes: Cyprinidae), from Guangxi, China

Yahui Zhao · Jiahu Lan · Chunguang Zhang

Received: 21 September 2007 / Accepted: 28 January 2008 / Published online: 21 February 2008

Abstract A new species of the cavefish genus *Sinocyclocheilus* is described based on six specimens collected from a subterranean river in Du'an County in the Guangxi Zhuang Autonomous Region of southern China. *Sinocyclocheilus brevibarbatus* sp. nov. can be distinguished from all congeners by the following combination of characteristics: a completely scaled body with well-developed eyes and a strongly-humped back; an almost straight lateral line possessing 49–51 perforated scales; seven soft rays on the dorsal fin; a relatively large eyeball (3.3–5.7% of SL) and short maxillary and rictal barbels (1.7–5.7% and 4.5–7.0% of SL, respectively).

Keywords *Sinocyclocheilus* · Cyprinidae · New species · China

Introduction

The species diversity of cavefishes is very rich in China. Most Chinese cavefishes belong to two genera, *Sinocyclocheilus* (Cyprinidae) and *Triplophysa* (Balitoridae) (Romero and Paulson 2001; Zhao and Zhang 2006). The genus *Sinocyclocheilus*, established by Fang in 1936, is endemic to China, and only distributed in a narrow karstric area of the Yun-gui Plateau and adjacent regions in southern China, including southern Guizhou, eastern Yunnan and northern Guangxi Provinces. At present, more than 50 *Sinocyclocheilus* species are recognized, making the genus one of the most species-rich genera of cyprinid fishes in China. Interestingly, almost all known species of *Sinocyclocheilus* are cave dwellers, although some species do not present typical characters of troglobitic fishes.

In 2001, six specimens of a thus far undescribed population of a cavefish were collected from an underground river in the northern part of the Guangxi Zhuang Autonomous Region in southern China. After comparison to specimens of all closely-related species, it was determined that the six new specimens should be recognized as a new species belonging to the genus *Sinocyclocheilus*.

Materials and methods

Type and comparative specimens for this study are deposited in the Animal Museum of the Institute of Zoology, the Chinese Academy of Sciences (ASIZB, institutional abbreviation is according to the list in Leviton et al. 1985). Specimens of the following

Y. Zhao · C. Zhang (✉)
Institute of Zoology, Chinese Academy of Sciences,
Beijing 100101, People's Republic of China
e-mail: fish@ioz.ac.cn

J. Lan
Aquatic Bureau of Du'an,
Anyang Town, Du'an County,
Guangxi 530700, People's Republic of China

species, which are most similar to the new species, were examined for comparison: *Sinocyclocheilus microphthalmus* Li, 1989, *Sinocyclocheilus altishoulderus* (Li and Lan 2002), and *Sinocyclocheilus jiuxuensis* (Li and Lan 2003). Information on specimen identities is given in the "Comparative materials" paragraph.

Measurements were taken point to point with a digital calliper to 0.1 mm precision. Morphometric and meristic characters were selected according to the methods described in Zhao et al. (2006). Osteological characters were observed on soft X-ray photographs following the methods described in Zhao et al. (2002, 2006). Vertebrae were counted excluding the Weberian complex. Morphometric and meristic information was analyzed using SYSTAT Version 10 (Wilkinson 2001).

Sinocyclocheilus brevibarbatus sp. nov. (Fig. 1; Table 1).

Holotype ASIZB 74228, 131.8 mm standard length (SL), Zhujiang (Pearl) River Basin: Hongshuihe River, Gaoling Town, Du'an County, Guangxi, P.R. China; 4 June 2001, J. H. Lan.

Paratypes ASIZB 74229, 69.8 mm SL, data as for holotype; ASIZB 93900–93904 (five specimens), 125.7–168.3 mm SL, Locality as for hototype, J. H. Lan.

Diagnosis The new species is distinguished from all congeners by the following combination of characteristics: Body completely scaled excluding head; back strongly humped; eye well-developed; lateral line almost straight, possessing 49–51 perforated scales; seven soft rays on dorsal fin; relatively large eye-ball (3.3–5.7% of SL) and short maxillary and rictal barbels (1.7–5.7% and 4.5–7.0% of SL, respectively).

Fig. 1 *Sinocyclocheilus brevibarbatus* sp. nov., holotype, ASIZB 74229, 69.8 mm SL

The new species is most similar to *S. microphthalmus*, *S. altishoulderus*, and *S. jiuxuensis* in terms its strongly-humped back and general body shape. Additionally, the distribution areas of these four species are also close to each other (Fig. 2). But *S. microphthalmus* and *S. altishoulderus* have eight soft dorsal rays whereas *S. brevibarbatus* sp. nov. has seven. *S. jiuxuensis* has a smaller eye-ball (2.3–3.3% of SL) and longer maxillary and rictal barbels (5.1–11.0% and 8.0–11.3% of SL, respectively) than *S. brevibarbatus*.

Description General body features are shown in Fig. 1. Meristics and proportional measurements as percentages of standard length are listed in Table 1.

Body compressed. Dorsal profile deeply convex while ventral profile slightly concave, tapering gradually toward anal fin. The greatest body depth in front of, not exactly at dorsal fin insertion.

Head rather small and compressed, relatively depressed in front of eyes. Snout duckbilled with a node on top, anterior tip rounded. Nostrils close to snout tip and one third of distance from snout tip to anterior margin of orbit. Anterior nostrils possessing a rim with a fleshy flap forming a complete tube; posterior nostril elliptical. Mouth curved and slightly inferior, with a projecting upper jaw. Two pairs of barbels; maxillary barbel insertion in front of anterior nostril, barbels short, not extending beyond posterior nostril; rictal barbels short, not extending beyond vertical through posterior margin of eye. Eye moderate, rounded. Gill opening large, operculum membrane not connected at isthmus. Joint of dentary-angulars close to each other at isthmus. Gill rakers triangular, well-developed, eight to nine in first gill arch (epibranchial with two, one specimen three, ceratobranchial with six).

Pectoral-fin insertion in vertical through posterior margin of operculum; pectoral-fin long, developed, extending beyond pelvic fin insertion. Pelvic fin insertion midway between pectoral-fin insertion and anus; pelvic fin long, almost reaching to anus (three-fourths of distance from pelvic fin insertion to anal fin insertion). Dorsal fin origin midway between snout tip and caudal-fin base, slightly closer to caudal fin, posterior to pelvic fin insertion. Last unbranched ray

Table 1 Meristics and morphometrics of *Sinocyclocheilus brevibarbatus* sp. nov. and the related species *S. jiuxuensis*, *S. altishoulderus* and *S. microphthalums*

	Sinocyclocheilus brevibarbatus sp. nov.					*Sinocyclocheilus jiuxuensis*					*Sinocyclocheilus altishoulderus*					*Sinocyclocheilus microphthalmus*				
	n	Min	Max	Mean	SD	*n*	Min	Max	Mean	SD	*n*	Min	Max	Mean	SD	*n*	Min	Max	Mean	SD
Dorsal fin	6	7	7	7		6	7	7	7		10	8	8	8		14	8	8	8	
Anal fin	6	5	5	5		6	5	5	5		10	5	5	5		14	5	5	5	
Pectoral fin	6	14	15	15		6	13	15	15		10	13	16	15		14	12	12	12	
Pelvic fin	6	8	9	9		6	7	8	8		10	7	8	7		14	7	7	7	
Total vertebrae	4	37	38	38		4	34	37	36											
Predorsal vertebrae	4	6	7	7		4	7	7	7											
Caudal vertebrae	4	20	21	21		4	19	21	20											
Gill rakers	6	8	9	8		6	8	10	9		10	9	10	9		14	9	12	11	
Lateral line scales	6	49	51	50		6	42	51	46		10	42	52	48		14	48	57	52	
Standard length	6	69.8	168.3	132.9	34.5	5	78.2	136.1	116.1	22.2	10	71.9	108.5	88.2	13.3	12	78.2	189.7	141.3	37.0
In % of standard length																				
Body depth	6	28.1	36.8	32.6	3.2	5	29.0	36.5	33.2	2.8	10	31.3	34.8	33.1	1.3	12	23.5	38.7	34.2	4.4
Predorsal length	6	58.5	62.1	60.2	1.4	5	59.1	62.0	60.1	1.1	10	54.9	63.3	59.2	2.5	12	51.6	57.4	53.6	1.8
Dorsal-fin base length	6	13.3	14.9	14.0	0.6	5	14.1	17.4	15.3	1.3	10	12.9	16.0	14.8	1.1	12	14.8	19.7	17.5	1.7
Dorsal fin length	6	20.0	24.7	21.9	1.7	5	23.2	24.5	23.8	0.5	10	22.0	29.2	24.6	2.0	12	22.2	29.8	25.9	2.3
Preanal length	6	72.6	77.0	74.0	1.6	5	68.9	76.8	72.2	3.1	10	69.8	75.5	72.9	1.9	12	59.8	73.1	69.1	3.5
Anal-fin base length	6	6.8	9.7	8.6	1.0	5	8.7	9.9	9.6	0.5	10	7.8	10.2	8.8	0.8	12	8.4	11.8	10.1	0.8
Anal fin length	6	13.8	19.2	17.0	2.2	5	18.5	21.6	20.2	1.1	10	18.3	23.1	20.7	1.5	12	18.2	24.0	21.1	1.9
Prepectoral length	6	30.9	34.2	32.1	1.2	5	29.5	32.8	30.9	1.3	10	27.1	32.2	29.8	1.8	12	24.6	29.6	26.4	1.6
Pectoral-fin base length	6	3.7	4.7	4.1	0.4	5	3.9	5.0	4.6	0.5	10	4.2	5.4	4.7	0.4	12	4.0	5.2	4.6	0.4
Pectoral fin length	6	22.2	28.0	25.4	1.9	5	25.6	28.2	26.8	1.3	10	25.3	28.8	27.2	1.2	12	25.5	32.9	28.4	2.3
Prepelvic length	6	48.1	51.3	50.0	1.2	5	48.1	52.6	50.3	1.7	10	47.8	51.8	49.5	1.2	12	25.8	48.3	44.1	5.9
Pelvic-fin base length	6	3.9	5.3	4.5	0.5	5	3.7	5.4	4.7	0.6	10	4.1	6.2	4.9	0.7	12	4.3	6.2	5.3	0.6
Pelvic fin length	6	18.7	20.7	19.5	0.7	5	17.1	21.3	20.0	1.7	10	18.7	23.2	20.4	1.3	12	19.7	26.1	23.0	2.0

Table 1 (continued)

	Sinocyclocheilus brevibarbatus sp. nov.					*Sinocyclocheilus jiuxuensis*					*Sinocyclocheilus altishoulderus*					*Sinocyclocheilus microphthalmus*				
	n	Min	Max	Mean	SD	*n*	Min	Max	Mean	SD	*n*	Min	Max	Mean	SD	*n*	Min	Max	Mean	SD
Caudal peduncle length	6	19.1	21.1	19.9	0.8	5	19.1	23.8	21.1	1.7	10	17.7	21.8	20.1	1.2	12	19.4	24.7	22.4	1.7
Caudal peduncle depth	6	11.1	13.5	12.5	0.9	5	11.1	13.6	12.2	1.0	10	11.5	14.2	12.8	1.0	12	10.6	14.5	13.4	1.2
Head length	6	30.5	34.1	31.5	1.3	5	29.6	32.4	30.9	1.1	10	27.8	32.6	30.7	1.6	12	24.8	30.7	26.6	1.9
Head depth	6	14.2	17.4	16.2	1.1	5	14.9	16.2	15.5	0.5	10	15.9	18.4	16.8	0.8	12	14.2	17.0	15.7	0.9
Head width	6	12.4	15.1	14.2	1.0	5	12.0	15.1	13.0	1.3	10	13.1	14.9	14.0	0.5	12	12.2	15.3	14.0	1.1
Snout length	6	9.1	10.1	9.5	0.4	5	9.1	10.4	9.7	0.5	10	8.1	12.0	9.4	1.1	12	7.6	10.3	8.8	0.7
Eye diameter	6	8.3	10.3	9.1	0.8	5	7.6	8.7	8.0	0.5	10	4.5	9.4	7.9	1.4	12	5.3	8.3	6.6	1.0
Eyeball diameter	6	3.3	5.7	4.1	0.9	5	2.3	3.3	2.8	0.4	10	1.9	4.7	3.3	0.9	12	0.8	3.0	2.0	0.6
Interorbital width	6	5.9	6.8	6.3	0.4	5	5.7	6.7	6.1	0.4	10	5.1	6.7	6.2	0.5	12	5.4	7.2	6.3	0.5
Prenostril length	6	4.9	5.7	5.2	0.3	5	4.3	5.1	4.6	0.3	10	3.7	5.6	4.7	0.6	12	4.1	5.4	4.7	0.5
Width between posterior nostrils	6	4.8	6.0	5.4	0.4	5	4.6	5.9	5.2	0.6	10	4.5	6.2	5.4	0.6	12	4.6	6.8	5.6	0.6
Upper jaw length	6	7.4	8.7	8.0	0.5	5	5.6	7.3	6.1	0.7	10	5.0	6.7	6.0	0.6	12	4.7	6.2	5.6	0.5
Lower jaw length	6	7.0	8.7	7.6	0.6	5	4.6	7.0	5.6	1.0	10	4.2	6.5	5.5	0.8	12	4.2	5.7	4.9	0.4
Mouth width	6	6.8	8.4	7.4	0.6	5	5.5	6.9	6.0	0.6	10	5.1	8.9	6.3	1.2	12	4.3	7.7	6.5	0.9
Maxilla barbel length	6	1.7	5.7	4.2	1.4	5	5.1	10.9	8.0	2.1	10	7.0	12.2	9.3	1.7	12	6.7	17.7	11.7	3.1
Rictal barbel length	6	4.5	7.0	5.6	1.0	5	8.0	11.3	9.4	1.2	10	6.3	11.5	9.4	1.7	12	9.0	17.4	12.6	2.4

of dorsal fin hard at base, softening toward tip, with serrations along posterior edge. Anal fin insertion midway between pelvic fin origin and caudal fin base, anal fin relatively long. Caudal fin bifurcate.

Lateral line almost straight from operculum to the caudal fin base. Body covered by moderately-small scales; lateral-line scales perforated and complete, larger than ones above and below lateral line. Lateral line perforated scale counts: 49(three specimens), 50 (one) and 51(two); scale row counts above lateral line: 14(four) and 15(two); scale row counts below lateral line: 9(one), 10(three) and 11(two); circum-peduncular scale counts: 24(three) and 26(three). Predorsal scales irregularly arranged, counts: 33 (two), 34(one) and 35(two). Pelvic-fin axillary scales present, generally two.

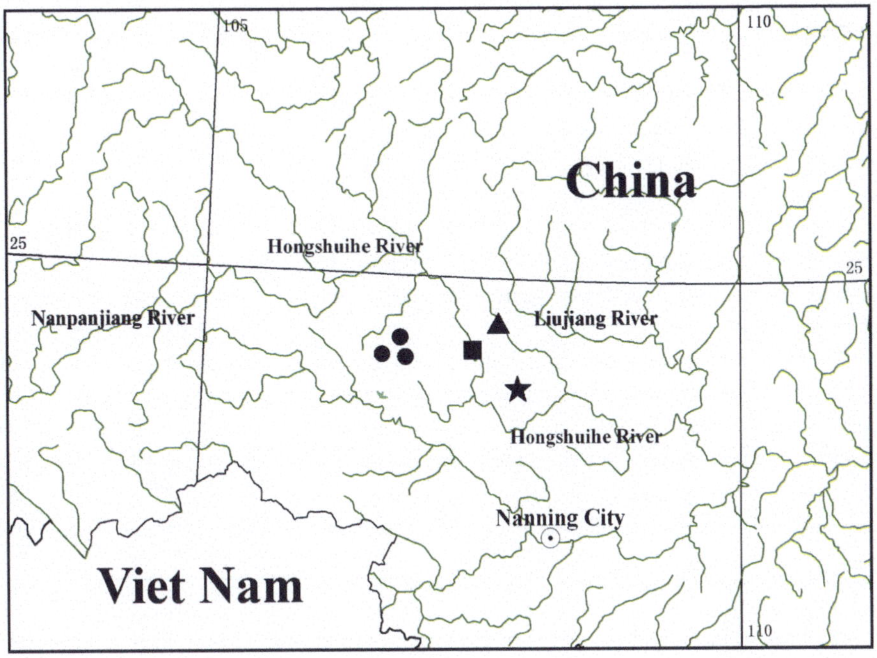

Fig. 2 The distribution of *Sinocyclocheilus brevibarbatus* sp. nov. and related species. *Filled star Sinocyclocheilus brevibarbatus* sp. nov., *filled circle S. microphthalmus*, *filled triangle S. jiuxuensis*, *filled square S. altishoulderus*

Osteological features

Total vertebrae 18+19=37 predorsal vertebrae seven. Pharyngeal teeth three rows, 2, 3, 4; 4, 3, 2. The first row of pharyngeal teeth strong, deeply compressed, with curved and pointed tips; the second row of pharyngeal teeth not well-developed, with very sharp tips.

Coloration in alcohol

Body generally brownish; abdomen light brownish. A few dark speckles on each side, but not very clear. Fins light yellowish.

Distribution

Known only from an underground river in the vicinity of Gaoling Town, Du'an County, Guangxi, China (Fig. 2). The underground river drains into the Hongshuihe River, which in turn flows into the Xijiang River, the longest tributary of the Pearl River Basin.

Etymology

The name of new species, *brevibarbatus*, means barbid fish with short barbels, as opposed to another species in the genus, *Sinocyclocheilus longibarbatus.*

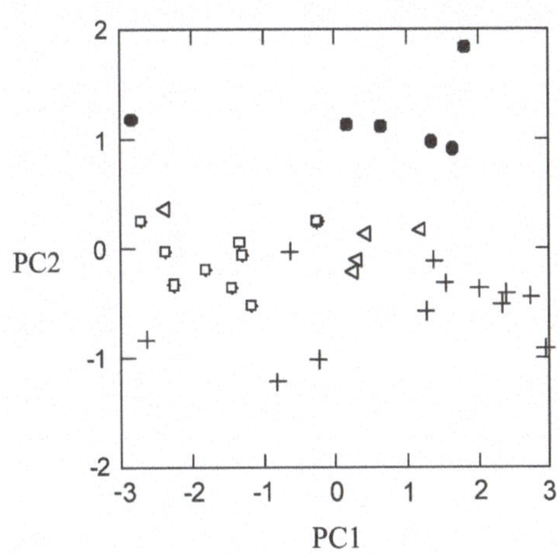

Fig. 3 Principal component analysis (first two dimensions) using the described morphometric values (Table 1)

Remarks

Species with humped backs constitute a special group within the genus *Sinocyclocheilus*. This group consists of at least 13 species including the new species recognized by the following combination of the characters: (1) a typically humped back; (2) long pectoral fin, extending beyond pelvic fin insertion; (3) smaller number of lateral line scales, generally less than 60 (Zhao 2006).

The new species is most similar in body shape to three other species in the humped-back group: *S. jiuxuensis*, *S. altishoulderus* and *S. microphthalmus*. Besides the differences between them mentioned in the "Diagnosis section", a principle component analysis of morphometric characters also showed that the new species is distinguished from other three by component II (Fig. 3), characterized by high loadings for length of barbels and eyeball diameter (Table 2).

Comparative materials

Sinocyclocheilus microphthalmus: ASIZB 73047–73048 (2), 78.2–99.5 mm SL, Zhujiang River basin: Hongshuihe River, Luolou Town, Lingyun County, Guangxi Province, 26 November 2001, C. G. Zhang, and Y. H. Zhao; Li-990417021 (1), 110.2 mm SL, Zhujiang River Basin: Hongshuihe River, Lingyun County, Guangxi Province, April 1999 W. X. Li,. *S. altishoulderus*: Li (species number lost, one of the

Table 2 Character loadings on principle components 1–5 for measurements from 6 specimens of the new species and 27 specimens of related species

	PC1	PC2	PC3	PC4	PC5
Standard length	0.3027	0.0103	0.0111	0.0041	0.0209
Body depth	0.3696	0.0099	−0.0314	0.0102	0.0073
Predorsal length	0.2792	0.0496	0.015	−0.0098	0.0091
Distance from head end to dorsal fin insertion (dorsal view)	0.3004	0.0693	0.0028	−0.0198	0.004
Dorsal-fin base length	0.3563	−0.0453	−0.0203	0.0491	−0.0036
Dorsal fin length	0.3022	−0.0363	−0.0192	0.0498	−0.0219
Preanal length	0.273	0.0315	0.0163	0.0051	0.0351
Anal-fin base length	0.3757	−0.0322	0.0158	0.024	0.0116
Anal fin length	0.2919	−0.0551	−0.0427	0.0345	−0.0272
Prepectoral length	0.2595	0.0711	0.0417	−0.0219	0.015
Pectoral-fin base length	0.3131	−0.0245	−0.0127	−0.0037	−0.0297
Pectoral fin length	0.3283	−0.018	−0.0013	0.0264	−0.0035
Prepelvic length	0.2521	0.0456	0.0317	−0.0133	0.0163
Pelvic-fin base length	0.3718	−0.0302	0.0082	0.0393	−0.0468
Pelvic fin length	0.3423	−0.0285	−0.0104	0.0502	−0.0209
Caudal peduncle length	0.3515	−0.0198	−0.0206	0.017	−0.0001
Caudal peduncle depth	0.3451	−0.0018	−0.0228	0.0208	0.0226
Head length	0.2465	0.0606	0.0349	−0.0248	0.0138
Head depth	0.2615	0.0177	0.0104	0.008	0.0147
Head width	0.3213	0.0152	0.0034	−0.0016	0.0095
Snout length	0.2826	0.0266	0.0354	−0.0174	0.0692
Eye diameter	0.2288	**0.1308**	0.0179	−0.0173	−0.0583
Interorbital width	0.2937	0.005	−0.0007	0.0118	0.0243
Prenostril length	0.3233	0.0496	0.0237	−0.019	0.0536
Width between posterior nostrils	0.3585	0.0167	0.0089	0.0147	−0.031
Upper jaw length	0.3097	0.1264	0.0645	−0.0525	−0.0211
Lower jaw length	0.2967	**0.1539**	0.0789	−0.0639	−0.0366
Mouth width	0.3723	0.0838	0.0209	−0.0048	−0.0316
Maxilla barbel length	0.3264	**−0.4334**	−0.0635	−0.1129	−0.0288
Rictal barbel length	0.3539	**−0.2999**	−0.0112	0.0007	0.0407
Eyeball diameter	0.2004	**0.3174**	−0.22	−0.0418	0.0132
Percent of Total Variance Explained %	80.6553	12.6783	2.0088	0.9797	0.6990

Five highest loadings on PC II indicated by boldface

paratypes), 96.7 mm SL, Zhujiang River Basin: Hongshuihe River, Taiping Town, Donglan County, Guangxi Province, 8 April 1989, W. X. Li, and J. H. Lan,; ASIZB 74214–74215 (2), 76.5–90.1 mm SL, Zhujiang River Basin: Hongshuihe River, Taiping Town, Donglan County, Guangxi Province, October 2001, C. G. Zhang, and Y. H. Zhao,. *Sinocyclocheilus jiuxuensis*: ASIZB 102259–102262 (4), 78.1–125.1 mm SL, Zhujiang River Basin: Hongshuihe River, Jiuxu Town, Jinchengjiang City, Guangxi, November 2001, J. H. Lan; Lan92060002, 92060006 (2), 117.1–136.1 mm SL, Zhujiang River Basin: Hongshuihe River, Jiuxu Town, Jinchengjiang City, Guangxi, June 1992, J. H. Lan.

Acknowledgement This study was funded by the National Natural Science Fund of China (NSFC-30470251 and NSFC-J0030092).

References

Fang PW (1936) *Sinocyclocheilus tingi*, a new genus and species of Chinese barbid fishes from Yunnan. Sin 7:588–593

Leviton AE, Gibbs RH, Heal E et al (1985) Standards in herpetology and ichthyology: part I. standard symbolic codes for institutional resource collections in herpetology and ichthyology. Copeia 1985(3):802–832

Li GL (1989) On a new fish of the genus *Sinocyclocheilus* from Guangxi, China. Acta Zootaxon Sin 14:123–126

Li WX, Lan JH (2002) A new genus and three new species of cyprinidae from Guangxi, China. J Zhanjiang Fish Coll 12:46–51

Li WX, Lan JH, Chen SY (2003) A new species of cave *Sinocyclocheilus* from Yunnan, China. J Guangxi Normal U 21:83–85

Romero A, Paulson KM (2001) It's wonderful hypogean life: a guide to the troglomorphic fishes of the world. Environ Biol Fish 62:13–41

Wilkinson L (2001) SYSTAT: the system for statistics, version 10.0. SPSS, Chicago, Illinois

Zhao YH (2006) An endemic cavefish genus *Sinocyclocheilus* in China—species diversity, systematics, and zoogeography (Cypriniformes: Cyprinidae). Dissertation, Chinese Academy of Sciences

Zhao YH, Zhang CG (2006) Cavefishes: concept, diversity and research progress. Biodivers Sci 14:451–460

Zhao YH, Zhang CG, Peng JX (2002) Methods of making transparent skeleton and X-ray photography of fish (in Chinese with English abstract). In: Lin Q, Zhang CG, Jin HN (eds) Proceedings of the conference, China Biological Collections for the 21st Century. China Science and Technology Press, Beijing, pp 298–301

Zhao YH, Watanabe K, Zhang CG (2006) *Sinocyclocheilus donglanensis*, a new cavefish (Teleostei: Cypriniformes) from Guangxi China. Ichthyol Res 53:121–128

Environ Biol Fish (2009) 86:211–278
DOI 10.1007/s10641-009-9441-3

The Hypogean fishes of China

Aldemaro Romero · Yahui Zhao · Xiaoyong Chen

Received: 7 October 2008 / Accepted: 6 January 2009 / Published online: 15 April 2009

Abstract China has 92 described species of hypogean (cave and artesian) fishes. That is nearly one third of all the described hypogean fish species (299), more than any other country. Of all Chinese hypogean fishes 56 species show troglomorphisms, i.e., adaptations that have been correlated to the hypogean environment such as reduction and/or loss of eyes, pigmentation, and the gas bladder. Additionally, two other characters seem to be unique to some Chinese hypogean species: presence of a horn-like structure and hyperdevelopment of the dorsal protuberance similar to a humpback. Despite the fact that the first written account of a cave fish was for species found in China in 1540 (Romero 2001; and Introduction to this special volume), almost all the new descriptions have taken place in the last 20 years mostly in papers written in Chinese and/or in journals of difficult access outside China. This paper summarizes all the knowledge we have on the hypogean fishes of China and puts them in context regarding all the hypogean fishes in the world.

Keywords Hypogean fishes · Biodiversity · Systematics · Taxonomy · Geographical distribution; convergent evolution

Aldemaro Romero and Yahui Zhao have contributed equally to this article.

A. Romero (✉)
Department of Biological Sciences,
Arkansas State University,
P.O. Box 599, Jonesboro, AR 72467, USA
e-mail: aromero@astate.edu

Y. Zhao
Key Laboratory of Zoological Systematics and Evolution,
Institute of Zoology,
Chinese Academy of Sciences,
Datun Road, Chaoyang District,
Beijing 100101, People's Republic of China

X. Chen
Kunming Institute of Zoology, Chinese Academy of Sciences,
No.32 Jiaochang Donglu,
Kunming 650223 Yunnan, People's Republic of China

Introduction

There are about 28,000 fish species in the world (Nelson 2006). More than 4,600 species have been reported for China (including Taiwan) (Wu et al. 1999). Of all fish species 299 have been reported for the hypogean (cave, phreatic) environment. Of those, 92 species belonging to three families are found in China. Furthermore, most of the hypogean fishes that have been described in the last two decade are from Asia, particularly from China. However, most of the literature on those species has been published in Chinese language in Chinese journals, with limited availability in western libraries, making access to that information difficult for many international researchers. This paper summarizes all known information about Chinese hypogean species whether they show troglomorphisms or not.

Troglomorphism is a term proposed by Christiansen (1962) to indicate the presence (or lack thereof) of a character that is considered related to the hypogean environment, such as the reduction or loss of the visual apparatus and pigmentation. Romero and Green (2005) listed other characters considered as troglomorphisms, including reduction or total loss in the number of scales and gas (swim) bladder. We have included the horn-like found among some hypogean Chinese fish species as a possible troglomorphic characteristic. Although the function of this structure is still unknown, it is found only among some hypogean species of the genus *Sinocyclocheilus*. An analogous structure has been reported for *Kustus gulliveri* (Perciformes: Kurtidae). In that species this structure called a 'hook' is found only on males and is used to carry fertilized eggs (Berra and Humphrey 2002). The horn we report for some species of *Sinocyclocheilus* is found in both sexes. Li et al. (1997) hypothesized that this structure could function as a protective organ but no there is as yet no confirmation of this. The humpback profile found among some hypogean *Sinocyclocheilus* species also has an unknown function. However, it is interesting to note that both characters may have a similar function because some species present a horn-like appearance formed by the humpback.

Methods

We collected all available information on hypogean fishes published up to September 2008 following the methodology described in Romero and Paulson (2001). We grouped the species by family following Nelson's (2006) systematic order, with a general summary of their biology. Given that the phylogenetic relationships with the rest of the family are uncertain for most genera and species, we list genera within families, as well as species within genera, alphabetically. We numbered the species to facilitate the reading of the maps (Fig. 1). Most of the information for each species comes from the original description; however, in the case of *Sinocyclocheilus* species one of us rechecked the original information for accuracy

a

b

c

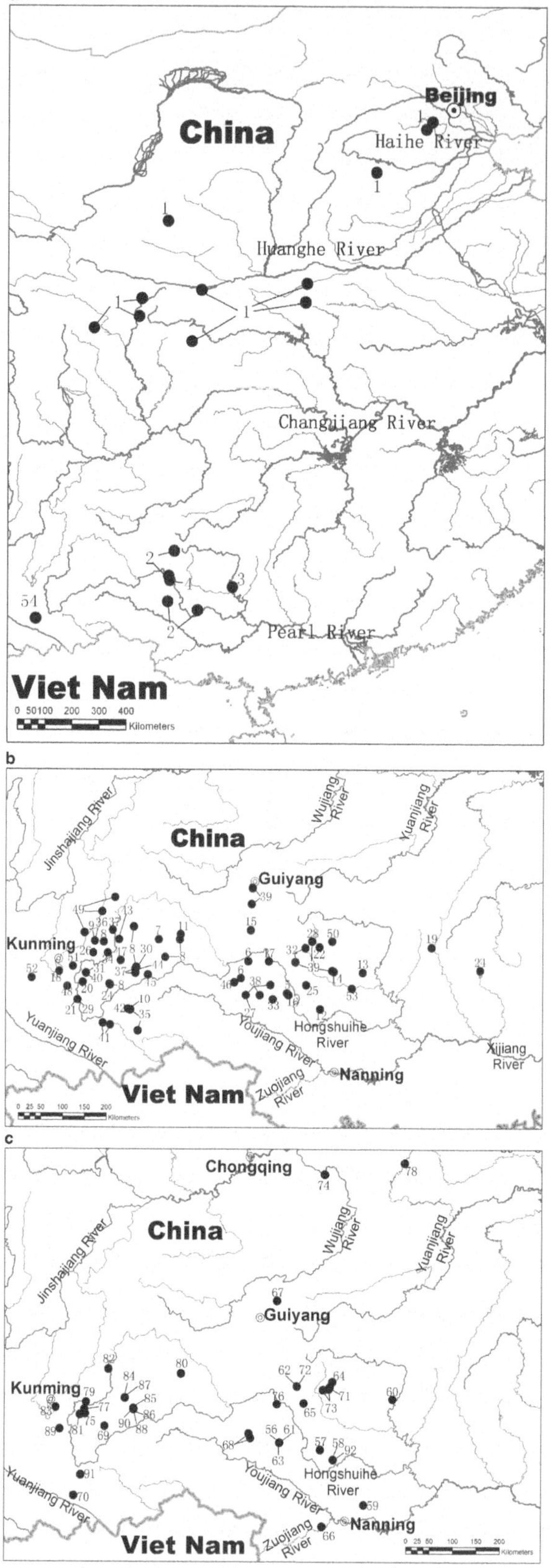

◀ **Fig. 1** Distribution map of all Chinese hypogean fishes. (**a**) Distribution of hypogean fishes from Cyprinidae (excluding the genus *Sinocyclocheilus*). (**b**) Distribution of the species from the genus *Sinocyclocheilus*. (**c**) Distribution of hypogean fishes from Cobitidae and Balitoridae

(Zhao 2006). Common name(s) are based on Wu et al. (1999) or from the original description. Etymology is given only for those names that were created to identify the hypogean and/or troglomorphic species and/or populations. Style for referring to etymologies follows Brown (1956) and Gotch (1995). Major synonyms are given only if they represent, might represent, or have represented a source of confusion. Historical information for species found both in the hypogean and the epigean environment refers only to hypogean populations. The remarks section aims to summarize information about troglomorphic features (reduction in eyes, pigmentation, and scales), behavioral features and phylogenetic relationships.

We created the maps accompanying this article with Arc View software, with the Mercator projection, which does create some distortion in the area of landmasses towards the poles. Since these maps exclude much of the polar landmasses and since troglomorphic fishes are not found at high latitudes, this is not much of an issue. We determined latitudes and longitudes of troglomorphic fish distributions using one of several methods: sometimes latitude and longitude information was found in the article describing the species; sometimes a description of the location was given (with varying degrees of accuracy and reproducibility), and latitude and longitude were determined using atlases and other sources. Some articles gave no more than a sketchy description of the location making the latitudes and longitudes for these few species rather vague. The latitude and longitude information included in the article is the same as that used in the creation of the maps. Where 'ca.' appears before the latitude and longitude, it means that it was determined using an atlas and thus reflects a lesser degree of accuracy. The reader should, therefore, take several things into consideration when using these maps. The maps are on a national scale and are intended to do little more than provide a general idea of where these species are located while giving a reference to the species numbers as described in the article. Some GIS information was directly from GPS by authors in the field. That is also one way to get latitudes and longitudes in the paper. The nature of the methods used and the scale of the maps mean that the dots are not always (though in some cases are accurate to the tenth of a second) on the *exact* location.

Results

Family: Cyprinidae—minnows or carps

The Cyprinidae is the largest family of freshwater fishes with nearly 2,500 species distributed throughout Eurasia, Africa, and North America. This overall number of cyprinid species may be artificially high due to lack of clarity in species status of many populations (Nelson 2006). Even among the hypogean species the phylogenetic distinctions may be exaggerated. For example, it is quite possible that *Barbopsis devecchii*, *Phreatichthys andruzzii*, *Caecobarbus geertsi* and *Iranocypris typhlops* should all be assigned to the genus *Barbus* (Ercolini et al. 1982). A very close genetic relationship between the first two is further supported by allozyme studies (Sbordoni et al. 1996). There are 96 species of hypogean cyprinids worldwide and 54 hypogean species have been reported for China.

Most cyprinids have scales to some degree. Nothing in their morphology, physiology or behavior seems to support the idea of a family being 'preadapted' to the hypogean environment. Thus, the large number of hypogean representatives for the family (96) seems to be just the result of the fact that this is a highly speciose family.

The vast majority of the 54 species of Chinese hypogean cyprinids have been discovered in since the 1980's. Of all the Chinese hypogean cyprinid species, 34 show one or more troglomorphisms. The presence of hypogean cyprinids that are hypogean but non-troglomorphic is not unusual. For example, Burr et al. (2001) reported normally eyed and pigmented individuals of *Cyprinus carpio* and *Pimephales promelas* for Mystery Cave, Perry County, Missouri. Trajano et al. (2002) found several individuals of epigean cyprinid species of the genera *Danio*, *Tor*, *Garra*, and *Propuntius* in caves in Thailand.

There are 51 species of the genus *Sinocyclocheilus* associated with the hypogean environment (this paper), 32 of them show troglomorphisms. Due to the isolation by different caves and subterranean rivers, the genus *Sinocyclocheilus* represents very high species diversity and suggests rapid speciation (Zhao and Zhang 2006).

Genus: *Onychostoma* Günther, 1896

Of the 16 recognized species of this Asian genus, only one (*O. macrolepis*) is found in caves where it is reported to hibernate (sic) during winter. This is the only hypogean species in China found north of the Changjiang River (Zhang 1986).

1. ***Onychostoma macrolepis*** (Bleeker, 1871) (Fig. 2). **Common name:** Multi-scaled sharp-jaw barbel (E, translation from Chinese); Taishan red scale fish (E, translation from Chinese). **Etymology:** *Onycho* (Gr) claw, hoof, after the sharp lower jaw; *stoma* (Gr) mouth in reference to the genus having sharp lower jaw; *macro* (Gr) huge, big; *lepis* (Gr) scale, after its large scales. **Major synonyms:** *Barbus tamusuiensis shansiensis*; *Gymnostomus macrolepis* (original combination); *Scaphesthes macrolepis*; *Varicorhinus shansiensis; Varicorhinus macrolepis*. **Ecological classification:** Troglophile. **History:** The cave-dwelling behavior of this species has been known for centuries by local people. For instance, every year many people will wait at the spring of the subterranean water where this fish hibernates to catch the fish during the night of Grain Rain (April 20th), one of 24 solar terms in China. The fish will always come out from the cave and look for the food at these days of Grain Rain period (April 20th–May 5th). **Morphological characteristics:** Mouth ventral, transverse; snout with lateral grooves extending to corners of mouth; lower lip separated from lower jaw, anterior border of lower jaw exposed, with a sharp horny sheath; branched dorsal ray 8; last dorsal ray slender, not serrated; lateral line scales more than 50, predorsal scales more than 18. Maximum standard length: 156 mm; **Troglomorphic characters:** None known. **Distribution:** Beijing City: Fangshan District: Jumahe River (ca. 39°34′ N, 115°42′ E); Hebei Province: Yixian County (ca. 39°20′ N, 115°30′ E); Shanxi Province: Niangziguan (ca. 37°57′ N, 113°53′ E); Shaanxi Province: Lueyang County (ca. 33°20′ N, 106°10′ E); Zhouzhi County (ca. 34°10′ N, 108°12′ E); Zhenba County (ca. 32°32′ N, 107°53′ E); Gansu Province: Wenxian County (ca. 32°58′ N, 104°41′ E); Liangdang County (ca. 33°54′ N, 106°15′ E); Henan Province: Luanchuan (ca. 33°47′ N, 111°36′ E), Luoning (ca. 34°23′ N, 111°39′ E); Shandong Province: Tai'an city (ca. 36°22′ N, 117°05′ E) (Fig. 3). **Habitat and ecology:** This species has been observed entering springs of karstic caves during winter. This hibernation-like activity lasts from late October until late April, spanning, thus, about 6 months. **Food and feeding:** Algae (mostly diatoms such as *Cymbella*, *Synedra*, *Fragilaria*, *Gomphonema*, and *Micrasterias*), insect larvae of Trichoptera, Plecoptera, Chironomidae, and Simuliidae, and organic material. **Reproduction and development:** Their reproductive activities start within 2–4 weeks after the fish come out of caves and continue for about 2 months with about 1 month of peak activities. According to measurements in the Hei He Basin, the

Fig. 2 *Onychostoma macrolepis*, collected in Heihe River, Shannxi Province in 2004. (Photo by Zhao, Y.)

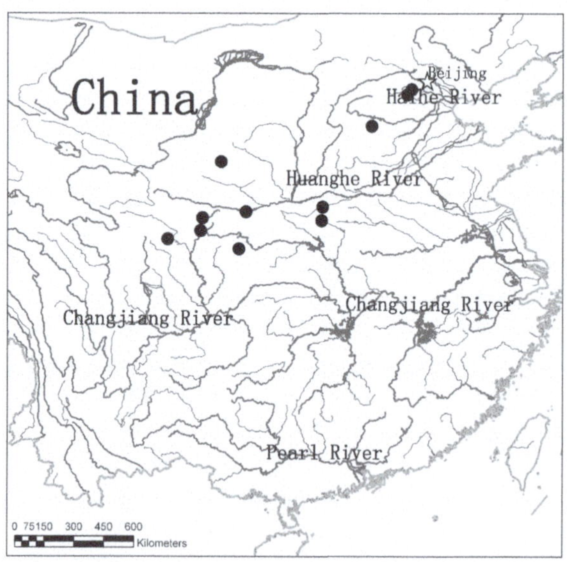

Fig. 3 The distribution of *Onychostoma macrolepis*

minimum water temperature for spawning is 16°C (Zhang 1986). **Other behavior:** The fish feed on algae on the surface of the gravel using their sharp lower jaw. **Conservation status:** It is classified as 'protected' in Hubei Province (Institute of Hydrobiology 2002, p. 90).[1] **Major threats:** Environmental impacts due to road construction and pollution; Excessive collecting for commercial purposes. **Conservation plans:** No governmental plan is directed to this species. **Phylogenetic relationships:** Individuals found in the hypogean environment do not seem to differ morphologically from the epigean form. Genetic analyses are yet to be conducted to see if there is any other differentiation. **Additional remarks**: The fish has been recorded as one of five best delicious fish from as early as the Tang Dynasty (618–907). The Chinese poet, Li Bai (701–762), even mentioned this fish in one of his poems. The fish has been produced by aquaculture in Shandong Province in recent years.

Genus ***Sinocrossocheilus*** Wu, 1977

The ten species of *Sinocrossocheilus* are endemic to the central and eastern Yunnan-Guizhou Plateau of southern China, in both the Pearl River drainage and the Yangtze River System. These fishes are found in both epigean and hypogean waters although the three species reported for hypogean environments show no troglomorphic characters (Su et al. 2003; Chen et al. 2006).

2. *Sinocrossocheilus bamaensis* (Fang, 1981) (Fig. 4). **Common name:** Bama tassel-lip barbel (E, translation from Chinese). **Etymology:** *Sino* (L) from China; *krossoi* (Gr) tassel; *chelos* (Gr) lips; *bamaensis* (L) after one locality of its distribution, Bama County in the Guangxi Zhuang Autonomous Region. **Major synonyms:** *Crossocheilus bamaensis*, original combination (Fang 1981); *Pseudocrossocheilus bamaensis* (Zhang and Chen 1997, Zhu 2006). **Ecological classification:** Troglophile. **History:** Type specimens were collected in 1974 (Fang 1981). **Morphological characteristics:** D iii, 8 A iii, 5 P i, 13–15 V i, 8. Body elongated; prefringe of rostral cap split into 5–6 fimbriations with fleshy papillae; papillate band very narrow; 2–3 rows horny tubercles confined to snout tip and never reaching nostril; fleshy papilla sparsely

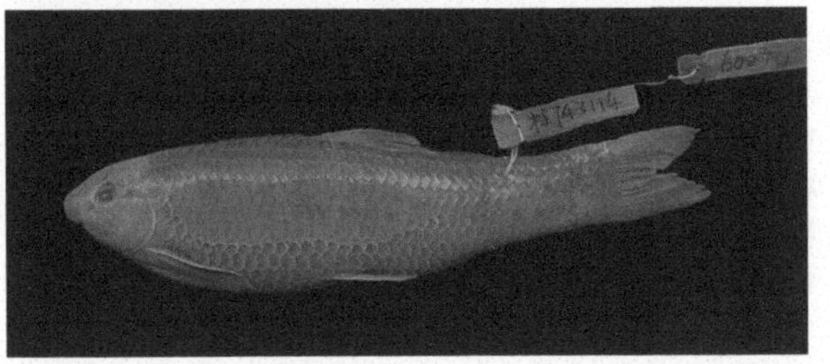

Fig. 4 *Sinocrossocheilus bamaensis*, holotype. (Photo by Ye, E. and Zhao, Y.)

arranged in rostral cap and lower lip; ventral fin not reaching anus; intestine very slender; interorbital wide; calvaria high, a deep groove between snout tip and calvaria; maxillary barbels longer than eye diameter; vertebrae 39–40. Maximum standard length: 116 mm (Fang 1981). Upper part of sides and dorsum of body brown and black; lower part of sides and ventral part of body yellow and brown; sides with a conspicuous black blotch on upper part of pectoral fin origin; prefringe of rostral cap and lower lip gray and white; outer margin of dorsal fin with a black band, two lateral margins of caudal fin black (Su et al. 2003). **Troglomorphic characters:** None known. **Distribution:** China: Guangxi Zhuang Autonomous Region: Bama County: Jiazhuan town (ca. 24°12′ N, 107°07′ E); Tian'e County (ca. 25°01′ N, 107°09′ E); Du'an County (ca. 23°55′ N, 108°06′ E); Guizhou Province: Pingtang County (ca. 25°49′ N, 107°20′ E) (Fig. 5). **Habitat and ecology:** It inhabits

Fig. 5 Distribution of *Sinocrossocheilus bamaensis*

[1] http://www-wds.worldbank.org/servlet/WDSContentServer/WDSP/IB/2002/06/12/000094946_02

clefts between rocks in fast-flowing sections of streams. During spring and summer, it leaves the caves to find food at both dawn and dusk. In winter, it spends all the time in caves (Su et al. 2003). **Food and feeding:** Algae (Fang 1981). **Reproduction and development:** Between May and June, the species moves out into the floodwaters to reproduce (Fang 1981). **Other behavior:** Unknown. **Conservation status:** Undetermined. **Major threats:** No survey has been conducted to assess its possible threats. **Conservation plans:** None. **Phylogenetic relationships:** The genus belongs to the subfamily Labeoninae.

3. *Sinocrossocheilus liuchengensis* (Liang et al. 1987) (Fig. 6). **Common name:** Liucheng tassel-lip barbel (E, translation from Chinese). **Etymology:** *liuchengensis* (L) after one locality of its distribution, Liucheng County in Guangxi, Zhuang Autonomous Region. **Major synonyms:** *Crossocheilus liuchenensis*, original combination (Liang et al. 1987); *Pseudocrossocheilus liuchengensis* (Zhang and Chen 1997). **Ecological classification:** Troglophile. **History:** Type specimens were collected in 1982 (Liang et al. 1987). **Morphological characteristics:** D iii, 8 A iii, 5 P i, 13–15 V i, 8. Body elongated; mouth cap deeply arched; rostral cap split into 11–13 fimbriations with fleshy papillae; horny tubercles absent from snout tip and sides; maxillary barbels much longer than rostral ones and eye diameter; predorsal scales irregularly arranged; ventral fin never reaching anus; vertebrae 40. Maximum standard length: 119 mm (Zhu 2006). Coloration of preserved specimens (in alcohol) is yellowish, ventral whitish gray; a row of black blotches (9–10) along lateral line; all fins without pigment, translucent (Su et al. 2003). **Troglomorphic characters:** None known. **Distribution:** China: Guangxi Zhuang Autonomous Region: Liucheng County (ca. 24°40′ N, 109°15′ E) (Fig. 7). **Habitat and ecology:** Subterranean stream. **Food and feeding:** Unknown. **Reproduction and development:** Unknown. **Other behavior:** None known. **Conservation status:** Undetermined. **Major threats:** No survey has been conducted to assess its possible threats. **Conservation plans:** None. **Phylogenetic relationships:** The genus belongs to the subfamily Labeoninae.

Fig. 7 Distribution of *Sinocrossocheilus liuchengensis*

Fig. 6 *Sinocrossocheilus liuchengensis*. Cited from Zhou and Zhang, 2006. (Photo by Lan, J.)

4. *Sinocrossocheilus megalophthalmus* (Chen et al. 2006) (Fig. 8). **Common name:** Big eye tassel-lip barbel (E, translation from Chinese). **Etymology:** *megas* (Gr) big; *ophthalmus* (Gr) eyes. **Major synonyms:** None. **Ecological classification:** Troglo-

Fig. 8 *Sinocrossocheilus megalophthalmus*, holotype, standard length 49.1 mm. (Photo by He, M.)

phile. **History:** Type specimens were collected in 1993 (Chen et al. 2006). **Morphological characteristics:** D iii, 7 A iii, 5 P i, 12–13 V i, 7. Two rows of pharyngeal teeth, no black spot above midpoint of pectoral fin; predorsal scales 15–16; bigger eye, eye diameter 2.5–3.1 in head length; rostral barbel reaching between anterior and posterior nostrils or reaching anterior edge of eye; maxillary barbel reaching between anterior edge of eye and midpoint of eye or reaching between midpoint of eye and posterior edge of eye; gill rakers 13; pelvic fin reach anus. Maximum standard length: 49 mm (Chen et al. 2006). Coloration of preserved specimens (in alcohol) is light brownish, ventral light yellowish, with a black stripe above lateral line, but not very clear; dorsal fin brownish, other fins light yellowish with no spot (Chen et al. 2006). **Troglomorphic characters:** None known. **Distribution:** China: Guangxi Zhuang Autonomous Region: Tian'e County: Banmo town (ca. 24°53′ N, 107°11′ E) (Fig. 9). **Habitat and ecology:** Subterranean stream. **Food and feeding:** Unknown. **Reproduction and development:** Unknown. **Other behavior:** None known. **Conservation status:** Undetermined. **Major threats:** No survey has been conducted to assess its possible threats. **Conservation plans:** None. **Phylogenetic relationships:** The species has close relationships with *Sinocrossocheilus microstomatus* and *S. labiatus* by having two rows of pharyngeal teeth (Chen et al. 2006).

Fig. 9 Distribution of *Sinocrossocheilus megalophthalmus*

Genus: *Sinocyclocheilus* Fang, 1936

Sinocyclocheilus is a highly diversified cyprinid genus with more than 50 species described so far. This genus is endemic to the provinces of Guizhou and Yunan and the Guangxi Zhuang Autonomous Region and is always found in karstic areas rich in caves and groundwater systems. Some of the hypogean species show clear troglomorphisms while others do not. Some of the morphological characters associated with species found in caves are quite unusual such as a horn on the nape and a humpbacked profile. The adaptive significance of such features is unknown. Molecular genetic data, complemented with morphological information, was utilized to construct a phylogenetic tree incorporating all known species (Zhao 2006). *Sinocyclocheilus* was found to be a monophyletic group with four clades, given the appellations *jii*, *angularis*, *cyphotergous* and *tingi*, from the four representative species respectively.

5. *Sinocyclocheilus altishoulderus* (Li and Lan, 1992) (Fig. 10). **Common name:** High-shoulder golden-line barbel (E, translation from Chinese). **Etymology:** *Sino* (L) from China; *cyclo* (Gr) circle; *cheilus* (Gr) lip; *alti* (L) high depth; *shoulderus* (L) shoulder. **Major synonyms:** *Anchicyclocheilus altishoulderus*, original combination. **Ecological classification:** Troglobite. **History:** First collected on 8 April 1989 by Li. **Morphological characteristics:** D: iii, 8; A: iii, 5; P: i, 13–16; V: i, 7–8. Body humpbacked; mouth sub-inferior and arched; eye small; pectoral fin long beyond pelvic fin insertion; lateral scale not very obviously larger than neighbor scales; lateral line complete and almost straight, possessing 42–52 scales; scale rows above and below lateral line 16–

Fig. 10 *Sinocyclocheilus altishoulderus*, one of type specimens, Li number lost, standard length 96.7 mm. (Photo by Zhao, Y.)

17, 10–11 respectively; caudal vertebrae 20–21. Maximum standard length: 96.7 mm (Zhao et al. 2006b). Live coloration is whitish and semi-transparent. Coloration of preserved specimens (in alcohol) is light yellowish, back darker, all fins light yellowish. **Troglomorphic characters:** Microphthalmic, depigmented, humpbacked. **Distribution:** China: Guangxi Zhuang Autonomous Region: Donglan County: Taiping Town (ca 24°22′ N, 107°22′ E, 550 m above sea level or meters above sea level) (Fig. 11). **Habitat and ecology:** A subterranean stream at about 10 to 20 m inside the mouth of a small cave located around 50 to 100 m up from the bottom of a 4 km2 lowland depression in a mountainous region. During the rainy season the subterranean water will flow out and make the lower depression look like a small lake. The species is sympatric with *Sinocyclocheilus donglanensis* (Zhao et al. 2006b). **Food and feeding:** Unknown. **Reproduction and development:** Unknown. **Other behavior:** Unknown. **Conservation status:** Undetermined. One of us (YZ) thinks that it should be categorized as VU (Vulnerable). **Major threats:** Exploitation of the underground water resource, solid waste and pesticide pollution, and general landscape alterations. **Conservation plans:** The Agriculture Ministry of China is organizing the new List of National Protected Aquatic Animals. One of us (YZ) participated in that plan and recommended that the entire genus be included.

Fig. 11 Distribution of *Sinocyclocheilus altishoulderus*

Phylogenetic relationships: This species belongs to the *cyphotergous* lineage.

6. *Sinocyclocheilus anatirostris* Lin and Luo, 1986 (Fig. 12). **Common name:** Duck-billed golden-line barbel (E, translation from Chinese). **Etymology:** *anatis* (L) duck-like; *rostris* (L) beak. **Major synonyms:** *Sinocyclocheilus guangxiensis* (Zhou and Li 1998a), *S. albeoguttatus* (Zhou and Li 1998b). **Ecological classification:** Troglobite. **History:** First collect in March 1983 (Lin and Luo 1986). **Morphological characteristics:** D: iii, 8; A: iii, 5; P: i, 12–13; V: i, 6–8. Body naked; snout duckbilled; blind; mouth sub-inferior and curved; lateral line almost straight, lateral line pores 33–42; pectoral fin long, beyond pelvic fin insertion. Maximum standard length: 129 mm (Shi et al. 1994). Living individuals lack pigmentation, being semi-transparent. **Troglomorphic characters:** No externally visible eyes, depigmented, scaleless, reduced gas bladder. **Distribution:** China: Guangxi Zhuang Autonomous Region: Lingyun and Leye counties. The fish is distributed in different branches of Bailang Subterranean River (ca. 24°40′–25 °00′ N, 106°26′–106°35′ E) (Fig. 13). **Habitat and ecology:** First found in two caves at about 30 m from their entrance at 1.5 m in depth. Water temperature 17.1–18.5°C, pH 6.0–6.4 (Shi et al. 1994). **Food and feeding:** Stomach contents include plant and insect remains (Shi et al. 1994). **Reproduction and development:** External fertilization and are non-guarder (Riehl and Baensch 1991). **Other behavior:** Unknown. **Conservation status:** VU in IUCN RL 2003; VU in China's Species Red List (Wang and Xie 2004). **Major threats:** Environmental degradation and Excessive collecting (Wang and Xie 2004). **Conservation plans:** None exist for this species. **Phylogenetic relationships:** It belongs to the *angularis* lineage.

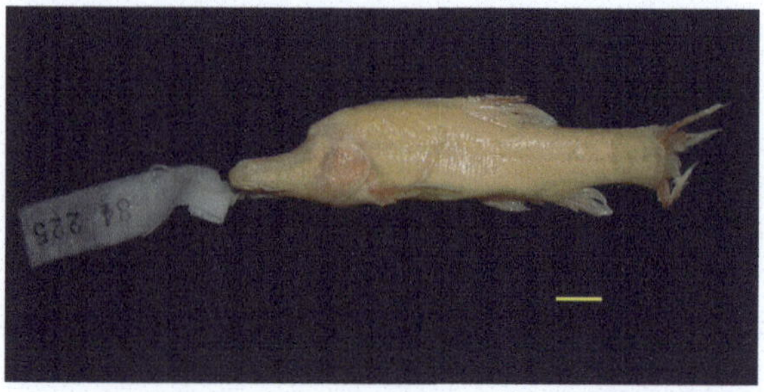

Fig. 12 *Sinocyclocheilus anatirostris*, holotype, IHB 84VII225, standard length 88.4 mm. (Photo by Zhao, Y.)

Fig. 13 Distribution of *Sinocyclocheilus anatirostris*

Fig. 15 Distribution of *Sinocyclocheilus angularis*

7. *Sinocyclocheilus angularis* Zheng and Wang, 1990 (Fig. 14). **Common name:** Angled golden-line barbel (E, translation from Chinese). **Etymology:** *angularis* (L) because at the end of the posterior portion of the head there is a horn forming an angle. **Major synonyms:** None. **Ecological classification**: Troglobite. **History:** First collected in June 1986 (Zheng and Wang 1990). **Morphological characteristics:** D: iii, 7 A: iii, 5 P: i, 15–18 V: i, 8–10. Simple horn on nape, not forked; snout duckbilled; mouth sub-inferior and curved; eye small; 2 pairs of barbels, short; pectoral fin long, reaching beyond pelvic fin insertion; scales small, embedded; lateral line complete, possessing 37–42 scales. Maximum standard length: 60.1 mm (Zhao 2006). Coloration of preserved specimens (in alcohol) light brownish, dorsum darker, many small black spots above lateral line; a black speckle on the dorsal fin base; all fins gray-whitish. **Troglomorphic characters:** Microphthalmic, scales embedded. **Distribution:** China: Guizhou Province: Panxian County: Cave located at 25°26′ N, 104°45′ E, at 1,540 m above sea level (Fig. 15). **Habitat and ecology:** Karst cave. **Food and feeding:** Unknown. **Reproduction and development:** Reproductive period is from June to July. **Other behavior:** Unknown. **Conservation status:** VU in IUCN RL 2003; VU in China's Species Red List (Wang and Xie 2004). **Major threats:** Environment disturbances and Excessive collecting (Wang and Xie 2004). **Conservation plans:** None. **Phylogenetic relationships:** It belongs to the *angularis* lineage.

Fig. 14 *Sinocyclocheilus angularis*, paratype, IHB 12209032-860219, standard length 59.8 mm. (Photo by Zhao, Y.)

8. *Sinocyclocheilus angustiporus* Zheng and Xie, 1985 (Fig. 16). **Common name:** Small gill-opening

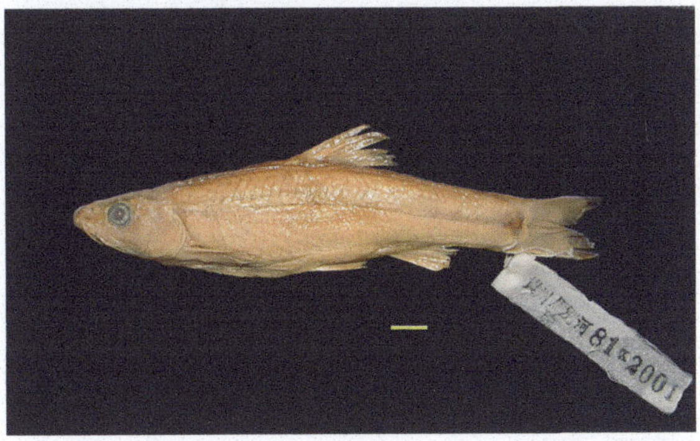

Fig. 16 *Sinocyclocheilus angustiporus*, holotype, IHB12209016-81X2001, standard length 104.0 mm. (Photo by Zhao, Y.)

golden-line barbel (E, translation from Chinese). **Etymology:** *angust* (L) narrow, small; *porus* (L), hole, opening. **Major synonyms:** *Sinocyclocheilus aluensis* (Li et al. 2005a). **Ecological classification:** Troglophile. **History:** Type specimens were collected in 1981 (Zheng and Xie 1985). **Morphological characteristics:** D iii, 7 A iii, 5 P i, 13–16, i, 7–8. Body elongated, scaled; mouth sub-inferior; 2 pairs of barbels, moderate long, rictal barbel beyond posterior edge of eye; dorsal-fin insertion opposite to anal-fin; upper corner of gill opening under horizontal line of superior edge of eye. Maximum standard length: 138.1 mm (Zhao 2006). Live coloration is gold. Coloration of preserved specimens (in alcohol) is yellow-brownish, back darker, many spots on sides, all fins light brownish. **Troglomorphic characters:** None known. **Distribution:** China: Guizhou Province: Xingyi (ca. 25°05′ N, 104°53′ E) city, Xingren (ca. 25°26′ N, 105°11′ E) and Zhenfeng (ca. 25°23′ N, 103°38′ E) counties; Yunnan Province: Luoping (ca. 24°53′ N, 104°18′ E), Fuyuan (ca. 25°41′ N, 104°15′ E) and Luxi (ca. 24°32′ N, 103°46′ E) counties (Fig. 17). **Habitat and ecology:** Karstic cave. The fish has been captured from the body of water around the cave exit. **Food and feeding:** Unknown. It can be fed in captivity with pet fish food. **Reproduction and development:** Unknown. **Other behavior:** In aquaria individuals hide in a small artificial cave (dark environment) (observation by YZ). **Conservation status:** Undetermined. **Major threats:** Extraction of water from the cave by the locals. **Conservation plans:** None. **Phylogenetic relationships:** It belongs to the *tingi* lineage.

Fig. 17 Distribution of *Sinocyclocheilus angustiporus*

Fig. 18 *Sinocyclocheilus anophthalmus*, holotype, KIZ865949, standard length 99.8 mm. (Photo by Zhao, Y.)

9. *Sinocyclocheilus anophthalmus* Chen and Chu, 1988 (in Chen et al. 1988b) (Fig. 18). **Common name:** Blind golden-line barbel (E, translation from Chinese). **Etymology:** *an* (Gr) without; *opthalmus* (Gr) eyes. **Major synonyms:** None. **Ecological classification:** Troglobite. **History:** First collected in 1986 (Chen et al. 1988b). **Morphological characteristics:** D iii, 7; A iii, 5; P i, 15–18; V ii, 8–9. Gill rakers on the first gill arch 7–8. Pharyngeal teeth in 3 rows with counts of 2, 3, 4–4, 3, 2. Body almost naked, lateral line complete, slightly curved, with indistinct scales and 46–59 easily-counted pores. Blind, either eyeless or with small eye remains. Snout duck-billed. Barbels 2 pairs. Last unbranched ray of dorsal fin hard at base, softening toward tip, with serrations along posterior edge. Pectoral fin long, reaching pelvic fin insertion. Maximum standard length: 113.3 mm (Zhao 2006). Live coloration is mostly semi-transparent with some shades of gray. **Troglomorphic characters:** Eyes not externally visible, or highly reduced, highly reduced pigmentation, few scales covering only parts of the body. **Distribution:** China: Yunnan Province: Yiliang County: cave (25°03′ N, 103°23′ E) (Fig. 19). **Habitat and ecology:** Subterranean stream in a limestone cave, located on a hillside around 20 m higher than a neighboring stream. **Food and feeding:** Insect remains have been found in stomach contents. **Reproduction and development:** External fertilization, non-guarder (Riehl and Baensch 1991). Spawning season is in late May or early June. **Other**

Fig. 19 Distribution of *Sinocyclocheilus anophthalmus*

behavior: Unknown. **Conservation status:** Rare in China's Red Data Book of Endangered Animals, Pisces (Yue and Chen 1998); Vulnerable in China's Species Red List, Vol. 1 Red List (Wang and Xie 2004); second grade protected animal category in Yunnan Province. **Major threats:** Habitat degradation and Excessive collecting (Yue and Chen 1998). **Conservation plans:** No official plan aiming at this species. **Phylogenetic relationships:** It belongs to the *tingi* lineage.

10. *Sinocyclocheilus aquihornes* Li and Yang, 2007 (in Li et al. 2007) (Fig. 20). **Common name:** Eagle-mouth-like horn golden-line barbel (E, translation from Chinese). **Etymology:** *aquila* (L) eagle; *horn* (Anglo Saxon) horn. **Major synonyms:** None. **Ecological classification:** Troglobite. **History:** First collected in 12 May 2006 (Li et al. 2007). **Morphological characteristics:** D iii, 7; A iii, 5; P i, 9; V i, 6. Gill rakers on the first gill arch 8. Pharyngeal teeth in 3 rows with counts of 2, 3, 4–4, 3, 2. Body elongated and naked. Lateral line complete, straight. Snout duck-billed. Blind. Mouth sub-inferior. Two pairs of barbels. Last unbranched ray of dorsal fin hard at base, softening toward tip, with serrations along posterior edge. Pectoral fin long, beyond pelvic fin insertion. Maximum standard length: 62 mm (Li et al. 2007). Live coloration is semi-transparent, pinkish (due to the visible blood vessels), vertebrae and entrails visible, a white speckle on each cheek; preserved specimens whitish, opaque, all fins lack coloration. **Troglomorphic characters:** Eyeless, depigmented, scaleless, horn. **Distribution:** China: Yunnan Province: Qiubei County (ca. 24°02′ N, 104° 11′ E) (Fig. 21). **Habitat and ecology:** Subterranean river in a cave. **Food and feeding:** Unknown. **Reproduction and development:** Unknown. **Other behavior:** Unknown. **Conservation status:** Undetermined, but only one specimen has been collected so far. **Major threats:** No survey has been conducted to assess its possible threats. **Conservation plans:**

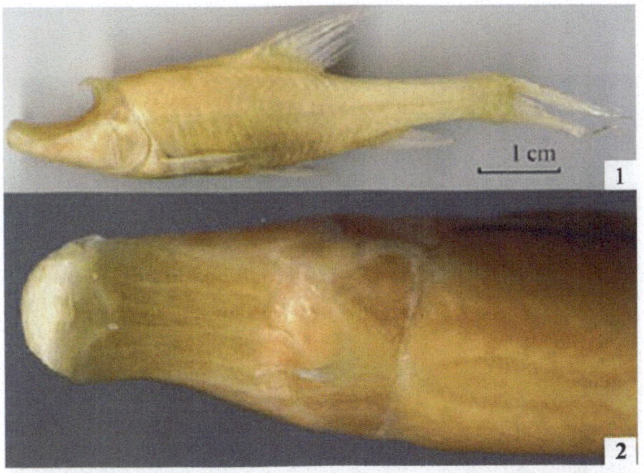

Fig. 20 *Sinocyclocheilus aquihornes*, holotype, Li 060512001, standard length 62 mm. (Cited from Li et al. 2007, photocopy) 1. Profile 2. Dosal view

Fig. 21 Distribution of *Sinocyclocheilus aquihornes*

None. **Phylogenetic relationships:** It belongs to the *angularis* lineage.

11. ***Sinocyclocheilus bicornutus*** Wang and Liao, 1997 (Fig. 22). **Common name:** Double-horned golden-line barbel (E, translation from Chinese). **Etymology:** *bi* (L) two, double; *cornutus* (L) horned, having a horn. **Major synonyms:** *Ceratobarbus biangularis* (Wang and Liao 1995). **Ecological classification:** Troglobite. **History:** Type specimens were collected in 1990 (Wang and Liao 1997). **Morphological characteristics:** D iii, 7 A iii, 5 P i, 13–16 V i, 6–7. Forked horn; snout duck-billed; mouth sub-inferior; eye very small; 2 pairs of barbels, maxillary barbel beyond posterior edge of eye, rictal barbel reaching anterior part of operculum; pectoral fin long, beyond pelvic fin insertion; scale small, embedded; lateral line scale 37–44. Maximum standard length: 100.1 mm (Zhao 2006). Living fish are whitish and semi-transparent, 7–8 irregular black spots along lateral line. **Troglomorphic characters:** Microphthalmic, depigmented, scales embedded, horn. **Distribution:** China: Guizhou Province: Xingren County: Gaowu town (ca. 25°32′ N, 105°12′ E) (Fig. 23). **Habitat and ecology:** No information available. **Food and feeding:** Unknown. **Reproduction and development:** Unknown. **Other behavior:** Unknown. **Conservation status:** Undetermined, but it should be categorized as VU (Vulnerable) at a minimum. **Major threats:** Local people extract saltpeter or other rocks and sometimes pollute the subterranean water. Also road construction is changing the original karstic landscape. **Conservation plans:** None. **Phylogenetic relationships:** It belongs to the *angularis* lineage.

Fig. 23 Distribution of *Sinocyclocheilus bicornutus*

12. ***Sinocyclocheilus brevibarbatus*** (Zhao et al. 2009a) (Fig. 24). **Common name:** Short-barbeled golden-line barbel (E, translation from Chinese). **Etymology:** *brevis* (L) short; *barbatus* (L) bearded. **Major synonyms:** None. **Ecological classification:** Troglobite. **History:** Collected for the first time on 4 June 2001. **Morphological characteristics:** D iii, 7 A iii, 5 P i, 14–15 V i, 8–9. Body humpbacked; scaled; snout slightly duckbilled; mouth sub-inferior; normal eyes; joint of dentary-angulars close to each other at isthmus; lateral line straight, possessing 49–51 scales; short maxillary and rictal barbels. Maximum standard length: 168.3 mm (Zhao et al. 2009a). Coloration of

Fig. 22 *Sinocyclocheilus bicornutus*, paratype, IHB12209043-9o5o241, standard length 98.8 mm. (Photo by Zhao, Y.)

Fig. 24 *Sinocyclocheilus brevibarbatus*, holotype, ASIZB74229, standard length 69.8 mm. (Photo by Zhao, Y.)

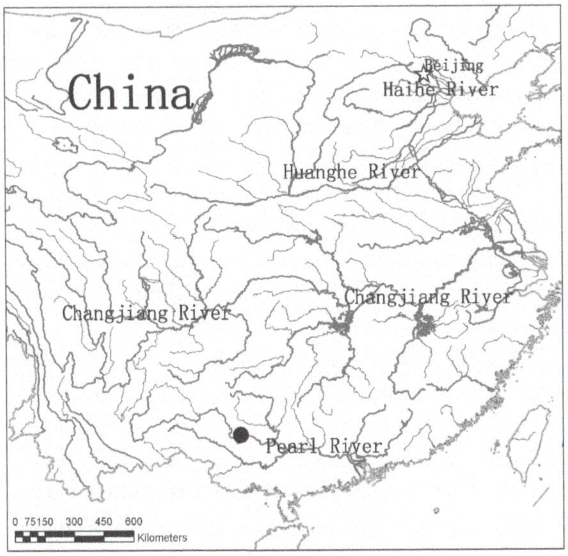

Fig. 25 Distribution of *Sinocyclocheilus brevibarbatus*

preserved specimens (in alcohol) is brownish, ventral part light yellowish, all fins light yellowish. **Troglomorphic characters:** Humpbacked. **Distribution:** China: Guangxi Zhuang Autonomous Region: Du'an County: Gaoling town (ca. 24°03′ N, 108°02′ E) (Fig. 25). **Habitat and ecology:** Karst cave. **Food and feeding:** Unknown. **Reproduction and development:** Unknown. **Other behavior:** Unknown. **Conservation status:** Undetermined. **Major threats:** No survey has been conducted to assess its possible threats. **Conservation plans:** None. **Phylogenetic relationships:** It belongs to the *cyphotergous* lineage.

13. *Sinocyclocheilus brevis* Lan and Chen, 1992 (in Chen and Lan 1992) (Fig. 26) **Common name:** Short-body golden-line barbel (E, translation from Chinese). **Etymology:** *brevis* (L) short. **Major synonyms:** None. **Ecological classification:** Troglophile. **History:** Collected in August 1987. **Morphological characteristics:** D iii, 7 A iii, 5 P i, 15–16 V i, 9–10. A little humpbacked, but not obvious; lateral line scales 51–56, scale rows above and below lateral 19–21, 11 respectively; rictal barbels long, reaching posterior edge of preopercle. Maximum standard length: 96.5 mm (Zhao 2006). Coloration of preserved specimens (in alcohol) light yellowish, all fins gray-whitish. **Troglomorphic characters:** None known except possibly being a little humpbacked. **Distribution:** China: Guangxi Zhuang Autonomous Region: Luocheng County (ca. 24°47′ N, 108°53′ E)

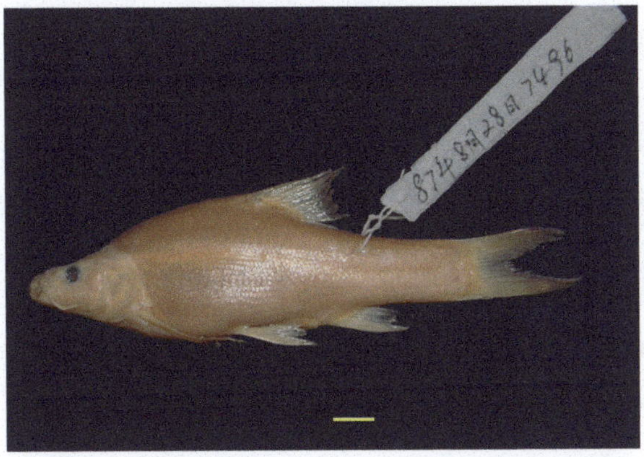

Fig. 26 *Sinocyclocheilus brevis*, holotype, IHB12209033-87087496, standard length 96.5 mm. (Photo by Zhao, Y.)

(Fig. 27). **Habitat and ecology:** Subterranean stream. **Food and feeding:** Unknown. **Reproduction and development:** Unknown. **Other behavior:** Unknown. **Conservation status:** It should be considered rare based on the field observations by one of us (YZ). **Major threats:** No survey has been conducted to assess its possible threats. **Conservation plans:** None. **Phylogenetic relationships:** It belongs to the *cyphotergous* lineage.

14. *Sinocyclocheilus broadihornes* Li and Mao, 2007 (Fig. 28). **Common name:** Broad horn golden-line barbel (E, translation from Chinese); horned fish (E, translation from Chinese). **Etymology:** *broadi*

Fig. 27 Distribution of *Sinocyclocheilus brevis*

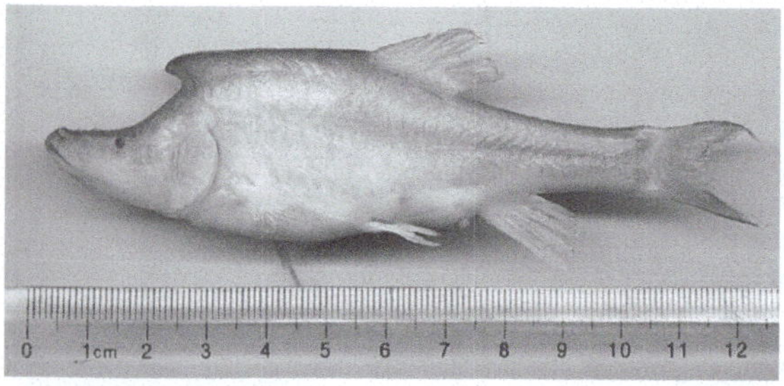

Fig. 28 *Sinocyclocheilus broadihornes*, holotype, Li 060412001, standard length 106 mm. (Cited from (Li and Mao 2007), photocopy)

(Anglo Saxon) broad; *horn* (Anglo Saxon) horn. **Major synonyms:** None. **Ecological classification:** Troglobite. **History:** This fish was first found in the 1980s by local people. Type specimens were collected on 12 April 2006 (Li and Mao 2007). **Morphological characteristics:** D iii, 6–7 A ii, 5 P i, 12–13 V i, 5–6. Horn forward on nape; snout duckbilled; mouth inferior; barbels well developed; almost naked, only lateral line scales present, 35–37. Maximum standard length: 106 mm (Li and Mao 2007). Live coloration is light gray-whitish. **Troglomorphic characters:** Microphthalmic, depigmented, reduced number of scales, horn. **Distribution:** China: Yunnan Province: Shilin County: Shilin town: Bat Cave (ca. 24°48′ N, 108°18′ E) at 1,750 m above sea level. (Fig. 29). **Habitat and ecology:** Bat Cave is a huge karst cave with a 400 m^2 subterranean lake. The water is very clear (Li and Mao 2007). **Food and feeding:** Unknown. **Reproduction and development:** Unknown. **Other behavior:** It does not respond to intense illumination, but is sensitive to water vibrations (Li and Mao 2007). **Conservation status:** Undetermined. **Major threats:** No survey has been conducted to assess its possible threats. **Conservation plans:** None. **Phylogenetic relationships:** It belongs to the *angularis* lineage.

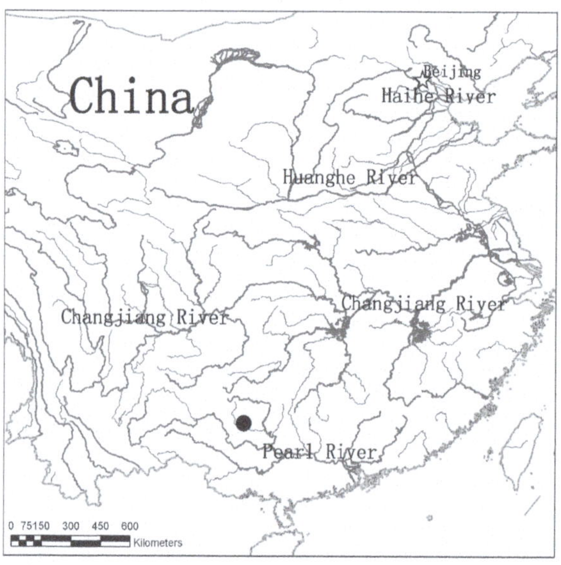

Fig. 29 Distribution of *Sinocyclocheilus broadihornes*

15. ***Sinocyclocheilus cyphotergous*** (Dai, 1988) (Fig. 30). **Common name:** Humpbacked golden-line barbel (E, translation from Chinese). **Etymology:** *cypho* (Gr) curved, convexed; *terg(o)us* (L) back, because of its humpback. **Major synonyms:** *Gibbibarbus cyphotergous*, original combination. **Ecological classification:** Troglobite. **History:** Type specimen was collected in 1986 (Dai 1988). **Morphological characteristics:** D iii, 8 A iii, 5 P i, 13–15 V i, 8. Humpbacked with a sarcous horn on back; eyes small, degenerate; pectoral fin short, not reaching pelvic fin base; lateral line straight, possessing 53 scales; scales small, embedded. Maximum standard length: 122 mm (Wang et al. 1995). Coloration of preserved specimens (in alcohol) brownish, back darker, no speckles, all fins light brownish. **Troglomorphic characters:** Microphthalmic or degenerated eyes, depigmented, scales small and embedded, horn, humpbacked. **Distribution:** China: Guizhou Province: Luodian County: cave at Daxiao Dong, Dajing village (ca. 25°37′ N, 107°05′ E) (Fig. 31). **Habitat and ecology:** Found at 300 m from the entrance of the cave. **Food and feeding:** Unknown. **Reproduction and development:** Unknown. **Other behavior:**

Fig. 30 *Sinocyclocheilus cyphotergous*, IHB12209040, standard length 116.2 mm (Photo by Zhao, Y.)

Fig. 31 Distribution of *Sinocyclocheilus cyphotergous*

Unknown. **Conservation status:** Endangered suggested based on the field observations by one of us (YZ). **Major threats:** No survey has been conducted to assess its possible threats. **Conservation plans:** None. **Phylogenetic relationships:** It belongs to the *cyphotergous* lineage.

16. *Sinocyclocheilus donglanensis* (Zhao et al. 2006a) (Fig. 32). **Common name:** Donglan golden-line barbel (E, translation from Chinese). **Etymology:** *donglanensis* (L) after the name of the type locality, Donglan County. **Major synonyms:** None. **Ecological classification:** Troglophile. **History:** Type specimens were collected in 2002 (Zhao et al. 2006a). **Morphological characteristics:** D iii, 7 A iii, 5 P i, 14–15 V i, 9. A completely scaled body with well-developed eyes; a curved lateral line possessing 57–64 scales; 8–9 pre-dorsal vertebrae; 8–9 gill rakers; joint of dentary-angulars not close to each other at the isthmus. Maximum standard length: 123.9 mm (Zhao et al. 2006a). Coloration of preserved specimens is brownish, abdomen light grayish; an indistinct black stripe running from posterior border of operculum along body midline to caudal fin base. Pectoral, dorsal and caudal fins dark grayish, pelvic and anal fins light yellowish. **Troglomorphic characters:** None known. **Distribution:** China: Guangxi Zhuang Autonomous Region: Donglan County: Taiping Town (ca. 24°20′ N, 107°24′ E) (Fig. 33). **Habitat and ecology:** The locality is the same as for the type specimens of *Sinocyclocheilus altishoulderus* (Zhao et al. 2006a). **Food and feeding:** Unknown. **Reproduction and development:** Unknown. **Other behavior:** Unknown. **Conservation status:** Undetermined. **Major threats:** Exploitation of the underground water resource, solid waste and pesticide pollution, and general landscape alterations. **Conservation plans:** None. **Phylogenetic relationships:** It belongs to the *cyphotergous* lineage.

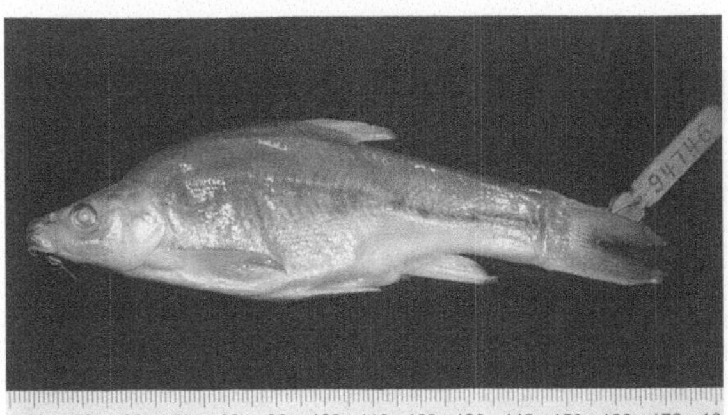

Fig. 32 *Sinocyclocheilus donglanensis*, holotype, ASIZB94746, standard length 98.1 mm. (Photo by Zhao, Y.)

17. *Sinocyclocheilus furcodorsalis* (Chen et al. 1997) (Fig. 34). **Common name:** crossed-fork back golden-line barbel (E, translation from Chinese). **Etymology:** *furco* (L) forked; *dorsalis* (L) on the back, after the bifurcation on the anterior part of the horn. **Major synonyms:** *Sinocyclocheilus tianeensis*

Fig. 33 Distribution of *Sinocyclocheilus donglanensis*

Fig. 34 *Sinocyclocheilus furcodorsalis*, ASIZB73164, standard length 69.4 mm. (Photo by Zhao, Y.)

(Li et al. 2003d). **Ecological classification:** Troglobite. **History:** Type specimens were collected in September 1993 (Chen et al. 1997). **Morphological characteristics:** D iii, 7 A iii, 5 P i, 14–15 V i, 7. Eyeless. Horn forked on tip; snout duck-billed; blind; mouth sub-inferior; scales small with most of them being embedded; pectoral fin long, beyond pelvic fin insertion. Maximum standard length: 88.0 mm (Zhao 2006). Live coloration is whitish and semi-transparent. Coloration of preserved specimens (in alcohol) is white-yellowish. **Troglomorphic characters:** No externally visible eyes, depigmented, scales reduced in size and embedded. **Distribution:** China: Guangxi Zhuang Autonomous Region: in an underground stream in Tian'e County (24°58′ N, 107°02′ E) (Fig. 35). The species is sympatric with *Triplophysa tianeensis*. **Habitat and ecology:** No information available. **Food and feeding:** Unknown. **Reproduction and development:** Unknown. **Other behavior:** Unknown. **Conservation status:** Endangered, suggested based on the field observations by one of us (YZ). **Major threats:** No survey has been conducted to assess its possible threats. **Conservation plans:** None. **Phylogenetic relationships:** It belongs to the *angularis* lineage.

Fig. 35 Distribution of *Sinocyclocheilus furcodorsalis*

18. *Sinocyclocheilus grahami* (Regan, 1904) (Fig. 36). **Common name:** Dianchi golden-line barbel (E, translation from Chinese). **Etymology:** *grahami* (L) after John Graham, who provided the specimens. **Major synonyms:** *Barbus grahami* (original combination), *Percocypris grahami* (Wu 1963), *Sinocyclocheilus hei*, *Sinocyclocheilus guanduensis*, *Sinocyclocheilus huanglongdongensis* (Xiao et al. 2004). **Ecological classification:** Troglophile. **History:** Original description did not mention the collection date of the type specimens. In the past, it was one of the most commercially exploited fish in Lake Dianchi. **Morphological characteristics:** D iii, 7 A iii, 5 P i, 15–17 V i, 8–9. Body elongated; pelvic fin insertion posterior to vertical line of dorsal fin insertion; most of scales embedded, lateral line scales 60–74; rictal barbel not reaching to posterior edge of preopercle. Maximum standard length: 130.1 mm (Zhao 2006). Coloration of preserved specimens (in alcohol) is brownish, abdomen lighter. **Troglomorphic characters:** None known although most scales are embedded. **Distribution:** China: Yunnan Province: Lake Dianchi and its connected subterranean waters (24°48′ N, 102°43′ E) (Fig. 37). **Habitat and ecology:** The Dianchi Lake, with many subterranean streams, is 370 km^2 in total surface, with a 4.4 m in depth on average, and located at 1,885 m above sea level. This is the eight largest lake in China and the

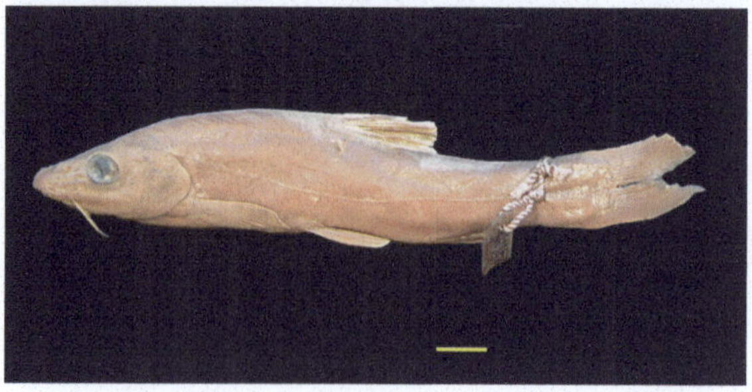

Fig. 36 *Sinocyclocheilus grahami*, ASIZB03496, standard length 112.6 mm. (Photo by Zhao, Y.)

Fig. 37 Distribution of *Sinocyclocheilus grahami*

largest in the Yunnan Province. **Food and feeding:** Aquatic insects, zooplankton (Yue and Chen 1998). **Reproduction and development:** During the spawning season it swims into the karst caves with springs on the lake beach to spawn and hatch (Yue and Chen 1998). There were two successful artificial breeding experiments in 2007. The mean fertilization rate was 73% and the mean hatching rate 36%. After 10 days 95% of the young fish survived and they reached 8–12 mm in length (Yang et al. 2007). **Other behavior:** Unknown. **Conservation status:** It was listed as a Second Class State Protected Animal in 1989. Endangered in China's Red Data Book (Yue and Chen 1998). Endangered in China's Species Red List (Wang and Xie 2004). **Major threats:** Anthropogenic impacts including the reclamation of land around the lake, water pollution, introduction of exotic species, and overfishing for commercial purposes. **Conservation plans:** No plans. **Phylogenetic relationships:** It belongs to the *tingi* lineage.

19. *Sinocyclocheilus guilinensis* (Zhao et al. 2009b) (Fig. 38). **Common name:** Guilin golden-line barbel (E, translation from Chinese). **Etymology:** *guilinensis* (L) after the collection locality, Guilin City. **Major synonyms:** *Sinocyclocheilus jii* (Zhang and Dai 1992). **Ecological classification:** Troglophile. **History:** First collected scientifically in 1982. **Morphological characteristics:** D iii, 7 A iii, 5 P i, 13–15 V i, 8. A completely scaled body with well-developed eyes; last unbranched ray of dorsal fin soft, without serration; scale row counts above and below lateral line 19–20 and 11–12 respectively; circum-peduncular scale counts 34–36; gill rakers 8–11; pre-dorsal vertebrae 8–9. Maximum standard length: 136.5 mm (Zhao et al. 2009b). **Troglomorphic characters:** None known. **Distribution:** China: Guangxi Zhuang Autonomous Region: Guilin city (ca. 25°17′ N, 110°17′ E) (Fig. 39). **Habitat and ecology:** Subterranean stream, karst cave, and deep well. **Food and feeding:** Unknown. **Reproduction and development:** Unknown. **Other behavior:** Unknown. **Conservation status:** Undetermined. **Major threats:** This species is found very near to Guilin City, which is heavily visited by tourists and which is being rapidly developing. **Conservation plans:**

Fig. 38 *Sinocyclocheilus guilinensis*, holotype, ASIZB113753, standard length 107.3 mm. (Photo by Zhao, Y.)

Fig. 39 Distribution of *Sinocyclocheilus guilinensis*

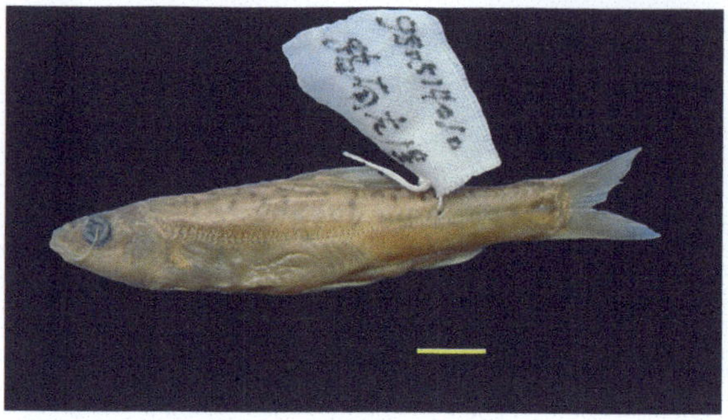

Fig. 40 *Sinocyclocheilus guishanensis*, paratype, Li980514005, standard length 64.9 mm. (Photo by Zhao, Y.)

None. **Phylogenetic relationships:** It belongs to the *jii* lineage.

20. ***Sinocyclocheilus guishanensis*** Li, 2003 (in Li et al. 2003b) (Fig. 40). **Common name:** Guishan golden-line barbel (E, translation from Chinese). **Etymology:** *guishanensis* (L) after the type locality, Guishan town. **Major synonyms:** None. **Ecological classification:** Troglophile. **History:** Type specimens were collected in 1998 (Li et al. 2003b). **Morphological characteristics:** D iii, 7 A iii, 5 P i, 15–18 V i, 8–9. Body elongated and scaled; pectoral fin short, not reaching to insertion of pelvic fin base, which is opposite to dorsal fin insertion; gill rakers on first gill arch 5–8; upper corner of gill open below horizontal line of upper edge of eye. Maximum standard length: 109.1 mm (Zhao 2006). **Troglomorphic characters:** None known. **Distribution:** China: Yunnan Province: Shilin County: Guishan town (24°35′ N, 103°32′) (Fig. 41). **Habitat and ecology:** Subterranean stream but also in the water near the cave mouth, but never far more than 10 m. **Food and feeding:** Unknown. **Reproduction and development:** Unknown. **Other behavior:** Unknown. **Conservation status:** Undetermined, but should classified as rare as suggested based on the field observations by one of us (YZ). **Major threats:** No survey has been conducted to assess its possible threats. **Conservation plans:** None. **Phylogenetic relationships**: It belongs to the *tingi* lineage.

Fig. 41 Distribution of *Sinocyclocheilus guishanensis*

21. ***Sinocyclocheilus huaningensis*** Li, 1998 (Fig. 42). **Common name:** Huaning golden-line barbel (E, translation from Chinese). **Etymology:** *huaningensis* (L) after the name of the type locality, Huaning County. **Major synonyms:** None. **Ecological classification:** Troglophile. **History:** First collected in 1992 (Li et al. 1998). **Morphological characteristics:** D iii, 7 A iii, 5 P i, 16–18 V i, 8–9. Body elongated and scaled; pectoral fin reaching to insertion of pelvic fin base; dorsal fin insertion posterior to pelvic fin insertion; lateral line curved, possessing 58–63 scales. Maximum standard length: 92.2 mm (Zhao 2006). Live coloration is golden with small black spots. **Troglomorphic characters:** None known. **Distribution:** China: Yunnan Province: Huaning County: Panxi town (ca. 24°14′ N, 103° 06′) (Fig. 43). **Habitat and ecology:** Exit of subterranean stream. **Food and feeding:** Unknown. **Repro-**

Fig. 42 *Sinocyclocheilus huaningensis*, ASIZB79228, standard length 92.2 mm. (Photo by Zhao, Y.)

Fig. 43 Distribution of *Sinocyclocheilus huaningensis*

duction and development: Unknown. **Other behavior:** Unknown. **Conservation status:** Undetermined. It should be classified as Rare, based on the field observations by one of us (YZ). **Major threats:** No survey has been conducted to assess its possible threats. **Conservation plans:** None. **Phylogenetic relationships:** It belongs to the *tingi* lineage.

22. ***Sinocyclocheilus hugeibarbus*** Li and Ran, 2003 (in Li et al. 2003c) (Fig. 44). **Common name:** Super-long barbeled golden-line barbel (E, translation from Chinese). **Etymology:** *hugei* (Anglo Saxon) huge, big; *barba* (L) beard. **Major synonyms:** None. **Ecological classification:** Troglobite. **History:** First collected in May 1995 (Li et al. 2003c). **Morphological characteristics:** D iii, 7 A iii, 5 P i, 15–16 V i, 7–8. Body humpbacked; barbels long, rictal barbel beyond posterior edge of opercle; lateral line curved, possessing 65–82 scales. Maximum standard length: 200.5 mm (Zhao 2006). Live coloration is whitish, semi-transparent. **Troglomorphic characters:** Depigmented, humpbacked. **Distribution:** China: Guizhou Province: Libo County: Dongtang town and Bantanba village (24°14′ N, 103°06′ E) (Fig. 45). **Habitat and ecology:** Subterranean stream. **Food and feeding:** Unknown. **Reproduction and development:** Unknown. **Other behavior:** Unknown. **Conservation status:** Undetermined. It should be classified as Rare as suggested based on the field observations by one of us (YZ). **Major threats:** No survey has been conducted to assess its possible threats. **Conservation plans:** None. **Phylogenetic relationships:** It belongs to the *cyphotergous* lineage.

Fig. 44 *Sinocyclocheilus hugeibarbus*, paratype, Li200210003, standard length 200.5 mm. (Photo by Zhao, Y.)

23. ***Sinocyclocheilus hyalinus*** Chen and Yang, 1994 (in Chen et al. 1994) (Fig. 46). **Common name:** Translucent golden-line barbel (E, translation from Chinese); hyaline fish (E, translation from Chinese). **Etymology:** *hyalinus*, from *hyalinos* (Gr) glass, after the transparency of the fish. **Major synonyms:** None. **Ecological classification:** Troglobite. **History:** Two specimens were collected in June 1991. This may have been the first troglomorphic fish in China ever reported in writing (Romero 2001). **Morphological characteristics:** D iii, 7 A iii, 5 P i, 12–13 V i, 6–7. Body naked; blind; horned; pectoral fin long, beyond

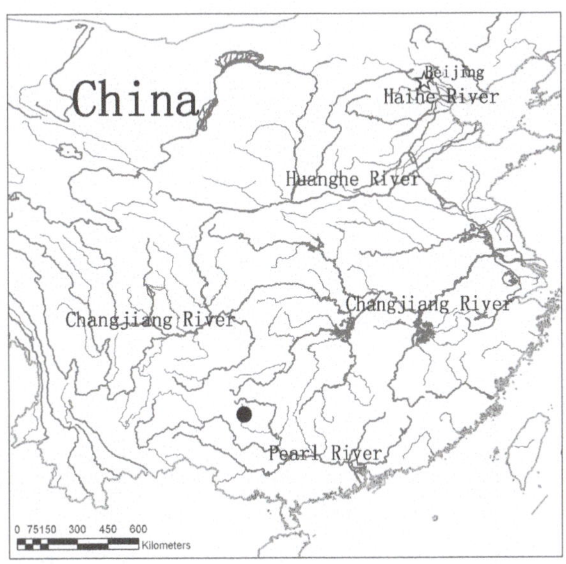

Fig. 45 Distribution of *Sinocyclocheilus hugeibarbus*

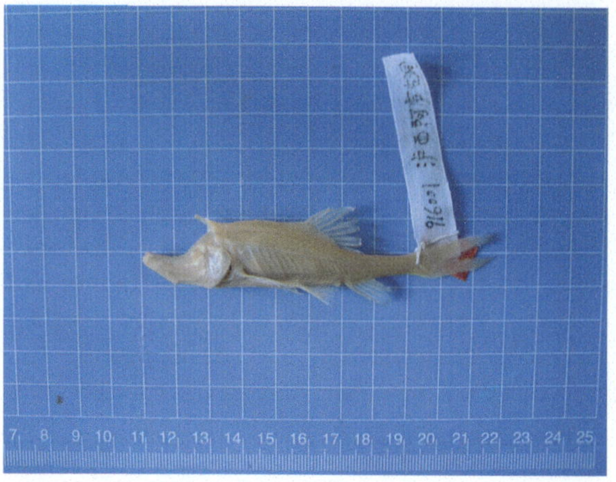

Fig. 46 *Sinocyclocheilus hyalinus*, holotype, KIZ916001, standard length 79.9 mm. (Photo by Zhao, Y.)

pelvic fin insertion; pharyngeal teeth 1 3 4: 4 3 1. Maximum standard length: 87 mm (Chen et al. 1994). Live coloration is whitish and semi-transparent. **Troglomorphic characters:** No externally visible eyes, depigmented, scaleless, horn. **Distribution:** China: Yunnan Province: Luxi County: Alu limestone caves (24°33′ N, 103°45′ E) (Fig. 47). **Habitat and ecology:** In an underground stream (Yusun River) of a cave of 625 m in length at an elevation of 1,712 m above sea level (Chen et al. 1994). **Food and feeding:** Feeds on microscopic animals and aquatic insects (Chen et al. 1994). **Reproduction and development:** Unknown. **Other behavior:** Unknown. **Conservation status:** VU in China's Species Red List (Wang and Xie 2004). **Major threats:** Excessive collecting and environmental degradation (Wang and Xie 2004). Alu Cave has been developed to a popular tourist site. **Conservation plans:** None. **Phylogenetic relationships:** It belongs to the *angularis* lineage.

Fig. 47 Distribution of *Sinocyclocheilus hyalinus*

24. *Sinocyclocheilus jii* Zhang and Dai, 1992 (Fig. 48). **Common name:** Ji golden-line barbel (E, translation from Chinese). **Etymology:** *Jii* (L) after Cunshan Ji, who described unofficially another species *S. guilinensis*. **Major synonyms:** None. **Ecological classification:** Troglophile. **History:** Type specimens were collected in 1992 (Zhang and Dai 1992). **Morphological characteristics:** D iii, 7 A iii, 5 P i, 14–15 V i, 8. Body scaled; last unbranched ray of dorsal fin soft, no serration; scale rows above and below lateral line 27–29, 15–17 respectively; scale around caudal peduncle 46–50; gill rakers on first gill 6–7. Maximum standard length is 123.6 mm. Live coloration is brownish with a black stripe along (a little above) lateral line. **Troglomorphic characters:** None known. **Distribution:** China: Guangxi Zhuang Autonomous Region: Fuchuan County (ca. 24°50′ N, 111°17′ E) (Fig. 49). **Habitat and ecology:** In the vicinity of a karst cave (Zhang and Dai 1992). **Food and feeding:** Unknown. **Reproduction and development:** Unknown. **Other behavior:** Unknown. **Conservation status:** Undetermined but it should be classified as Endangered as suggested based on the field observations by one of us (YZ). **Major threats:** No survey has been conducted to assess its possible threats. **Conservation plans:** None. **Phylogenetic relationships:** It belongs to the *jii* lineage.

25. *Sinocyclocheilus jiuxuensis* Li and Lan, 2003 (in Li et al. 2003a) (Fig. 50). **Common name:** Jiuxu

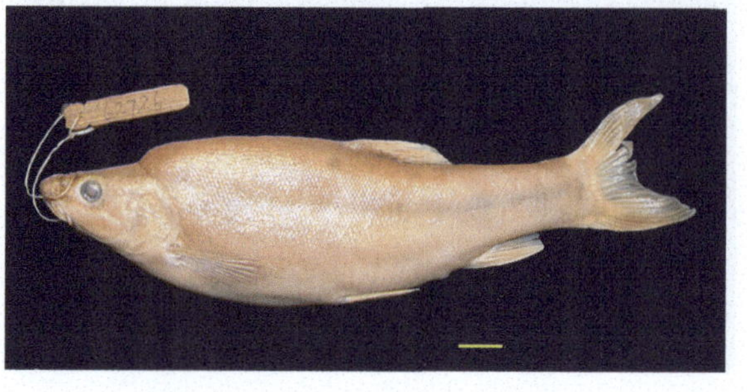

Fig. 48 *Sinocyclocheilus jii*, holotype, ASIZB62726, standard length 123.6 mm. (Photo by Zhao, Y.)

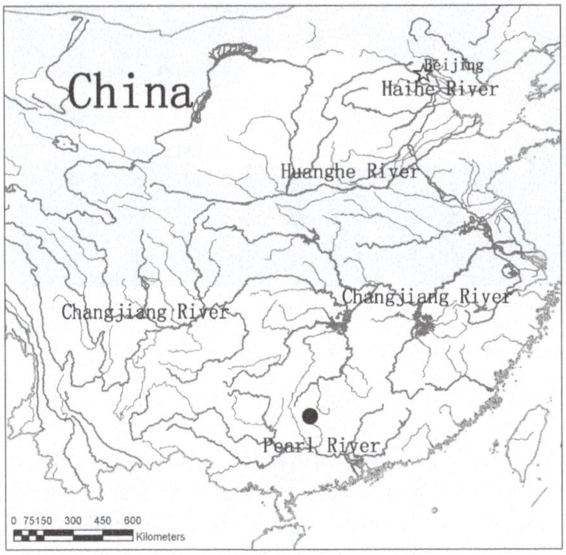

Fig. 49 Distribution of *Sinocyclocheilus jii*

Fig. 51 Distribution of *Sinocyclocheilus jiuxuensis*

golden-line barbel (E, translation from Chinese). **Etymology:** *Jiuxuensis* (L) after the name of the type locality, Jiuxu town. **Major synonyms:** None. **Ecological classification:** Troglobite. **History:** Type specimens were collected in 2002 (Li et al. 2003a). **Morphological characteristics:** D iii, 7 A iii, 5 P i, 13–15 V i, 7–8. Body scaled; eyes small; humpbacked; lateral line straight, possessing 42–51 scales; barbels short. Maximum standard length: 136.1 mm. Live coloration is whitish and semi-transparent. **Troglomorphic characters:** Microphthalmic, depigmented, humpbacked. **Distribution:** China: Guangxi Zhuang Autonomous Region: Jinchengjiang city: Jiuxu town (24°32′ N, 107°45′ E) (Fig. 51). **Habitat and ecology:** Subterranean stream. **Food and feeding:** Unknown. **Reproduction and development:** Unknown. **Other behavior:** Unknown. **Conservation status:** Undetermined but it should be classified as Rare, based on the second author's (YZ) anecdotal field observations. **Major threats:** No survey has been conducted to assess its possible threats. **Conservation plans:** None. **Phylogenetic relationships:** It belongs to the *cyphotergous* lineage.

26. *Sinocyclocheilus lateristritus* Li, 1992 (Fig. 52). **Common name:** Side-stripe golden-line barbel (E, translation from Chinese). **Etymology:** *lateris* (L) brick, tile; *striatus* (L) having stripes. **Major synonyms:** None. **Ecological classification:** Troglophile. **History:** Type specimens were collected in 1986 (Li 1992). **Morphological characteristics:** D iii, 7 A iii, 5 P i, 15–19 V i, 8–9. Body elongated; last unbranched dorsal ray hard with serration; pelvic fin

Fig. 50 *Sinocyclocheilus jiuxuensis*, ASIZB102260, standard length 125.1 mm. (Photo by Zhao, Y.)

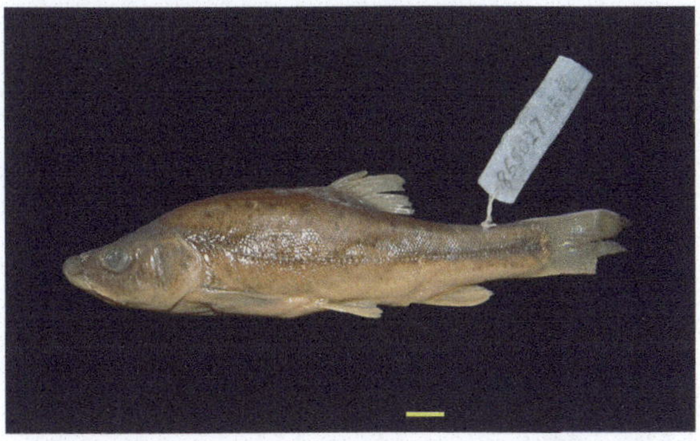

Fig. 52 *Sinocyclocheilus lateristritus*, holotype, IHB12209036-865027, standard length 120.0 mm. (Photo by Zhao, Y.)

insertion posterior to vertical line of dorsal fin insertion; rictal short, not beyond to post edge of preopercle. Maximum standard length: 120.0 mm (Zhao 2006). Live coloration is yellow-brownish, a black stripe along lateral line. **Troglomorphic characters:** None known. **Distribution:** China: Yunnan Province: Luliang County: Fanghua town (26°10′ N, 103°43′) (Fig. 53). **Habitat and ecology:** Mouth of subterranean stream, which has been built into a pond, locates in a deserted residential area 1,879 m above sea level. **Food and feeding:** Unknown. **Reproduction and development:** Unknown. **Other behavior:** Unknown. **Conservation status:** Undetermined. It should be classified as Rare, based on the field observations by one of us (YZ). **Major threats:** Original exit of the subterranean stream has been replaced by several artificial pounds. **Conservation plans:** None. **Phylogenetic relationships:** It belongs to the *tingi* lineage.

27. *Sinocyclocheilus lingyunensis* Li, Xiao and Luo, 2000 (in Li et al. 2000b) (Fig. 54). **Common name:** Lingyu golden-line barbel (E, translation from Chinese). **Etymology:** *lingyunensis* (L) after the type locality, Lingyu County. **Major synonyms:** None. **Ecological classification:** Troglobite. **History:** Holotype collected on 18 April 1999. **Morphological characteristics:** D iii, 7 A iii, 5 P i, 14 V i, 8–9. Body a little humpbacked, not obvious; scaled; pectoral fin long, beyond pelvic fin insertion; lateral line scales 54–57; gill raker on first gill arch 10. Maximum standard length: 110.2 mm (Zhao 2006). Living individuals are white-pinkish, semi-transparent. **Troglomorphic characters:** Depigmented. **Distribution:** China: Guangxi Zhuang Autonomous Region: Lingyun County: Sicheng Town: Shadong Cave (24°20′ N, 1,063°32′ E) (Fig. 55). **Habitat and ecology:** Subterranean stream, 470 m above sea level. It shares the habitat with two other troglobite fish species, *Sinocyclocheilus microphthalmus* and *Schistura lingyunensis* (Li et al. 2000b). **Food and feeding:** Unknown. **Reproduction and development:** Believed that the reproductive season begins after May. **Other behavior:** Unknown. **Conservation status:** None given but it should be classified as Endangered, based on the field observations by one of us (YZ). **Major threats:** Environmental impacts due to road

Fig. 54 *Sinocyclocheilus lingyunensis*, ASIZB 73038, standard length 84.3 mm. (Photo by Zhao, Y.)

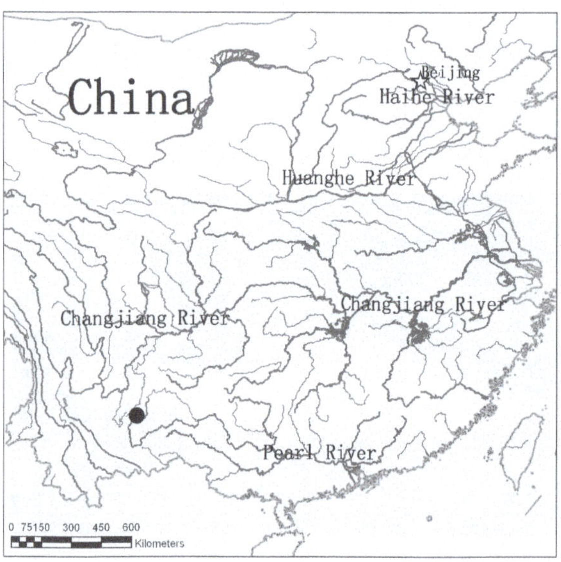

Fig. 53 Distribution of *Sinocyclocheilus lateristritus*

Fig. 55 Distribution of *Sinocyclocheilus lingyunensis*

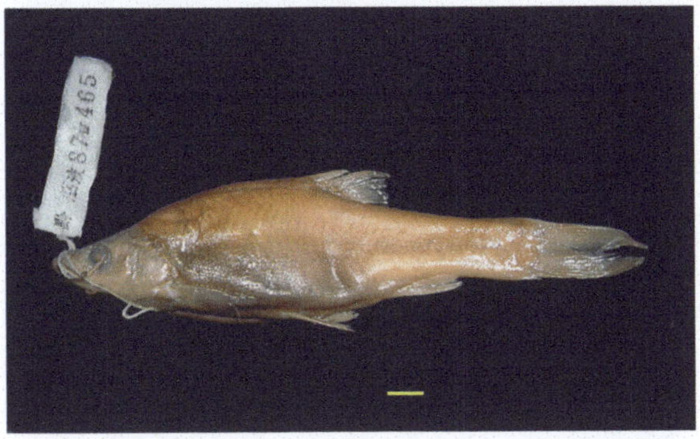

Fig. 56 *Sinocyclocheilus longibabarbus*, paratype, IHB87IV465, standard length 114.1 mm. (Photo by Zhao, Y.)

construction. **Conservation plans:** None. **Phylogenetic relationships:** It belongs to the *cyphotergous* lineage.

28. *Sinocyclocheilus longibarbatus* Wang and Chen, 1989 (Fig. 56). **Common name:** Long-barbeled golden-line barbel (E, translation from Chinese). **Etymology:** *longibarbatus* (L) after its long barbels. **Major synonyms:** None. **Ecological classification:** Troglobite. **History:** First collected in April 1984. **Morphological characteristics:** D iii, 7 A iii, 5 P i, 15–18 V i, 8–10. Body a little humpbacked; lateral line curved, possessing 63–75 scales; scale rows above and below lateral line 28–29, 15–18 respectively. Maximum standard length: 137 mm (Wang and Chen 1989). **Troglomorphic characters:** Microphthalmic, depigmented, humpbacked. **Distribution:** China: Guizhou Province: Libo County (25° 24′ N, 107°52′ E) (Fig. 57). **Habitat and ecology:** Karst cave, exit of subterranean stream (Wang and Chen 1989). **Food and feeding:** Unknown. **Reproduction and development:** Unknown. **Other behavior:** Unknown. **Conservation status:** None assigned but it should be classified as Rare, based on the field observations by one of us (YZ). **Major threats:** There is a National Park (Maolan Karst Forestry Natural Reserve) in Libo County. The main target for protection in this park is the forest, not the karst system. Inside the park the environment has been protected, but outside the park, the habitat is being rapidly affected by development. This fish can be found both inside and outside the protected area of the park. **Conservation plans:** None. **Phylogenetic relationships:** It belongs to the *cyphotergous* lineage.

29. *Sinocyclocheilus longifinus* Li, 1998 (in Li et al. 1998) (Fig. 58). **Common name:** Long fin golden-

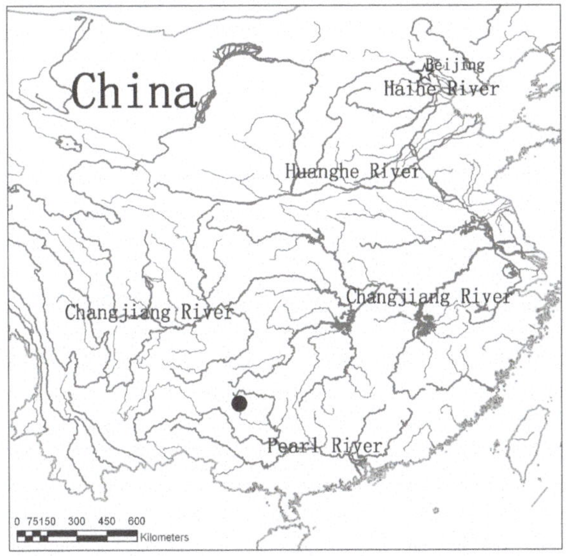

Fig. 57 Distribution of *Sinocyclocheilus longibabarbus*

line barbel; white golden-line barbel (E, translation from Chinese). **Etymology:** *longi* (L) long; *finus* (L) fin, after the long pectoral fin of the fish. **Major synonyms:** None. **Ecological classification:** Troglobite. **History:** First collected in 1991 (Li et al. 1998). **Morphological characteristics:** D iii, 7 A ii, 5 P i, 16 V i, 8. Body elongated; almost naked, only a few small scales on posterior part of caudal peduncle; mouth sub-superior; 2 pairs of barbels, well-developed; lateral line complete, straight, possessing 70–72 pores; pectoral fin long, reaching to pelvic fin insertion. Maximum standard length: 154 mm (Li et al. 1998). Live coloration is whitish, back grayish. Coloration of preserved specimens (in alcohol) is light grayish in the dorsal area, all fins lack coloration (Li et al. 1998). **Troglomorphic characters:** Depigmented, scaleless greatly reduced in size and number. **Distribution:** China: Guizhou Province: Huaning County: Panxi town (ca. 24°14′ N, 103°06′ E) (Fig. 59). **Habitat and ecology:** Exit of subterranean

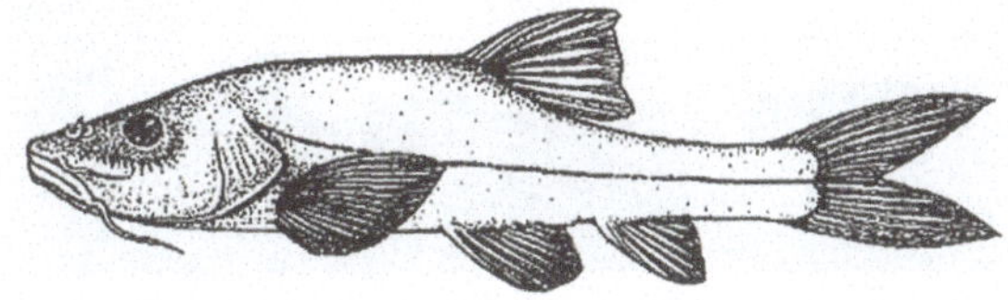

Fig. 58 *Sinocyclocheilus longifinus*, cited from (Li et al. 1998)

Fig. 59 Distribution of *Sinocyclocheilus longifinus*

Fig. 61 Distribution of *Sinocyclocheilus luopingensis*

stream, same locality as *Sinocyclocheilus huaningensis* (Li et al. 1998). **Food and feeding:** Unknown. **Reproduction and development:** Unknown. **Other behavior:** Unknown. **Conservation status:** At least Rare since only two specimens (the type specimens) have been collected so far, one in 1991, another in 1997. **Major threats:** This fish is known only from the Heilongtan Pool. This locality has been developed into a local park and opened for tourism. **Conservation plans:** None. **Phylogenetic relationships:** It belongs to the *tingi* lineage.

30. *Sinocyclocheilus luopingensis* Li and Tao, 2002 (in Li et al. 2002c) (Fig. 60). **Common name:** Luoping golden-line barbel (E, translation from Chinese). **Etymology:** *luopingensis* (L) after the name of the type locality, Luoping County. **Major synonyms:** None. **Ecological classification:** Troglobite. **History:** First collection took place on 1 August 1998 (Li et al. 2002c). **Morphological characteristics:** D iii, 7 A iii, 5 P i, 14 V i, 8. Body humpbacked, scales embedded; lateral line straight, possessing 57 scales; gill rakers on first gill arch 8. Maximum standard length: 130.8 mm (Zhao 2006). Live coloration is gold, back darker, all fins light yellowish. **Troglomorphic characters:** Scales embedded, humpbacked. **Distribution:** China: Yunnan Province: Luoping County (ca. 24°53′ N, 104°18′ E) (Fig. 61). **Habitat and ecology:** Subterranean stream. **Food and feeding:** Unknown. **Reproduction and development:** Unknown. **Other behavior:** Unknown. **Conservation status:** It should be classified as Endangered since only one specimen has been collected since 1996. **Major threats:** No survey has been conducted to assess its possible threats. **Conservation plans:** None. **Phylogenetic relationships:** It belongs to the *cyphotergous* lineage.

Fig. 60 *Sinocyclocheilus luopingensis*, holotype, Li980831001, standard length 130.8 mm. (Photo by Zhao, Y.)

31. *Sinocyclocheilus macrocephalus* Li, 1985 (Fig. 62). **Common name:** Big head golden-line barbel; big mouth oil fish (E, translation from Chinese). **Etymology:** *macros* (Gr) long, big; *cephal* (Gr) head. **Major synonyms:** None. **Ecological classification:** Troglophile. **History:** Type specimens were collected in 1966 (Li 1985). **Morphological**

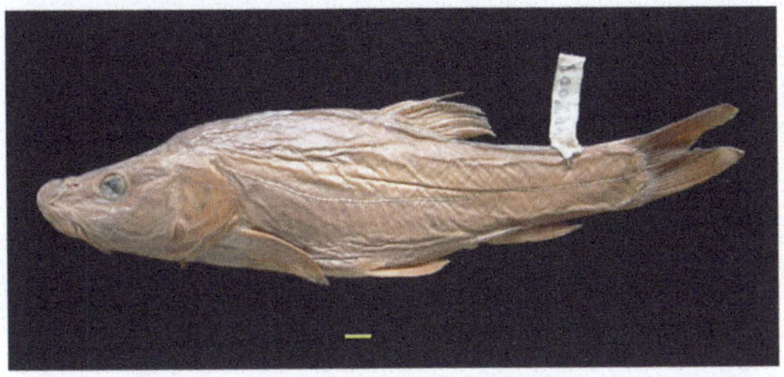

Fig. 62 *Sinocyclocheilus macrocephalus*, holotype, IHB12209012-662001, standard length 202.3 mm. (Photo by Zhao, Y.)

characteristics: D iii, 7 A iii, 5 P i, 15–17 V i, 8. Body elongated; mouth slightly superior; lateral line curved, possessing 63–72 scales; last unbranched dorsal ray strong with serration. Maximum standard length: 202.3 mm (Zhao 2006). Live coloration is golden. **Troglomorphic characters:** None known. **Distribution:** China: Yunnan Province: Shilin County: Heilongtan reservoir (ca. 24°46′ N, 103°16′ E) (Fig. 63). **Habitat and ecology:** Subterranean stream that now flows into an anthropogenic reservoir. **Food and feeding:** Unknown. **Reproduction and development:** Unknown. **Other behavior:** Unknown. **Conservation status:** None assigned but it should be classified as Vulnerable, based on the field observations by one of us (YZ). **Major threats:** Environment degradation due to modifications of the subterranean streams containment into a small reservoir. **Conservation plans:** None. **Phylogenetic relationships:** It belongs to the *tingi* lineage.

Fig. 63 Distribution of *Sinocyclocheilus macrocephalus*

Fig. 64 *Sinocyclocheilus macrolepis*, holotype, IHB12209035-87IV457, standard length 93.3 mm. (Photo by Zhao, Y.)

32. *Sinocyclocheilus macrolepis* Wang and Chen, 1989 (Fig. 64). **Common name:** Large scale golden-line barbel (E, translation from Chinese). **Etymology:** *macros* (Gr) long, big; *lepis* (Gr) scale. **Major synonyms:** None. **Ecological classification:** Troglophile. **History:** Type specimens were collected in 1987 (Wang and Chen 1989). **Morphological characteristics:** D iii, 7 A iii, 5 P i, 12–15 V i, 7–10. Body elongated; last unbranched ray of dorsal fin soft, without serration; lateral line scales 53–56; scale row above and below lateral line 12–14, 5–6 respectively. Maximum standard length: 122 mm (Wang and Chen 1989). Coloration of preserved specimens (in alcohol) is dark brownish, lower lighter. **Troglomorphic characters:** None known. **Distribution:** China: Guizhou Province: Libo County: Wangmeng town (ca. 25°16′ N, 107°44′ E); Guangxi Zhuang Autonomous Region: Nandan County (ca. 24°59′ N, 107°32′ E) (Fig. 65). **Habitat and ecology:** Exit of subterranean stream (Wang and Chen 1989). **Food and feeding:** Algae (Wang and Chen 1989) and probably insects too. **Reproduction and development:** Spawning season starts at the end of April. Egg diameter is 1.5–2 mm. **Other behavior:** Unknown. **Conservation status:** None assigned but it should be classified as Vulnerable, based on the field observations by one of us (YZ). **Major threats:** No survey has been conducted to assess its possible threats. **Conservation plans:** None. **Phylogenetic relationships:** It belongs to the *cyphoterous* lineage.

33. *Sinocyclocheilus macrophthalmus* Zhang and Zhao, 2001 (Fig. 66). **Common name:** Big eye golden-line barbel (E, translation from Chinese). **Etymology:** *macros* (Gr) long, big; *ophthalmus* (Gr)

Fig. 65 Distribution of *Sinocyclocheilus macrolepis*

Fig. 67 Distribution of *Sinocyclocheilus macrophthalmus*

eyes. **Major synonyms:** None. **Ecological classification:** Troglophile. **History:** The first collection was in 1999 (Zhang and Zhao 2001). **Morphological characteristics:** D iii, 7 A iii, 5 P i, 14–15 V i, 8. Body elongated, scaled; mouth inferior; gill rakers on first gill arch 10–11. Maximum standard length: 97.0 mm (Zhao 2006). Live coloration is light yellowish, semi-transparent, a dark spot on the caudal base. After being kept for several months under natural light, individuals become more pigmented. **Troglomorphic characters:** Depigmented. **Distribution:** China: Guangxi Zhuang Autonomous Region: Du'an County: Xia'ao town (ca. 24°15′ N, 107°05′ E) (Fig. 67). **Habitat and ecology:** Karst cave at 210 m above sea level. The species is sympatric with at least *Protocobitis typhlops* and *Oreonectes translucens*. **Food and feeding:** It feeds on water flea and pet fish food in captivity. **Reproduction and development:** Unknown. **Other behavior:** Unknown. **Conservation status:** None assigned but it should be classified as Vulnerable, based on the field observations by one of us (YZ). **Major threats:** Excessive collecting. **Conservation plans:** None. **Phylogenetic relationships:** It belongs to the *cyphoterous* lineage.

34. *Sinocyclocheilus macroscalus* (Li, 1994) (after Li 1992) (Fig. 68). **Common name:** Luliang golden-line barbel (E, translation from Chinese). **Etymology:** *macros* (Gr) long, big; *scalus* (L) scale. **Major synonyms:** *Anchicyclocheilus macrolepis, Sinocyclocheilus macrolepis*. **Ecological classification:** Troglophile. **History:** Type specimens were collected in 1986 (Li 1992). **Morphological characteristics:** D iii, 7 A iii, 5 P i, 16–17 V i, 9. Body elongated; mouth terminal; last unbranched dorsal ray strong with serration; lateral line scale the same size as neighbor-

Fig. 66 *Sinocyclocheilus macrophthalmus*, holotype, ASIZB70907, standard length 44.9 mm. (Photo by Zhao, Y.)

Fig. 68 *Sinocyclocheilus macroscalus*, holotype, IHB12209034-866007, standard length 136.9 mm

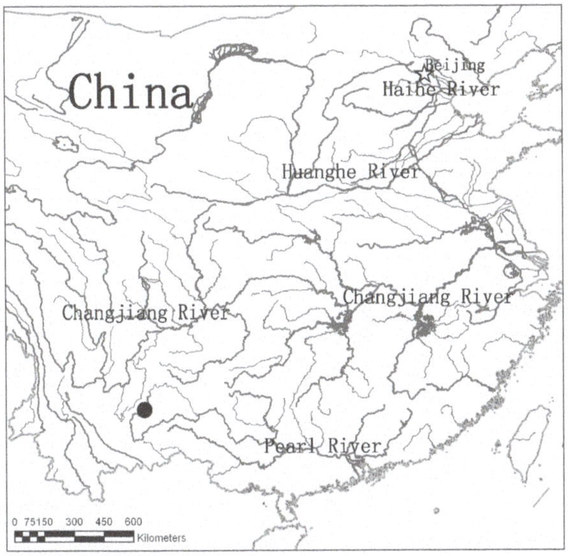

Fig. 69 Distribution of *Sinocyclocheilus macroscalus*

ing scales, lateral line curved. Maximum standard length: 167.6 mm (Zhao 2006). Live coloration is bright golden. **Troglomorphic characters:** None known. **Distribution:** China: Yunnan Province: Luliang County: Fanghua town (26°10′ N, 103°43′) (Fig. 69). **Habitat and ecology:** Exit of the subterranean stream. The species is sympatric with *Sinocyclocheilus lateristritus*. **Food and feeding:** Unknown. **Reproduction and development:** Unknown. **Other behavior:** Unknown. **Conservation status:** None assigned but it should be classified as Endangered as suggested based on the field observations by one of us (YZ). **Major threats:** Original exit of subterranean has been replaced by several artificial ponds. **Conservation plans:** None. **Phylogenetic relationships:** It belongs to the *tingi* lineage.

35. *Sinocyclocheilus maculatus* Li, 2000 (in Li et al. 2000c) (Fig. 70). **Common name:** Maculate golden-line barbel (E, translation from Chinese). **Etymology:** *maculates* (L) speckled, maculate. **Major synonyms:** None. **Ecological classification:** Troglobite. **History:** First specimen was collected in April 1987. Staff from KIZ, (Kunming Institute of Zoology, Chinese Academy of Sciences) first found the species but did not publish its description. Type specimens were collected in August of the same year. **Morphological characteristics:** D iii, 7 A iii, 5 P i, 14–15 V i, 7–8. Body a little humpbacked; naked; lateral line pore present, 81–88; gill rakers on first gill arch 14–17. Maximum standard length: 82.7 mm

Fig. 70 *Sinocyclocheilus maculatus*, holotype, Li870808001, standard length 82.7 mm. (Photo by Zhao, Y.)

(Zhao 2006). Live coloration is golden with dark gray speckles. Preserved coloration (in alcohol) is brownish with many dark brown speckles. **Troglomorphic characters:** Scaleless, slightly humpbacked. **Distribution:** China: Yunnan Province: Yanshan County (ca. 23°37′ N, 104°20′ E) and Qiubei County (ca. 24° 02′ N, 104°11′ E) (Fig. 71). **Habitat and ecology:** Mouth of subterranean stream (Li et al. 2000c). **Food and feeding:** Unknown. **Reproduction and development:** Unknown. **Other behavior:** Unknown. **Conservation status:** None has been assigned but it should be classified as Endangered, based on the field observations by one of us (YZ). **Major threats:** Environment change. **Conservation plans:** None.

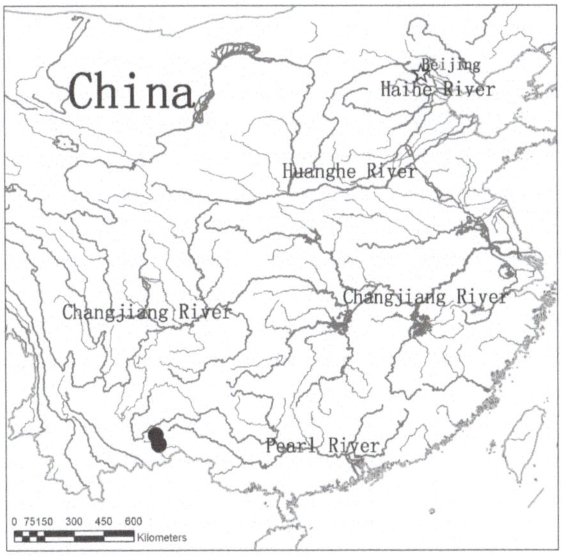

Fig. 71 Distribution of *Sinocyclocheilus maculatus*

Fig. 72 *Sinocyclocheilus maitianheensis*, holotype, IHB12209039-874001, standard length 90.0 mm. (Photo by Zhao, Y.)

Fig. 73 Distribution of *Sinocyclocheilus maitianheensis*

Phylogenetic relationships: It belongs to the *tingi* lineage.

36. *Sinocyclocheilus maitianheensis* Li, 1992 (Fig. 72). **Common name:** Maitianhe golden-line barbel (E, translation from Chinese). **Etymology:** *maitianheensis* (L) after the name of the type locality, Maitianhe River. **Major synonyms:** None. **Ecological classification:** Troglophile. **History:** First collection took place in 1997 (Li 1992). **Morphological characteristics:** D iii, 7 A iii, 5 P i, 14–15 V i, 9. Body elongated; scales embedded; pectoral fin short, not reaching to pelvic fin base; dorsal fin insertion posterior to vertical line of pelvic insertion; lateral line almost straight, possessing 70–82 scales. Maximum standard length: 147.8 mm (Zhao 2006). Coloration of preserved specimens (in alcohol) is light brownish; 5–7 black speckles on back side from dorsal to caudal fin base. **Troglomorphic characters:** Scales embedded. **Distribution:** China: Yunnan Province: Yiliang County: Maitianhe River (ca. 25°24′ N, 103°27′ E) (Fig. 73). **Habitat and ecology:** Exit of the subterranean stream. **Food and feeding:** Unknown. **Reproduction and development:** Unknown. **Other behavior:** Unknown. **Conservation status:** None has been assigned but it should be classified as Endangered as suggested based on the field observations by one of us (YZ). **Major threats:** Changes in the hydrological balance of the area due to dam construction and water extraction. **Conservation plans:** None. **Phylogenetic relationships:** It belongs to the *tingi* lineage.

37. *Sinocyclocheilus malacopterus* Chu and Cui, 1985 (Fig. 74). **Common name:** Soft ray golden-line barbel (E, translation from Chinese). **Etymology:** *malac* (Gr) soft; *ptero* (Gr) fin. **Major synonyms:** None. **Ecological classification:** Troglophile. **History:** First collection took place in 1977 (Chu and Cui 1985). **Morphological characteristics:** D iii, 7 A iii, 5 P i, 14–18 V i, 9. Body elongated; scales embedded; last unbranched ray of dorsal fin soft with serration. Maximum standard length: 172 mm (Chu and Cui 1985). Coloration of preserved specimens (In alcohol) is dorsally dark grayish, ventrally yellow brownish, black speckles in three rows on back and two on the sides (Chu and Cui 1985). **Troglomorphic characters:** Scales embedded. **Distribution:** China: Yunnan Province: Luoping County: Xinzhai town (ca. 24°46′ N 104°17′ E), Yangzhewo reservoir (ca. 24°51′ N 104°17′ E); Zhanyi County (ca. 25°37′ N 103°49′ E) (Fig. 75). **Habitat and ecology:** At the exit of a subterranean stream in a karstic cave. **Food and**

Fig. 74 *Sinocyclocheilus malacopterus*, paratype, KIZ775831, standard length 129.2 mm. (Photo by Zhao, Y.)

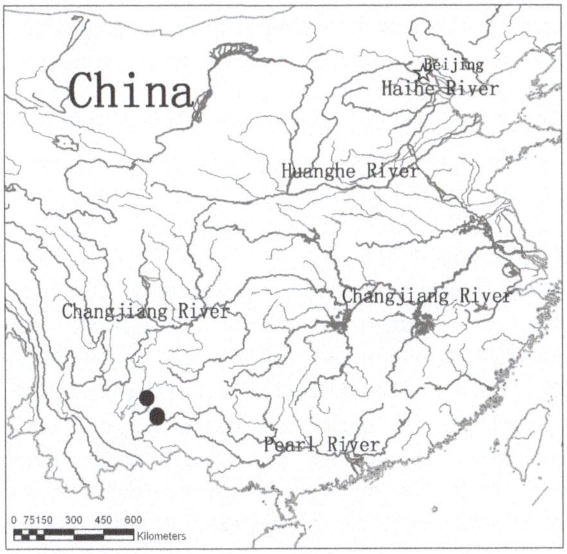

Fig. 75 Distribution of *Sinocyclocheilus malacopterus*

Fig. 77 Distribution of *Sinocyclocheilus microphthalmus*

feeding: Unknown. **Reproduction and development:** Unknown. **Other behavior:** Unknown. **Conservation status:** None assigned but it should be classifies as Rare based on the field observations by one of us (YZ). **Major threats:** The underground water is being used by local people in different ways. **Conservation plans:** None. **Phylogenetic relationships:** It belongs to the *tingi* lineage.

38. ***Sinocyclocheilus microphthalmus*** Li, 1989 (Fig. 76). **Common name:** Small eye golden-line barbel (E, translation from Chinese). **Etymology:** *micro* (Gr) small; *ophthalmus* (Gr) eyes. **Major synonyms:** *Anchicyclocheilus halfibindus*. **Ecological classification:** Troglobite. **History:** Li collected it in 1986, between April and May (Li 1989). **Morphological characteristics:** D iii, 8 A iii, 5 P i, 12 V i, 7. Body humpbacked; snout duck-billed; eye degenerated, forming small eye-dot; lateral line scaled 48–57, scale rows above and below lateral line 11–12, 7–9, respectively. Maximum standard length: 189.7 mm (Zhao 2006). Live coloration is whitish, somewhat transparent. **Troglomorphic characters:** Microphthalmic, depigmented, humpbacked. **Distribution:** China: Guangxi Zhuang Autonomous Region: Ling Yun County: Luo-lou town (24°20′ N 106°49′ E) and Sicheng Town: Shadong Cave (24°20′ N, 106°32′ E); Fengshan County: Fengcheng Town (24°32′ N 107° 02′ E) (Fig. 77). **Habitat and ecology:** At the exit of underground streams in karstic caves. In Lingyun (Sicheng Town), the species is sympatric with *Sinocyclocheilus lingyunensis* and *Schistura lingyunensis*. **Food and feeding:** Unknown. **Reproduction and development:** External, non-guarder (Riehl and Baensch 1991). **Other behavior:** Unknown. **Conservation status:** VU, in IUCN RL 2003; VU, in China's Species Red List (Wang and Xie 2004). **Major threats:** Excessive collecting and environmental degradation (Wang and Xie 2004). **Conservation plans:** None. **Phylogenetic relationships:** It belongs to the *cyphotergous* lineage.

Fig. 76 *Sinocyclocheilus microphthalmus*, ASIZB73047, standard length 99.5 mm. (Photo by Zhao, Y.)

39. ***Sinocyclocheilus multipunctatus*** Pellegrin, 1931 (Fig. 78). **Common name:** Speckled golden-line barbel (E, translation from Chinese). **Etymology:** *multi* (L) many; *punctatus* (L) having speckles or spots. **Major synonyms:** *Schizothorax multipunctatus*, original combination. **Ecological classification:**

Fig. 78 *Sinocyclocheilus multipunctatus*, ASIZB73000, standard length 188.6 mm. (Photo by Zhao, Y.)

Troglobite. **History:** Type specimens were collected by Père Cavaleris in 1913 (Pellegrin 1931). **Morphological characteristics:** D iii, 8 A iii, 5 P i, 15–18 V i, 8–9. Body humpbacked; scaled; mouth terminal; pectoral fin short, not reaching to pelvic fin base; lateral line straight, possessing 49–60 scales. Maximum standard length: 188.6 mm (Zhao 2006). Coloration of preserved specimens (in alcohol) is brownish, back darker and abdomen light brownish; small dark brown spots on sides above lateral line. **Troglomorphic characters:** Humpbacked. **Distribution:** China: Guangxi Zhuang Autonomous Region: Huanjiang County (26°08′ N, 106°39′ E) and Nandan County (24°49′ N, 108°16′ E), and Guizhou Province: Huishui (24°59′ N, 107°32′ E), Libo (25°24′ N, 107° 52′ E), Huaxi (26°27′ N, 106°40′ E) counties (Fig. 79). **Habitat and ecology:** Subterranean streams. **Food and feeding:** Unknown. **Reproduction and development:** Unknown. **Other behavior:** Unknown. **Conservation status:** None assigned but it should be classified as Rare, based on the field observations by one of us (YZ). **Major threats:** Environment degradation. **Conservation plans:** None. **Hylogenetic relationships:** It belongs to the *cyphotergous* lineage.

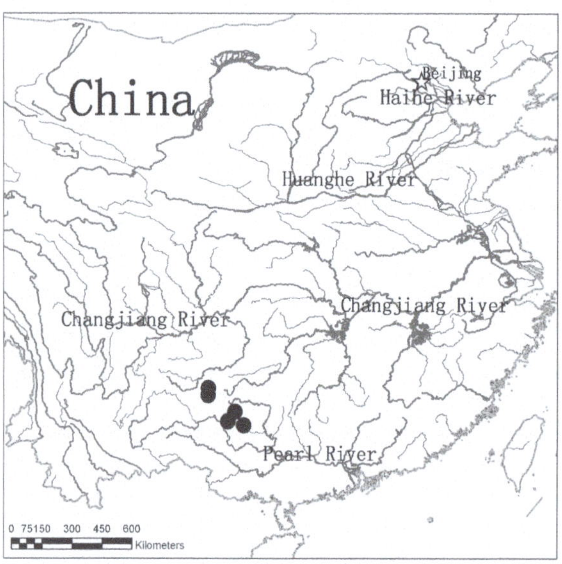

Fig. 79 Distribution of *Sinocyclocheilus multipunctatus*

40. *Sinocyclocheilus oxycephalus* Li, 1985 (Fig. 80). **Common name:** Cuspate head golden-line barbel; sharp mouth oil fish; Kuma fish (E, translation from Chinese). **Etymology:** *Oxy* (Gr) sharp, peaked, cuspate; *cephal* (Gr) head. **Major synonyms:** *Sinocyclocheilus lunanensis*. **Ecological classification:** Troglophile. **History:** Type specimens were collected in 1965 (Li 1985). **Morphological characteristics:** D iii, 7 A iii, 5 P i, 15–16 V i, 8–9. Body elongated and naked; snout acute; lateral line pores 63–72, pectoral fin short, not reaching to pelvic fin base. Maximum standard length: 135.8 mm (Zhao 2006). Live coloration is golden, abdomen light yellowish, some dark speckles on back and sides (above lateral line). **Troglomorphic characters:** Scaleless. **Distribution:** China: Yunnan Province: Shilin County: Heilongtan reservoir (ca. 24°46′ N, 103°16′ E) (Fig. 81). **Habitat and ecology:** Subterranean stream whose course has been modified to end into a reservoir. **Food and feeding:** Unknown. **Reproduction and development:** Unknown. **Other behavior:** Unknown. **Conservation status:** None has been assigned but it should be classified as Rare as suggested based on the field observations by one of us (YZ). **Major threats:** Environment degradation due to modifications in the subterranean streams and containment into a small

Fig. 80 *Sinocyclocheilus oxycephalus*, holotype, IHB12209013-652047, standard length 103.1 mm. (Photo by Zhao, Y.)

Fig. 81 Distribution of *Sinocyclocheilus oxycephalus*

reservoir. **Conservation plans:** None. **Phylogenetic relationships:** It belongs to the *tingi* lineage.

41. ***Sinocyclocheilus purpureus*** Li, 1985 (Fig. 82). **Common name:** Purple golden-line barbel; oil fish (E, translation from Chinese). **Etymology:** *purpureus* (L) purple, light reddish. **Major synonyms:** None. **Ecological classification:** Troglophile. **History:** Type specimens were collected in 1973 (Li 1985). **Morphological characteristics:** D iii, 7 A iii, 5 P i, 15–16 V i, 8–9. Body elongated and almost naked, lateral line scales present, 61–70. Maximum standard length: 97.6 mm (Zhao 2006). According to the original description live coloration is purplish, back darker, however one of us (YZ) could not see this latter feature when checking the type specimens.

Fig. 82 *Sinocyclocheilus purpureus*, holotype, IHB12209015-731004, standard length 97.6 mm. (Photo by Zhao, Y.)

Fig. 83 Distribution of *Sinocyclocheilus purpureus*

Troglomorphic characters: Scaleless. **Distribution:** China: Yunnan Province: Yanshan County: Pingyuanjie town (ca. 23°44′ N, 103°46′ E); Kaiyuan city: Zhongheying town (ca. 23°46′ N, 103°37′ E) (Fig. 83). **Habitat and ecology:** Subterranean stream, original environment has been modified. **Food and feeding:** Unknown. **Reproduction and development:** Unknown. **Other behavior:** Unknown. **Conservation status:** None has been assigned but it should be classified as Endangered, based on the field observations by one of us (YZ). **Major threats:** The underground water has been drawn for agricultural irrigation and the pool at the exit of the cave has been greatly reduced in size. **Conservation plans:** None. **Phylogenetic relationships:** It belongs to the *tingi* lineage.

42. ***Sinocyclocheilus qiubeiensis*** Li, 2002 (in Li et al. 2002a) (Fig. 84). **Common name:** Qiubei golden-

Fig. 84 *Sinocyclocheilus qiubeiensis*, holotype, Li990527002. standard length 103.3 mm. (Photo by Zhao, Y.)

line barbel (E, translation from Chinese). **Etymology:** *qiubeiensis* (L) after the name of the type locality: Qiubei County. **Major synonyms:** *Sinocyclocheilus jiuchengensis* (Li et al. 2002a). **Ecological classification:** Troglophile. **History:** Type specimens were collected in 1999 (Li et al. 2002a). **Morphological characteristics:** D iii, 7 A iii, 5 P i, 14–17 V i, 8–9. Body elongated; scales embedded; mouth terminal; lateral line curved; dorsal fin insertion posterior to vertical line of pelvic fin insertion; speckled; rictal barbel short, reaching to posterior edge of eye. Maximum standard length: 120.6 mm (Zhao 2006). Live coloration is golden, with some dark speckles on both sides. **Troglomorphic characters:** Scales embedded. **Distribution:** China: Yunnan Province: Qiubei County: Jinping town: Jiucheng village (ca. 24° 03′ N, 104°08′ E) (Fig. 85). **Habitat and ecology:** Exit of a subterranean stream. **Food and feeding:** Unknown. **Reproduction and development:** Unknown. **Other behavior:** Unknown. **Conservation status:** None has been assigned but it should be classified as Rare, based on the field observations by one of us (YZ). **Major threats:** Environmental degradation due to recreational use of the underground water for bathing and fishing. Carps have been introduced and excessive collecting of the cave fish has occurred. **Conservation plans:** None. **Phylogenetic relationships:** It belongs to the *tingi* lineage.

Fig. 86 *Sinocycocheilus qujingensis*, ASIZB78790, standard length 99.9 mm. (Photo by Zhao, Y.)

43. *Sinocyclocheilus qujingensis* Li, Mao and Lu, 2002 (in Li et al. 2002b) (Fig. 86). **Common name:** Qujing golden-line barbel (E, translation from Chinese). **Etymology:** *qujingensis* (L) after the name of the type locality: Qujing City. **Major synonyms:** None. **Ecological classification:** Troglophile. **History:** Type specimens were collected in 1991 (Li et al. 2002b). **Morphological characteristics:** D iii, 7 A iii, 5 P i, 15–18 V i, 6–8. Body elongated; scaled; lateral line almost straight, possessing 69–72 scales; mouth inferior; rictal barbel long, beyond posterior edge of preopercle. Maximum standard length: 100 mm (Zhao 2006). Live coloration is golden. **Troglomorphic characters:** None known. **Distribution:** China: Yunnan Province: Qujing city: Ciying town (ca. 25° 26′ N, 103°57′ E) (Fig. 87). **Habitat and ecology:**

Fig. 85 Distribution of *Sinocyclocheilus qiubeiensis*

Fig. 87 Distribution of *Sinocycocheilus qujingensis*

Exit of a subterranean stream. **Food and feeding:** Unknown. **Reproduction and development:** Spawning time takes place around May (Li et al. 2002b). **Other behavior:** Unknown. **Conservation status:** None has been assigned but it should be classified as Rare, based on the field observations by one of us (YZ). **Major threats:** The pool (including the exit of the subterranean flow) is very close to a populated area, so it is frequently affected by local people by water pollution, swimming, and agricultural water usage. **Conservation plans:** None. **Phylogenetic relationships:** It belongs to the *tingi* lineage.

44. *Sinocyclocheilus rhinocerous* Li and Tao, 1994 (Fig. 88). **Common name:** Rhinocerontic horn golden-line barbel (E, translation from Chinese). **Etymology:** *rhinokeros* (Gr) rhino. **Major synonyms:** None. **Ecological classification:** Troglobite. **History:** Type specimens were collected in 1993 (Li and Tao 1994). **Morphological characteristics:** D iii, 7 A iii, 5 P i, 11–14 V i, 6–7. Body with horn on nape, but simple; eyes small; scales degenerated, lateral line scales present, 35–40; pelvic fin short, not reaching to anus; gill rakers on first gill arch 8–10. Maximum standard length: 86.7 mm (Zhao 2006). Coloration of preserved specimens (in alcohol) is brownish. **Troglomorphic characters:** Microphthalmic, degenerated scales, presence of horn. **Distribution:** China: Yunnan Province: Luoping County: Xinzhai town (ca. 24°46′ N 104°17′ E) (Fig. 89). **Habitat and ecology:** Subterranean stream. **Food and feeding:** Unknown. **Reproduction and development:** Unknown. **Other behavior:** Unknown. **Conservation status:** VU, in China's Species Red List (Wang and Xie 2004). **Major threats:** Excessive collecting has taken place as well as excessive water withdrawl and pollution of the underground water (Wang and Xie 2004). **Conservation plans:** None. **Phylogenetic relationships:** It belongs to the *angularis* lineage.

Fig. 89 Distribution of *Sinocyclocheilus rhinocerous*

45. *Sinocyclocheilus robustus* Chen and Zhao, 1988 (Fig. 90). **Common name:** robust golden-line barbel (E, translation from Chinese). **Etymology:** *robustus* (L) strong. **Major synonyms:** None. **Ecological classification:** Troglophile. **History:** Type specimens were collected in 1980. **Morphological characteristics:** D iii, 7 A iii, 5 P i, 15 V i, 9. Body elongated and naked; pectoral fin short, not reach to pelvic fin base; last unbranched ray of dorsal fin strong with serration; dorsal fin insertion posterior to vertical line of pelvic fin insertion; lateral line curved, possessing 77 pores. Maximum standard length: 162.7 mm (Zhao 2006). Coloration of preserved specimen (in alcohol) is yellow-brownish, with some

Fig. 88 *Sinocyclocheilus rhinocerous*, ASIZB93907, standard length 78.2 mm. (Photo by Zhao, Y.)

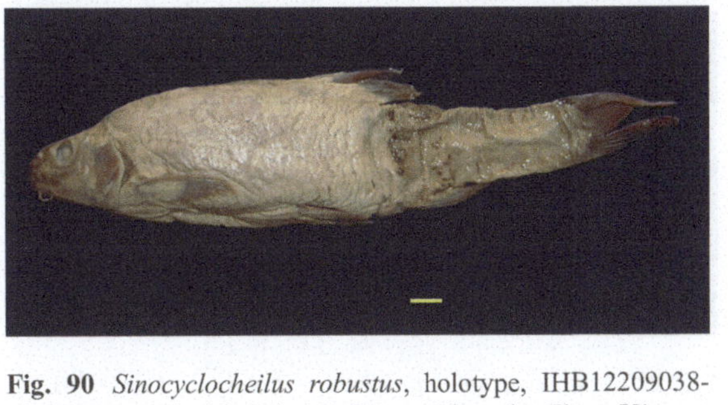

Fig. 90 *Sinocyclocheilus robustus*, holotype, IHB12209038-8001091, standard length 162.7 mm. (Photo by Zhao, Y.)

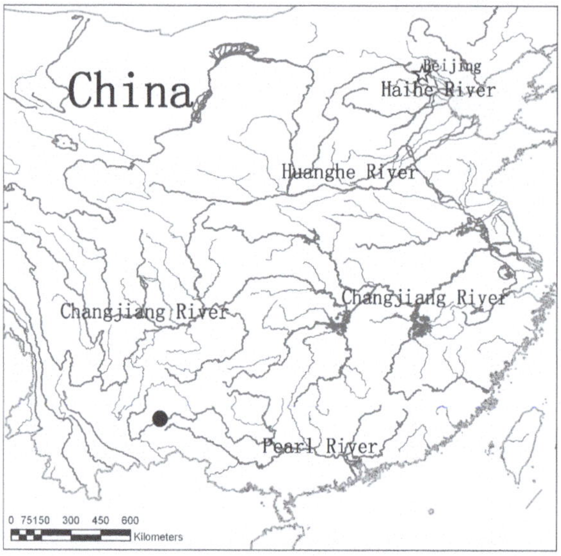

Fig. 91 Distribution of *Sinocyclocheilus robustus*

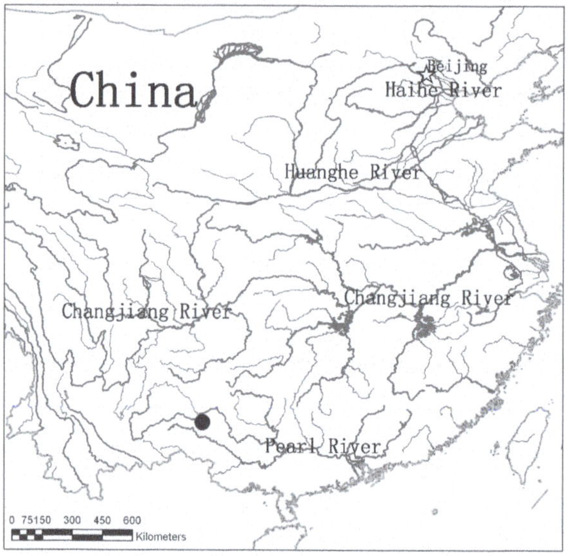

Fig. 93 Distribution of *Sinocyclocheilus tianlinensis*

dark spots (Chen et al. 1988a). **Troglomorphic characters:** Scaleless. **Distribution:** China: Guizhou Province: Xingyi City: Huangnihe River (ca. 24°44′ N 104°32′ E) (Fig. 91). **Habitat and ecology:** Unknown. **Food and feeding:** Unknown. **Reproduction and development:** Unknown. **Other behavior:** Unknown. **Conservation status:** None assigned but it should be classified as Endangered because only one specimen has been found since 1980. **Major threats:** No survey has been conducted to assess its possible threats. **Conservation plans:** None. **Phylogenetic relationships:** It belongs to the *tingi* lineage.

46. *Sinocyclocheilus tianlinensis* (Zhou et al. 2004) (Fig. 92). **Common name:** Tianlin golden-line barbel (E, translation from Chinese). **Etymology:** *tianlinensis* (L) after the name of the type locality, Tianlin County in Guangxi. **Major synonyms:** None.

Fig. 92 *Sinocyclocheilus tianlinensis*, holotype, ASIZB74125, standard length 93.0 mm. (Photo by Zhao, Y.)

Ecological classification: Troglobite. **History:** On 1 April 1998, *Guangxi Daily* reported that local people found blind fish in Tianlin. **Morphological characteristics:** D iii, 8 A iii, 5 P i, 12 V i, 7. Body elongated and naked; snout duck-billed; lateral line pores 41. Maximum standard length: 93.0 mm (Zhao 2006). Live coloration is whitish, semi-transparent. **Troglomorphic characters:** No externally visible eyes, depigmented, scaleless. **Distribution:** China: Guangxi Zhuang Autonomous Region: Tianlin County: Pingshan Town (ca. 24°35′ N, 106°18′ E) (Fig. 93). **Habitat and ecology:** The karstic cave located in the half-hill of the northwest side of Cenwanglaoshan Mountains, around 60 m from the nearest stream. **Food and feeding:** Unknown. **Reproduction and development:** Unknown. **Other behavior:** Unknown. **Conservation status:** None has been assigned but it should be classified as Endangered, based on the field observations by one of us (YZ). **Major threats:** Environment change. **Conservation plans:** None. **Phylogenetic relationships:** It belongs to the *angularis* lineage.

47. *Sinocyclocheilus tileihornes* Mao, Lu and Li, 2003 (in Mao et al. 2003) (Fig. 94). **Common name:** Tile-like horn golden-line barbel (E, translation from Chinese). **Etymology:** *tilei* (Anglo Saxon) for shingle; *hornes* (Anglo Saxon) for the horn of the fish. **Major synonyms:** None. **Ecological classification:** Troglobite. **History:** Type specimens were collected

Fig. 94 *Sinocyclocheilus tileihornes*, ASIZB78377, standard length 63.3 mm. (Photo by Zhao, Y.)

in 2002 (Mao et al. 2003). **Morphological characteristics:** D iii, 7 A iii, 5 P i, 12–14 V i, 6–7. Horn complex, tile-like; body scaleless; lateral line pores 35–37. Maximum standard length: 64.3 mm (Zhao 2006). Live coloration is light brownish, with some dark speckles. **Troglomorphic characters:** Microphthalmic, scaleless, horn. **Distribution:** China: Yunnan Province: Luoping County: A'gang Town (ca. 25°01′ N 103°59′ E) (Fig. 95). **Habitat and ecology:** A subterranean stream with several dolines. **Food and feeding:** Unknown. **Reproduction and development:** Unknown. **Other behavior:** Unknown. **Conservation status:** None assigned but it should be classified as Endangered, based on the field observations by one of us (YZ). **Major threats:** A farm has been developed around the exit of the subterranean flow. **Conservation plans:** None. **Phylogenetic relationships:** It belongs to the *angularis* lineage.

Fig. 95 Distribution of *Sinocyclocheilus tileihornes*

Fig. 96 *Sinocyclocheilus tingi*, ASIZB60227, standard length 109 mm. (Photo by Zhao, Y.)

48. *Sinocyclocheilus tingi* Fang, 1936 (Fig. 96). **Common name:** Fuxian golden-line barbel (E, translation from Chinese), boluoyu. **Etymology:** *tingi* (L) after a geologist, V. K. Ting, former general secretary of the Academia Sinica (Fang 1936). **Major synonyms:** *S. grahami tingi* (Fang 1936). **Ecological classification:** Troglophile. **History:** Type specimens were collected in 1934 (Fang 1936). **Morphological characteristics:** D iii, 7 A iii, 5 P i, 14–16 V i, 6–8. Body elongated, mouth terminal, lateral line scales 63–72; dorsal fin insertion posterior to vertical line of pelvic fin insertion. Maximum standard length: 195 mm (Chu and Cui 1989). Live coloration is golden. **Troglomorphic characters:** None known. **Distribution:** China: Yunnan Province: Lake Fuxian (ca. 24°30′ N 102°53′ E) (Fig. 97). **Habitat and ecology:** Lake at the exit of a subterranean stream.

Fig. 97 Distribution of *Sinocyclocheilus tingi*

Food and feeding: Unknown. **Reproduction and development:** Unknown. **Other behavior:** Unknown. **Conservation status:** None assigned but it should be classified as Rare, based on the field observations by one of us (YZ). **Major threats:** Excessive collecting, water pollution. **Conservation plans:** None. **Phylogenetic relationships:** It belongs to the *tingi* lineage.

49. *Sinocyclocheilus wumengshanensis* Li, Mao and Lu, 2003 (in Li et al. 2003b) (Fig. 98). **Common name:** Wumengshan golden-line barbel (E, translation from Chinese). **Etymology:** *wumengshanensis* (L) after the type locality: Wumenghshan Mountains in Yunnan Province. **Major synonyms:** None. **Ecological classification:** Troglophile. **History:** First collected in 1990. **Morphological characteristics:** D iii, 7 A iii, 5 P i, 15–18 V i, 8–10. Body elongated, pectoral fin long, reaching to pelvic fin base; barbel long, rictal barbel generally beyond posterior edge of preopercle; lateral line curved, possessing 71–81 scales. Maximum standard length: 100.5 mm (Zhao 2006). Live coloration is golden. **Troglomorphic characters:** None known. **Distribution:** China: Yunnan Province: Xundian County (ca. 25°34′ N 103°14′ E); Zhanyi County: Deze town (ca. 25°59′ N 103°36′ E); Xuanwei County: Xize town (ca. 26°16′ N 103°52′ E) (Fig. 99). **Habitat and ecology:** Subterranean streams. **Food and feeding:** Unknown. **Reproduction and development:** Unknown. **Other behavior:** Unknown. **Conservation status:** None has been assigned but it should be classified as Rare, based on the field observations by one of us (YZ). **Major threats:** No survey has been conducted to assess its possible threats. **Conservation plans:** None. **Phylogenetic relationships:** It belongs to the *tingi* lineage.

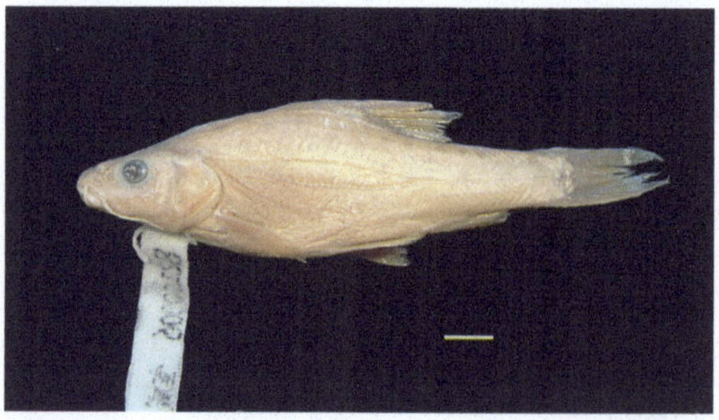

Fig. 98 *Sinocyclocheilus wumengshanensis*, KIZ82100006, standard length 92.8 mm. (Photo by Zhao, Y.)

Fig. 99 Distribution of *Sinocyclocheilus wumengshanensis*

50. *Sinocyclocheilus xunlensis* (Lan et al. 2004) (Fig. 100). **Common name:** Xunle golden-line barbel (E, translation from Chinese). **Etymology:** *xunlensis* (L) after the type locality: Xunle town. **Major synonyms:** None. **Ecological classification:** Troglobite. **History:** Type specimens were collected in May 2000. **Morphological characteristics:** D iii, 7 A iii, 5 P i, 11–13 V i, 7. Body scaled; snout duck-billed; blind; lateral line straight, possessing 41–48 scales; pectoral fin long, beyond pelvic fin base. Maximum standard length: 129 mm (Lan et al. 2004). Live coloration is whitish, semi-transparent. **Troglomorphic characters:** No externally visible eyes, depigmented. **Distribution:** China: Guangxi Zhuang Autonomous Region: Huanjiang County: Xunle Town (ca. 25°24′ N 108°16′ E) (Fig. 101). **Habitat and ecology:** A karstic cave. **Food and feeding:** Un-

Fig. 100 *Sinocyclocheilus xunlensis*, holotype, ASIZB73169, standard length 129.0 mm. (Photo by Zhao, Y.)

Fig. 101 Distribution of *Sinocyclocheilus xunlensis*

known. **Reproduction and development:** Unknown. **Other behavior:** Unknown. **Conservation status:** None assigned but it should be classified as Endangered as suggested based on the field observations by one of us (YZ). **Major threats:** No survey has been conducted to assess its possible threats. **Conservation plans:** None. **Phylogenetic relationships:** It belongs to the *angularis* lineage.

51. ***Sinocyclocheilus yangzongensis*** Tsü and Chen, 1977 (in Wu et al. 1977) (Fig. 102). **Common name:** Yangzong golden-line barbel (E, translation from Chinese). **Etymology:** *yangzongensis* (L) after the type locality: Lake Yangzonghai, Yunnan Province. **Major synonyms:** *Sinocyclocheilus grahami yangzongensis*. **Ecological classification:** Troglophile. **History:** First collecting took place in 1955 (Wu et al. 1977). **Morphological characteristics:** D iii, 7 A iii, 5 P i, 16 V i, 9. Body elongated; mouth

Fig. 102 *Sinocyclocheilus yangzongensis*, holotype, KIZ6351069, standard length 144.5 mm. (Photo by Zhao, Y.)

Fig. 103 Distribution of *Sinocyclocheilus yangzongensis*

superior; lateral line scales 71–81; pectoral fin short, not reaching to pelvic fin base. Maximum standard length: 150 mm (Shan et al. 2000). Live coloration is golden. **Troglomorphic characters:** None known. **Distribution:** China: Yunnan Province: Lake Yangzonghai (ca. 24°54′ N 103°00′ E) (Fig. 103). **Habitat and ecology:** Lake at the exit of a subterranean stream. **Food and feeding:** Unknown. **Reproduction and development:** Unknown. **Other behavior:** Unknown. **Conservation status:** None assigned but it should probably be classified as Endangered based on the fact that only a few individuals can be seen every year and yet they are subject to collecting and water pollution. **Major threats:** Excessive collecting and water pollution. **Conservation plans:** None. **Phylogenetic relationships:** It belongs to the *tingi* lineage.

52. ***Sinocyclocheilus yimenensis*** Li and Xiao, 2005 (in Li et al. 2005b) (Fig. 104). **Common name:** Yimen golden-line barbel (E, translation from Chinese). **Etymology:** *yimenensis* (L) after the type locality: Yimen County. **Major synonyms:** None. **Ecological classification:** Troglophile. **History:** First collecting took place in 1999 (Li et al. 2005b). **Morphological characteristics:** D iii, 7 A iii, 5 P i, 15 V i, 9. Body elongated and scaled; mouth ventral; gill rakers on first gill arch 5; dorsal fin insertion posterior to vertical line of pelvic fin insertion; rictal barbel beyond posterior edge of preopercle; lateral line curved, possessing 71–74 scales. Maximum standard length: 105 mm (Li et al. 2005b). Live

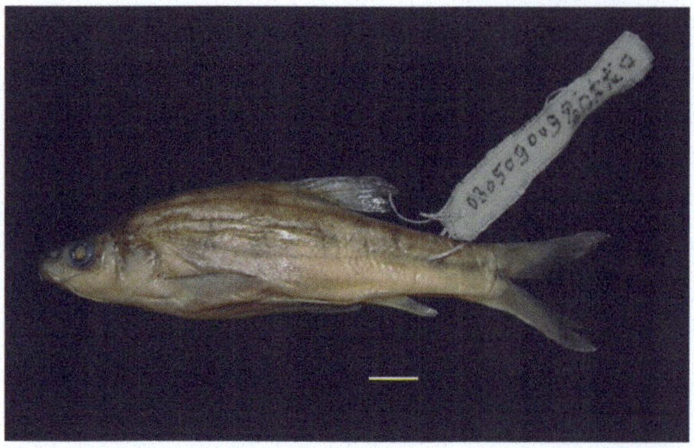

Fig. 104 *Sinocyclocheilus yimenensis*, paratype, Li030509009, standard length 92.5 mm. (Photo by Zhao, Y.)

coloration is golden, back darker, with some dark speckles. **Troglomorphic characters:** None known. **Distribution:** China: Yunnan Province: Yimen County (ca. 24°40′ N 102°09′ E) (Fig. 105). **Habitat and ecology:** Exit of a subterranean stream (Li et al. 2005b). **Food and feeding:** Unknown. **Reproduction and development:** Unknown. **Other behavior:** Unknown. **Conservation status:** Undetermined. **Major threats:** No survey has been conducted to assess its possible threats. **Conservation plans:** None. **Phylogenetic relationships:** It belongs to the *tingi* lineage.

53. *Sinocyclocheilus yishanensis* Li and Lan, 1992 (Fig. 106). **Common name:** Yishan goldenline barbel (E, translation from Chinese). **Etymology:** *yishanensis* (L) after the type locality: Yizhou (original name Yishan) City. **Major synonyms:** None. **Ecological classification:** Troglobite. **History:** First collecting took place in 1983. **Morphological characteristics:** D iii, 7 A iii, 5 P i, 13–15 V i, 7–9. Body a little humpbacked and scaled; last unbranched ray of dorsal fin soft, without serration; pectoral fin short, not reaching to pelvic fin base; predorsal vertebrae 6–7; gill rakers on first gill anch 9–11. Maximum standard length: 142 mm (Li and Lan 1992). Coloration of preserved specimens (in alcohol) is dark brownish, abdomen lighter. **Troglomorphic characters:** Humpbacked. **Distribution:** China: Guangxi Zhuang Autonomous Region: Yizhou city (ca. 24°28′ N 108°40′ E) (Fig. 107). **Habitat and ecology:** Original environment was a subterranean river. Now it has been changed to a reservior. **Food and feeding:** Unknown. **Reproduction and development:** Unknown. **Other behavior:** Unknown.

Fig. 106 *Sinocyclocheilus yishanensis*, ASIZB74145, standard length 98.3 mm. (Photo by Zhao, Y.)

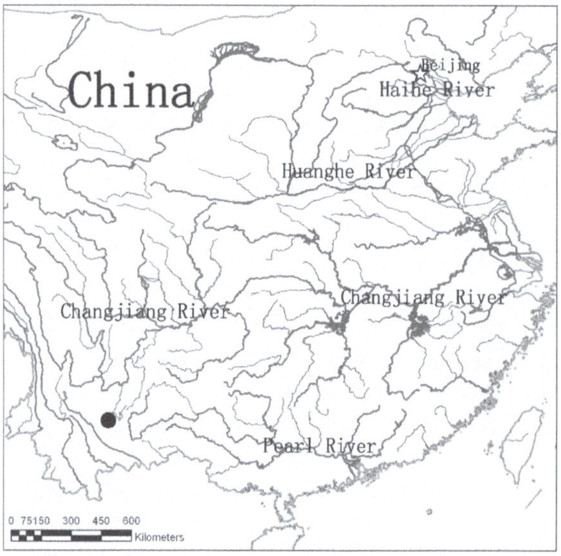

Fig. 105 Distribution of *Sinocyclocheilus yimenensis*

Fig. 107 Distribution of *Sinocyclocheilus yishanensis*

Fig. 108 *Typhlobarbus nudiventris*, holotype, standard length 46 mm. (Photo by He, M.)

Fig. 109 Distribution of *Typhlobarbus nudiventris*

Conservation status: None has been assigned but it should be classified as Rare a suggested based on the field observations by one of us (YZ). **Major threats:** Reservior. Irrigation. **Conservation plans:** None. **Phylogenetic relationships:** It belongs to the *jii* lineage.

Genus *Typhlobarbus* Chu and Chen, 1982

This is a monotypic genus.

54. *Typhlobarbus nudiventris* Chu and Chen, 1982 (Fig. 108). **Common name:** Blind naked-belly barbel (E, translation from Chinese). **Etymology:** *typhlo* (Gr) blind; *barbus* (L) carp; *nudi* (L) naked; *ventris* (L) ventral. **Major synonyms:** None. **Ecological classification:** Troglobite. **History:** First collection took place in 1976. First reported in 1979. Known from only three specimens (Chu and Chen 1982). **Morphological characteristics:** D ii, 8 A ii, 5 P i, 15 V ii, 8. Body elongated; mouth inferior and arched; eye degenerated; scaled, but anterior back and thorax-abdomen regions naked; lateral line scales 39–40; dorsal and anal spine soft. Maximum standard length: 45.5 mm (Shan et al. 2000). Live coloration is pale pinkish, semi-transparent. The branchial region is very reddish due to the blood circulation in that area. **Troglomorphic characters:** No externally visible eyes, partially scaleless. **Distribution:** China: Yunnan Province: Jianshui County: Yangjieba (ca. 23°39′ N, 102°46′ E) (Fig. 109). **Habitat and ecology:** The underground water in the karst cave is about 100 m below the surface at an altitude of 1,450 m above sea level. The water moves slowly (Chu and Chen 1982; Yue and Chen 1998). **Food and feeding:** Outer shells of chitinozoa have been found in feces suggesting that plankton is at least part of their diet (Chu and Chen 1982). **Reproduction and development:** External, non-guarder (Riehl and Baensch 1991). **Other behavior:** They support their bodies on the outer edge of pectoral and pelvic fins with the tip of the lower lobe of the caudal fin and the head raised upward. When swimming, they open their nostrils, lift the valve of their noses with their snouts extending forward and their bellies against the wall of the container and they again became still after swimming several rounds. They are very sensitive to vibration but show no response to light (Chu and Chen 1982; Yue and Chen 1998). **Conservation status:** VU in IUCN RL 2003, VU in China's Species Red List (Wang and Xie 2004), Rare in China's Red Data Book of Endangered Animals (Yue and Chen 1998) and a Second Class Protected Animal of Yunnan Province. **Major threats:** Water extraction by the locals. **Conservation plans:** None. **Phylogenetic relationships:** No known phylogenetic relationships except for the fact that it has been placed in the subfamily Barbinae. This species is superficially very similar to *Linichthys laticeps* (Zhang and Fang 2005).

Family: Cobitidae—loaches

This is a family composed of about 180 species of freshwater fishes from Eurasia and Morocco. They have a wormlike to fusiform body and are mostly bottom dwellers (Nelson 2006; pp. 146–147). Only one troglomorphic species has been described and it is from China.

Genus *Protocobitis* (Yang et al. 1994)

There are two species for this genus and both as hypogean.

55. ***Protocobitis polylepis*** (Zhu et al. 2008) (Fig. 110). **Common name:** None given**. Etymology**: *proto* (Gr) primitive; *cobitis* (L) a generic name for members of the subfamily Cobitinae; *poly* (Gr) many; lepis (Gr) scales. **Major synonyms:** None. **Ecological classification:** Troglobite. **History:** Two specimens collected on May 2006 by Sheng Zhang. **Morphological characteristics:** D ii, 6–7, A 2–5 P 1–7, V 1–4. Gill rakers 8. Blind. Scales covering the entire body except for head and abdomen. It lacks a bony gas bladder capsule, which may be a primitive character. Maximum standard length: 33.0 mm (Zhu et al. 2008). Pigmented. **Troglomorphic characters:** No externally visible eyes, absence of gas bladder. **Distribution:** China: Guangxi Zhuang Autonomous Region: Wuming County (ca. 23°10′ N, 108°17′ E) (Fig. 111). **Habitat and ecology:** Underground water. **Food and feeding:** Unknown. **Reproduction and development:** Unknown. **Other behavior:** Unknown. **Conservation status:** None has been assigned. **Major threats:** Unknown**. Plans:** None. **Phylogenetic relationships:** The genus may be the primitive sister-group of the whole subfamily Cobitinae (Yang et al. 1994).

56. ***Protocobitis typhlops*** (Yang et al. 1994) (Fig. 112). **Common name:** None given. **Etymology:** *proto* (Gr) primitive; *cobitis* (L) a generic name for members of the subfamily Cobitinae; *typhlos* (Gr) blind; *ops* (Gr) eye. **Major synonyms:** None. **Ecological classification:** Troglobite. **History:** Specimens collected in August 1991 by Lan Jia-Hu.

Fig. 110 *Protocobitis polylepis*, paratype KIZ20060002. (Photo by He, M.)

Fig. 111 Distribution of *Protocobitis polylepis*

Morphological characteristics: D iii, 7 A iii, 4–5 P i, 6–7, V i, 5. Gill rakers 7–8. Vertebrae 4 + 43. Blind. Scales rudimentary and only along the midline of the sides of the body. Lacks lateral line. It also lacks a bony gas bladder capsule, which may be a primitive character. Maximum standard length: 54.0 mm (Yang et al. 1994). Live fish pinkish and translucent, viscera and vertebrae clearly visible. **Troglomorphic characters:** No externally visible eyes, depigmented, scales reduced in number, size, and structure, absence of gas bladder. **Distribution:** China: Guangxi Zhuang Autonomous Region: Du'an County: Xia'ao town (24°15′ N, 107°05′ E) (Fig. 113). **Habitat and ecology:** In a cave at 210 m above sea level. The species is sympatric with *Sinocyclocheilus macrophthalmus* and *Oreonectes translucens*. **Food and feeding:** Unknown. **Reproduction and develop-**

Fig. 112 *Protocobitis typhlops*, ASIZB70026, standard length 39 mm. (Photo by Ye, E and Zhao, Y.)

Fig. 113 Distribution of *Protocobitis typhlops*

ment: Unknown. **Other behavior:** Swims slowly near the bottom. **Conservation status:** VU (IUCN). **Major threats:** Excessive collecting. **Plans:** None. **Phylogenetic relationships:** None known.

Family: Balitoridae—River loaches

This is another freshwater family of the Order Cypriniformes, with about 600 species from Eurasia. They have three or more pair of barbels. Some species are scaleless. Many tend to hide underneath rocks (Nelson 2006). They have 32 hypogean representatives of which 11 are troglobitic. Trajano et al. (2002) found numerous individuals of the genera *Homaloptera* and *Balitora* in caves in Thailand. In 1997 an epigean individual of the species *Balitora burmanica* was collected in a cave in Thailand (Trajano et al. 2002).

Genus *Heminoemacheilus* Zhu and Cao, 1987

Only two species have been described, both of them in caves and in a subterranean stream in Du'an County, Guangxi.

57. ***Heminoemacheilus hyalinus*** (Lan et al. 1996) (Fig. 114). **Common name:** translucent loach (E, translation from Chinese). **Etymology:** *hemi* (Gr) half, *nema* (Gr) filament, *cheilos* (Gr) lip; *hyalinus*, from *hyalinos* (Gr) of glass, referring to the transparency of the fish. **Major synonyms:** None. **Ecological classification:** Troglobite. **History:** Type specimens were collected in September 1994 (Lan et al. 1996). **Morphological characteristics:** D iii, 7–8 A iii, 4–5 P i, 11–12, V i, 5. Body almost naked; lateral line incomplete; blind; branched caudal rays 11–12, barbels short, not reaching to opercular, head small. Maximum standard length: 43.5 mm (Lan et al. 1996). Live coloration is translucent with viscera and vertebrae clearly visible. Coloration of preserved specimens (in alcohol) is whitish with no speckles. **Troglomorphic characters:** No externally visible eyes, depigmented, mostly scaleless. **Distribution:** China: Guangxi Zhuang Autonomous Region: Du'an County: Bao'an town (ca. 24°07′ N, 107°51′ E) (Fig. 115). **Habitat and ecology:** Subterranean stream. **Food and feeding:** Unknown. **Reproduction**

Fig. 114 *Heminoemacheilus hyalinus*, cited from Lan and Zhang 2006. (Photo by Lan, J.)

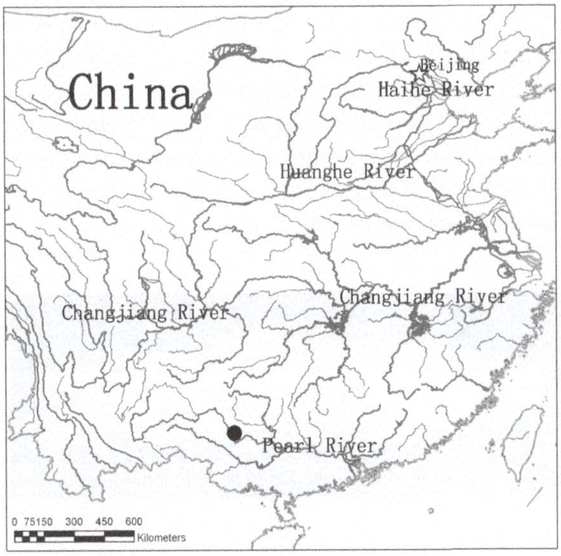

Fig. 115 Distribution of *Heminoemacheilus hyalinus*

and development: Unknown. **Other behavior:** Unknown. **Conservation status:** Undetermined. **Major threats:** No survey has been conducted to assess its possible threats. **Conservation plans:** None. **Phylogenetic relationships:** The genus is closely related to *Paranemacheilus*.

58. *Heminoemacheilus zhengbaoshani* Zhu and Cao, 1987 (Fig. 116). **Common name:** Zheng's loach (E, translation from Chinese). **Etymology:** *zhengbaoshani* (L) after Dr. Zheng Baoshan's contribution to Chinese ichthyology. **Major synonyms:** None. **Ecological classification:** Troglophile. **History:** Type specimens were collected in September 1974 (Zhu and Cao 1987). **Morphological characteristics:** D iii, 7–9 A iii, 5 P i, 11–12, V i, 6–7. Body elongated; mouth sub-inferior, lower lip with a slight median incision, upper jaw with a tooth like structure; barbels reaching operculum or beyond; dorsal fin origin slightly in advance of ventral, its origin being considerable nearer to caudal base than to snout tip; ventral not reaching vent; lateral line incomplete; scales small and head naked. Maximum standard length: 87 mm (Zhu and Cao 1987). Coloration of preserved specimens (in alcohol) is pale brown, somewhat lighter below, no markings (Zhu and Cao 1987). **Troglomorphic characters:** Possibly smaller scales and/or lack thereof on head. **Distribution:** China: Guangxi Zhuang Autonomous Region: Du'an County (ca. 23°56′ N, 108°05′ E) (Fig. 117). **Habitat and ecology:** Subterranean stream. **Food and feeding:** Unknown. **Reproduction and development:** Unknown. **Other behavior:** Unknown. **Conservation status:** Undetermined. **Major threats:** No survey has been conducted to assess its possible threats. **Conservation plans:** None. **Phylogenetic relationships:** The genus is closely related to *Paranemacheilus*. **Remarks:** It is the type species of the genus *Heminoemacheilus*.

Fig. 116 *Heminoemacheilus zhengbaoshani*, holotype, ASIZB61075 (Field No. 741802), standard length 85 mm. (Photo by Ye, E and Zhao, Y.)

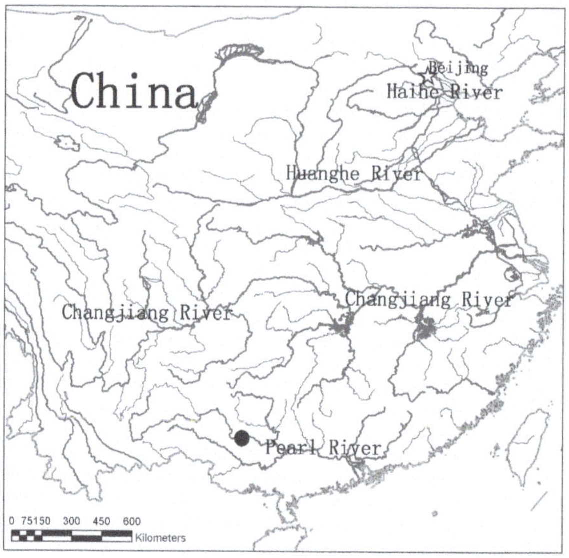

Fig. 117 Distribution of *Heminoemacheilus zhengbaoshani*

Genus *Oreonectes* Günther, 1868

There are seven known valid species for this genus. Except for type species, *Oreonectes platycephalus* and *O. polystigmus*, the rest of species are all cave dwellers. Interestingly, all cave fishes in the genus are all found in Guangxi, China. Classification of the genus is ambiguous and is still in need of further study.

59. *Oreonectes anophthalmus* Zheng, 1981 (Fig. 118). **Common name:** Eyeless flat-headed loach (E, translation from Chinese). **Etymology:** *Oreos* (Gr) mountain, *nekton* (Gr) swimmer; *an* (Gr) without; *opthalmus* (Gr) eyes. **Major synonyms:** *Nemacheilus anophthalmus* (Kottelat 1988). **Ecological classification:** Troglobite. **History:** Type specimens were collected in 1977. **Morphological characteristics:** D ii, 7 A ii, 5 P i, 10, V i, 4. Body

Fig. 118 *Oreonectes anophthalmus*, holotype, ASIZB60294 (Field No. Gui77001), standard length 41.3 mm (Photo by Ye, E. and Zhao, Y.)

naked; anterior section of body strong and slightly depressed, posterior part compressed; mouth inferior; 3 pairs of barbels, rostral one 2 pairs and maxillary one 1 pair; blind; dorsal fin opposite behind pelvic fin base; caudal fin rounded. Maximum standard length: 44 mm (Lan and Zhang 2006). Live coloration is semitransparent, fins transparent. **Troglomorphic characters:** No externally visible eyes, depigmented, scaleless. **Distribution:** China: Guangxi Zhuang Autonomous Region: Wuming County: Qifeng Shan (hill): Taiji Cave (ca. 23°06′ N, 108°40′ E) (Fig. 119). **Habitat and ecology:** Subterranean stream in karst cave. **Food and feeding:** Unknown. **Reproduction and development:** External reproduction, non-guarder (Breder and Rosen 1966). **Other behavior:** Unknown. **Conservation status:** Rare in China's Red Data Book of Endangered Animals, Pisces (Yue and Chen 1998); Vulnerable in China's Species Red List, Vol. 1 Red List (Wang and Xie 2004). **Major threats:** Excessive collecting. Environment change since the type locality has been changed to a local park. **Conservation plans:** None. **Phylogenetic relationships:** The species is more closely related to *O. platycephalus* by having similar dorsal fin position, caudal fin shape and numbers of dorsal, and anal fin rays (Lan et al. 1995).

60. *Oreonectes furcocaudalis* Zhu and Cao, 1987 (Fig. 120). **Common name:** Forked caudal-fin flat-headed loach (E, translation from Chinese). **Etymology:** *furca* (L) fork; *cauda* (L) tail. **Major synonyms:** None. **Ecological classification:** Troglobite. **History:** Type specimens were collected in 1983. **Morphological characteristics:** D iii, 8 A iii, 6 P i, 12–13, V i, 7. Body elongated, compressed; mouth inferior; eyes small; caudal fin forked; lateral line incomplete; very small scales on back behind dorsal fin. Maximum standard length: 72 mm (Lan and Zhang 2006). Coloration of preserved specimens (in alcohol) is grayish, back darker, light brownish, no speckles. **Troglomorphic characters:** Small eyes, **depigmented,** scales reduced in size and numbers. **Distribution:** China: Guangxi Zhuang Autonomous Region: Rongshui County (ca. 25°04′ N, 109°13′ E) (Fig. 121). **Habitat and ecology:** Mouth of a

Fig. 120 *Oreonectes furcocaudalis*, cited from Lan and Zhang 2006. (Photo by Lan, J.)

Fig. 119 Distribution of *Oreonectes anophthalmus*

Fig. 121 Distribution of *Oreonectes furcocaudalis*

subterranean stream. **Food and feeding:** Unknown. **Reproduction and development:** Unknown. **Other behavior:** Unknown. **Conservation status:** Undetermined. **Major threats:** No survey has been conducted to assess its possible threats. **Conservation plans:** None. **Phylogenetic relationships:** None known.

61. *Oreonectes microphthalmus* (Du et al. 2008) (Fig. 122). **Common name:** small eyes loach. **Etymology:** *micron* (Gr) small, and *phthalm* (Gr) eye. **Major synonyms:** None. **Ecological classification**: Troglobite. **History**: Du et al. (2008) were the first to mention this species. **Morphological characteristics**: D iii, 10; A ii, 7; Pi, 10; V i, 7; C 15. Body elongate; snout elongate; eyes reduced to a black pigment; body naked; cephalic lateral-line system with 2+2 supratemporal, 7 supraorbital, 3+ 0 infraorbital and 8 preoperculomandibular pores; lateral line incomplete, with 3 pores (Du et al. 2008). Live coloration is without pigments including the fins. **Troglomorphic characters:** Microphthalmic and degenerated, scaleless. **Distribution:** China: Guangxi Zhuang Autonomous Region: Du'an County (24 °15′ N 107 °05′ E) (Fig. 123). **Habitat and ecology:** Unknown. **Food and feeding:** Unknown. **Reproduction and development:** Unknown. **Other behavior:** Unknown. **Conservation status**: Undetermined. **Major threats:** No survey has been conducted to assess its possible threats. **Conservation plans:** None. **Phylogenetic relationships:** None known. Remarks: More studies are needed to determine the relationship with *Oreonectes translucens.*

62. *Oreonectes retrodorsalis* (Lan et al. 1995) (Fig. 124). **Common name:** Back dorsal-fin flat-headed loach (E, translation from Chinese). **Etymology:** *retro* (L) back, backwards, behind; *dorsalis* (L) back, dorsal, referring to the dorsal fin. **Major synonyms:** None. **Ecological classification:** Troglophile. **History:** Type specimens were collected in 1991 (Lan et al. 1995). **Morphological characteristics:** D iv, 7 A iv, 5 P i, 10–12, V i, 7. Body elongated, slightly compressed; head naked; mouth ventral; distance from dorsal fin insertion to snout tip obviously longer than to caudal fin base; caudal fin forked; lateral line incomplete. Maximum standard length: 56 mm (Lan and Zhang 2006). Coloration of preserved specimens (in alcohol) is grayish, ventral yellowish with a black stripe along midline of body sides. **Troglomorphic characters:** None known. **Distribution:** China: Guangxi Zhuang Autonomous Region: Nandan County: Liuzhai town: Longli Village (ca. 25°10′ N, 107°10′ E) (Fig. 125). **Habitat and ecology:** Exit of a subterranean stream. **Food and feeding:** Unknown. **Reproduction and development:** Unknown. **Other behavior:** Unknown. **Conservation status:** VU, in China's Species Red List (Wang and Xie 2004). **Major threats:** No survey

Fig. 122 *Oreonectes microphthalmus*, holotype KIZ20030001, standard length 39 mm. (Photo by He, M.)

Fig. 123 Distribution of *Oreonectes microphthalmus*

Fig. 124 *Oreonectes retrodorsalis*, cited from Lan and Zhang 2006. (Photo by Lan, J.)

Fig. 125 Distribution of *Oreonectes retrodorsalis*

has been conducted to assess its possible threats. **Conservation plans:** None. **Phylogenetic relationships:** The species is closely related to *O. platycephalus* and *O. anophthalmus* since the position of dorsal fin and the numbers of branched rays of anal fin are similar to the other two species (Lan et al. 1995).

63. ***Oreonectes translucens*** (Zhang et al. 2006) (Fig. 126). **Common name:** Transparent flat-headed loach (E, translation from Chinese). **Etymology:** *trans* (L) across; *lucens* (L) light. **Major synonyms:** None. **Ecological classification:** Troglobite. **History:** Yang et al. (1994) mentioned this species. **Morphological characteristics:** D iii, 8 A iii, 6 P i, 11, V i, 6. Body elongated; head slightly depressed; blind; well developed crests; dorsal fin insertion opposite pelvic fin insertion; lateral line incomplete; vertebrae 4+ 32. Maximum standard length: 45.8 mm (Zhang et al. 2006). Live coloration is absent. Coloration of preserved specimens (in alcohol) is grayish with no speckles. **Troglomorphic characters:** No externally visible eyes, depigmented, scaleless. **Distribution:**

Fig. 126 *Oreonectes translucens*, holotype, ASIZB94785, standard length 45.8 mm. (Photo by Zhang, Z.)

Fig. 127 Distribution of *Oreonectes translucens*

China: Guangxi Zhuang Autonomous Region: Du'an County: Xia'ao town (24°15′ N, 107°05′ E) (Fig. 127). **Habitat and ecology:** In a cave at 210 m above sea level The species is sympatric with *Sinocyclocheilus macrophthalmus* and *Protocobitis typhlops*. **Food and feeding:** Unknown. **Reproduction and development:** Unknown. **Other behavior:** Unknown. **Conservation status:** None assigned but it should be classified as Endangered, based on the field observations by one of us (YZ). **Major threats:** No survey has been conducted to assess its possible threats. **Conservation plans:** None. **Phylogenetic relationships:** None known. **Remarks:** It needs a further study on the relationship with *Oreonectes microphthalmus*.

Genus *Paracobitis* Bleeker, 1863

More than 20 species of this genus have been described. *Paracobitis* species are distributed from Middle Asia to China and south-east Asia. Three species have been found from subterranean streams of south-west China in recent years.

64. ***Paracobitis maolanensis*** (Li et al. 2006) (Fig. 128). **Common name:** Maolan blind loach (E, translation from Chinese). **Etymology:** *maolanensis* (L) after the type locality, Maolan Karst Forestry Natural Reserve. **Major synonyms:** None. **Ecological classification:** Troglobite. **History:** The holotype was collected on 3 January 2005. **Morphological**

Fig. 128 *Paracobitis maolanensis*, holotype, Li050103001, standard length 46 mm. Cited from Li et al. 2006

characteristics: D iii, 8 A ii, 5 P i, 11, V i, 6. Body elongated, naked; mouth inferior; barbels 3 pairs, not well-developed; dorsal fin insertion opposite pelvic fin insertion; caudal fin deeply forked. Maximum standard length: 46 mm (Li et al. 2006). Live coloration is semi-transparent with viscera and blood vessel of caudal peduncle clearly visible. Coloration of preserved specimens (in alcohol) is whitish. **Troglomorphic characters:** No externally visible eyes, depigmented, scaleless. **Distribution:** China: Guizhou Province: Libo County: Maolan Karst Forestry Natural Reserve (a national park) (25°23′ N, 108°04′ E) (Fig. 129). **Habitat and ecology:** A karstic cave. **Food and feeding:** Unknown. **Reproduction and development:** Unknown. **Other behavior:** Unknown. **Conservation status:** Undetermined. Only one specimen has been found so far, so it should be classified as at least Rare. **Major threats:** No survey has been conducted to assess its possible threats. **Conservation plans:** None. **Phylogenetic relationships:** None known. **Remarks:** From description of Li et al. (2006), it is probably a species of *Triplóphysa* and distinct from *T. longibarbatus*. Because it is sympatric with *T. longibarbatus* and has similar morphology, more studies are needed to determine its systematic status.

Fig. 129 Distribution of *Paracobitis maolanensis*

65. ***Paracobitis posterodarsalus*** Li, Ran and Chen, 2006 (in Ran et al. 2006) (Fig. 130). **Common name:** Poster dorsal-fin blind loach (E, translation from Chinese). **Etymology:** *post* (L) after; *darsum* (L) back; for the insertion of the dorsal fin behind the vertical line of the pelvic fin insertion. **Major synonyms:** None. **Ecological classification:** Troglobite. **History:** Only one specimen (the holotype) has been found so far. It was collected on 5 October 2004 (Ran et al. 2006). **Morphological characteristics:** D iii,6 A ii, 4 P i, 13, V i, 5. Body elongated; naked; mouth inferior; barbels 3 pairs, well-developed; blind; dorsal fin insertion opposite behind pelvic fin insertion; caudal fin deeply forked, upper lobe longer than lower one. Maximum standard length: 53 mm (Ran et al. 2006). Live coloration is semi-transparent. Coloration of preserved specimens (in alcohol) is whitish. **Troglomorphic characters:** No externally visible eyes, depigmented, scaleless. **Distribution:** China: Guangxi Zhuang Autonomous Region: Nandan County (ca. 24°59′ N, 107°32′ E) (Fig. 131). **Habitat and ecology:** A karstic cave. **Food and feeding:** Unknown. **Reproduction and development:** Unknown. **Other behavior:** Unknown. **Conservation status:** Undetermined, but because only one specimen has been found so far it should be classified as at least Rare. **Major threats:** No survey has been conducted to assess its possible threats. **Conservation plans:** None. **Phylogenetic relationships:** None known. **Remarks:** Li et al. (2006) proposed three diagnoses to differentiate *Paracobitis posterodarsalus* from *P. longibarbatus*. The first one, relative position of dorsal and pelvic fins, is actually not a prominent difference. The other two are fin ray counts of dorsal

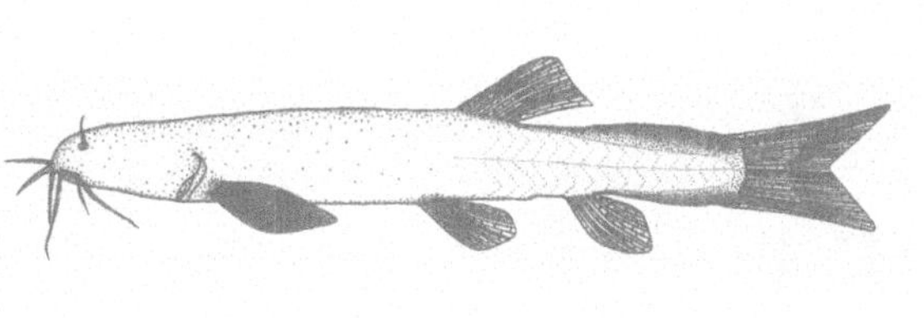

Fig. 130 *Paracobitis posterodarsalus*, cited from Ran et al. 2006

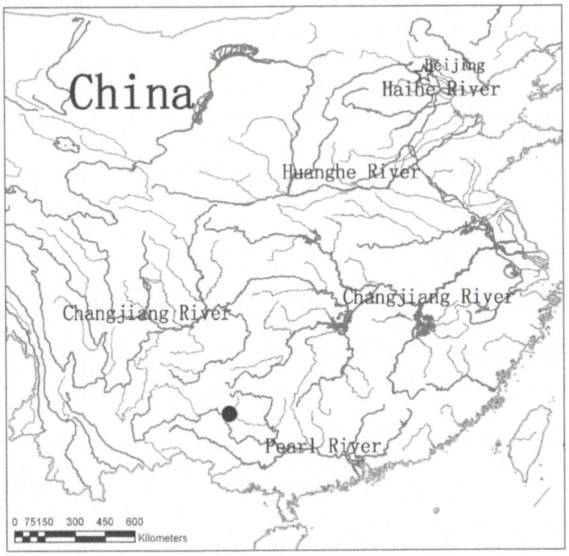

Fig. 131 Distribution of *Paracobitis posterodarsalus*

and anal fins. From Fig. 1 of Li et al. (2006), the branched dorsal-fin ray is at least 7 vs. 6 in the diagnosis and the branched anal-fin ray is at least 6 vs. 4 in the diagnosis. Therefore, the accuracy of fin-ray counts in this paper is problematic. We suggest that *Paracobitis posterodarsalus* might be a junior synonym of *Triplophysa longibarbatus*.

Genus *Paranemacheilus* Zhu, 1983

This is a monotypic genus from Guangxi.

66. *Paranemacheilus genilepis* Zhu, 1983 (Fig. 132). **Common name:** Unknown. **Etymology:** *para* (Gr) close to; *nema* (Gr) line; *geni* (Gr) from genys, meaning cheek, *lepis* (Gr) scales. **Major synonyms**: There seems to be a misspelling in the way the genus was originally described. It should be *Paranemacheilus* since the original intent was to relate it to the genus *Nemacheilus*. According to

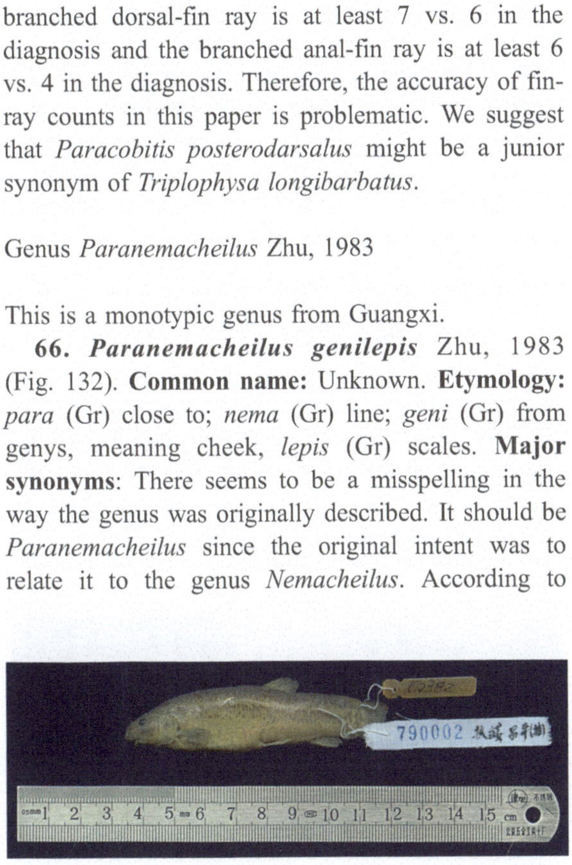

Fig. 132 *Paranemachilus genilepis*, holotype ASIZB60382, standard length 79 mm. (Photo by Ye, E.)

Article 32.5.1. of the International Code of Zoological Nomenclature (Fourth Edition) 'If there is in the original publication itself, without recourse to any external source of information, clear evidence of an inadvertent error, such as a *lapsus calami* or a copyist's or printer's error, it must be corrected. Incorrect transliteration or latinization, or use of an inappropriate connecting vowel, are not to be considered inadvertent errors.' Therefore we believe that the correct spelling for this genus should be *Paranemacheilus* not *Paranemachilus*. **Ecological classification**: Troglophile. **History:** First specimen was collected in September 1979. **Morphological characteristics**: D iii, 7–8; A iii,5; P i, 11–13; V i, 6–7. Body robust, slightly elongate and compressed, completely covered with minute scales. Suborbital spine absent. Lateral line incomplete. Head slightly depressed, naked above, cheeks scaly. Nostrils close together, anterior one tubular. Barbels 6, 4 rostral and 2 maxillary. Coloration of the dorsum of preserved specimens (in formalin) is brownish with many irregular speckles and spots; some irregular short, vertical dark stripes along midline of each side from the gill opening to the caudal fin base. In some male specimens, it becomes a long horizontal stripe. **Troglomorphic characters:** None. **Distribution:** China: Guangxi Zhuang Autonomous Region: Fusui County: subterranean river of Changping (ca. 22°42′ N, 107°53′ E) (Fig. 133). **Habitat and ecology**:

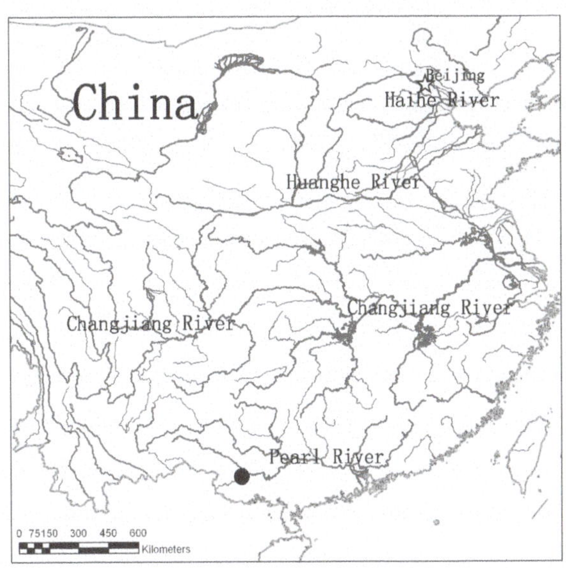

Fig. 133 Distribution of *Paranemachilus genilepis*

Unknown. **Food and feeding:** Unknown. **Reproduction and development:** Unknown. **Other behavior:** Unknown. **Conservation status**: Undetermined. **Major threats:** No survey has been conducted to assess its possible threats. **Conservation plans:** None. **Phylogenetic Relationship:** None known.

Genus *Schistura* McClelland, 1838

This is a genus represented by about 200 species of which five are hypogean.

67. ***Schistura dabryi microphthalmus*** Liao and Wang, 1997 (in Liao et al. 1997) (Fig. 134). **Common name:** Small eye Dabry's loach (E, translation from Chinese). **Etymology:** *schistura* (Gr) or skhizeîn = to split; *oura* (Gr) tail as an allusion to the forked tail; *micro* (Gr) small; *ophthalmus* (Gr) eyes. **Major synonyms:** None. **Type:** Troglobite. **History:** Type specimens were collected in 1991 (Liao et al. 1997). **Morphological characteristics:** D iii, 8 A ii, 5 P i, 9 V i, 7. Body elongated; naked; anterior and posterior nostrils situate closely together; mouth inferior; barbels 3 pairs; eye present, very small; dorsal fin insertion in front of vertical line of pelvic fin insertion; caudal fin forked; lateral line complete. Maximum standard length: 62 mm (Liao et al. 1997). Live coloration is grayish with irregular speckles on both sides. **Troglomorphic characters:** Microphthalmic. **Distribution:** China: Guizhou Province: Weng'an County: Hawk Cave (ca. 26°53′ N, 107°02′ E), Yangtze River drainage (Fig. 135). **Habitat and ecology:** A karstic cave. **Food and feeding:** Unknown. **Reproduction and development:** Unknown. **Other behavior:** Unknown. **Conservation status:** Undetermined. **Major threats:** No survey has been conducted to assess its possible threats. **Conservation plans:** None. **Phylogenetic relationships:** The species is most similar to *Triplophysa napanjiangensis*, which used to be *Schistura dabryi napanjingensis*, as a subspecies of *Schistura dabryi*. The systematics and taxonomy on this species needs to be reviewed.

Fig. 135 Distribution of *Schistura dabryi microphthalmus*

68. ***Schistura lingyunensis*** Liao and Luo, 1997 (in Liao et al. 1997) (Fig. 136). **Common name:** Lingyun loach (E, translation from Chinese). **Etymology:** *lingyunensis* (L) after its type locality, Lingyun County in Guangxi. **Major synonyms:** None. **Type:** Troglobite. **History:** Type specimens were collected in 1992 (Liao et al. 1997). **Morphological characteristics:** D iii, 6–7 A iii, 5 P i, 9 V i, 5–7. Body elongated; naked. Head slightly depressed; anterior and posterior nostrils situated closely together; mouth inferior; barbels 3 pairs, well-developed; eye very small, present as a black dot. Dorsal fin insertion in front of vertical line of pelvic fin insertion; anal fin close to anus, not reaching to caudal fin base, caudal fin forked. Lateral line

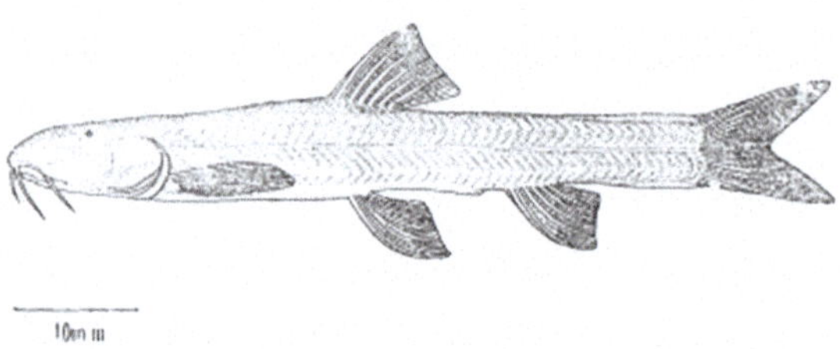

Fig. 134 *Schistura dabryi microphthalmus*, cited from Liao et al. 1997

Fig. 136 *Schistura lingyunensis*, cited from Lan and Zhang 2006. (Photo by Lan, J.)

incomplete, stops below insertion of dorsal fin. Maximum standard length: 62 mm (Liao et al. 1997). Live coloration is mostly transparent, operculum red due to blood circulation (Liao et al. 1997). **Troglomorphic characters:** Eyes degenerated, completely depigmented, scaleless. **Distribution:** China: Guangxi Zhuang autonomous region: Lingyun county: Guancang village (ca. 24°25′ N, 106°30′ E) and Sicheng Town: Shadong Cave (24°20′ N, 1063°32′ E) (Fig. 137). **Habitat and ecology:** Subterranean streams, belonging to Nanpanjiang River basin. In Shadong Cave (Sand Cave), the species is sympatric with *Sinocyclocheilus microphthalmus* and *Sinocyclocheilus lingyunensis* (Li et al. 2000b). **Food and feeding:** Unknown. **Reproduction and development:** Unknown. **Other behavior:** Unknown. **Conservation status:** Undetermined. **Major threats:** No survey has been conducted to assess its possible threats. **Conservation plans:** None. **Phylogenetic relationships:** The species is most similar to *Schistura incerta*, which has bigger eyes and a complete lateral line.

Genus *Triplophysa* Rendahl, 1933

The species of the genus *Triplophysa* are widely distributed in Qinghai-Tibet Plateau and its related areas, 107 species have been described so far. Some of them are cave dwellers. At least 11 cavefish species from the genus are found in south-west China.

Fig. 137 Distribution of *Schistura lingyunensis*

Fig. 138 *Triplophysa aluensis*, holotype, KIZ20006007, standard length 74 mm. (Photo by Ms. He, M.)

69. *Triplophysa aluensis* Li and Zhu, 2000 (Fig. 138). **Common name:** Alu plateau loach (E, translation from Chinese). **Etymology:** *Triplos* (Gr) thrice; *physa* (Gr) tube; *alunensis* (L) after the name of the type locality, Alu Cave in Luxi County, Yunnan Province. **Major synonyms:** None. **Ecological classification:** Troglobite. **History:** The only specimen collected so far was in June 2000. **Morphological characteristics:** D iii, 7 A iii, 5 P i, 9, V i, 6. Body elongated; naked; mouth inferior; barbels 3 pairs, well-developed; eye degenerated; dorsal fin insertion slightly in front of vertical line of pelvic fin insertion; caudal fin forked; lateral line present, straight. Maximum standard length: 74 mm (Li and Zhu 2000). Live coloration is light grayish, back darker, two speckles on caudal fin. **Troglomorphic characters:** Microphthalmic, depigmented, scaleless. **Distribution:** China: Yunnan Province: Luxi County: Alu limestone caves (24°33′ N, 103°45′ E) (Fig. 139). **Habitat and ecology:** In an underground stream

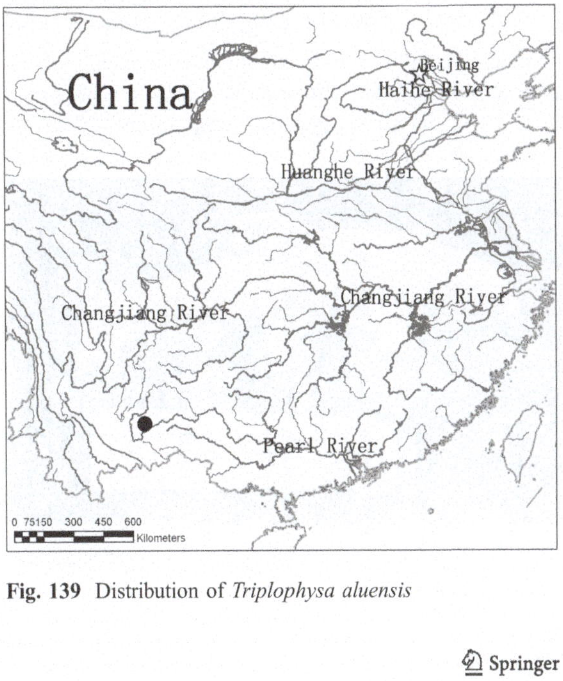

Fig. 139 Distribution of *Triplophysa aluensis*

(Yusun River) of a cave of 625 m in length at an elevation of 1,713 m above sea level **Food and feeding:** Feeds on microscopic animals including aquatic insects. **Reproduction and development:** Unknown. **Other behavior:** Unknown. **Conservation status:** Undetermined. **Major threats:** Environmental degradation. Alu Cave has been developed as a popular tourist site. **Conservation plans:** None. **Phylogenetic relationships:** None known. **Remarks**: The fin-ray count in Li and Zhu (2000) is not accurate and some additional characters have been described elsewhere (see Chen and Yang 2005).

70. ***Triplophysa gejiuensis*** (Chu and Chen, 1979) (Fig. 140). **Common name:** Gejiu blind plateau loach (E, translation from Chinese). **Etymology:** *gejiuensis* (L) after the type locality, Gejiu City, Yunnan Province. **Major synonyms:** *Nemacheilus gejiuensis* (original combination) (Chu and Chen 1979), *Schistura gejiuensis* (Zhu 1989). **Ecological classification:** Troglobite. **History:** The first nine specimens were captured in March 1978. **Morphological characteristics:** D ii, 7–8 A ii, 4–6 P i, 9–10, V i, 5. Body elongated; naked; blind; 3 pairs of barbels, well-developed; mouth inferior; lateral line present, straight. Maximum standard length: 52.0 mm (Yang 1990). Live coloration is whitish, semi-transparent with all fins transparent. Coloration of preserved specimens (in alcohol) is milk-like whitish. **Troglomorphic characters:** No externally visible eyes, depigmented, scaleless. **Distribution:** China: Yunnan Province: Gejiu City: Bajiaojing 4 km from Kafang (23°16′ N, 103°09′ E) (Fig. 141). **Habitat and ecology:** In subterranean waters at a depth of 400 m underground. Water temperature 17°C, air temperature 20°C in cave, ground temperature 22°C in August (Chu and Chen 1979). **Food and feeding:** Feeds on microscopic animals and aquatic insects. **Reproduction and development:** External reproduction, non-guarder (Breder and Rosen 1966). **Other behavior:** Unknown. **Conservation status:** VU, in IUCN RL2003; VU, in China's Species Red List (Wang and Xie 2004), Rare, in China's Red Data Book of Endangered Animals (Yue and Chen 1998). **Major threats:** Original environment has been severely modified. **Conservation plans:** None. **Phylogenetic relationships:** More closely related to *T. shilinensis* by having 14–15 branched rays in the caudal fin, and the position of dorsal fin insertion (Chen et al. 1992).

Fig. 141 Distribution of *Triplophysa gejiuensis*

71. ***Triplophysa longibarbatus*** (Chen et al. 1998) (Fig. 142). **Common name:** Long-barbel blind loach

Fig. 140 *Triplophysa gejiuensis*, paratype, KIZ1978000983. (Photo by He, M.)

Fig. 142 *Triplophysa longibarbatus*, holotype, KIZ1995000636, standard length 68 mm. (Photo by He, M.)

(E, translation from Chinese). **Etymology:** *longi* (L) long; *barbatus* (L) barbels. **Major synonyms:** None. **Ecological classification:** Troglobite. **History:** The first three specimens were collected in March 1995 by a joint Slovenian-Chinese expedition (Chen et al. 1998). **Morphological characteristics:** D ii, 8 A ii, 6 P i, 10–11, V i, 5–6. Body elongated; naked; mouth inferior; blind; barbels well-developed; pectoral and pelvic fins long; adipose keel present, along dorsal and ventral edge of caudal peduncle; posterior margin of caudal fin strongly concave. Maximum standard length: 68 mm (Chen et al. 1998). Live coloration is transparent. Coloration of preserved specimens is whitish. **Troglomorphic characters:** No externally visible eyes, depigmented, scaleless. **Distribution:** China: Guizhou Province: Libo County, in a cave 13 km northeast of the town of Libo (ca. 25°15′ N, 108°00′ E) (Fig. 143). **Habitat and ecology:** The cave contains both streams and lakes with a sandy/rocky bottom. **Food and feeding:** Unknown. **Reproduction and development:** Unknown. **Other behavior:** Unknown. **Conservation status:** VU, in China's Species Red List (Wang and Xie 2004). **Major threats:** No survey has been conducted to assess its possible threats. **Conservation plans:** None. **Phylogenetic relationships:** It is related to *Paracobitis variegatus longidorsalis*, which occurs in the same basin of Xijiang (Chen et al. 1998). **Remarks:** A recent study (Du et al. 2008) confirmed *Paracobitis longibarbatus* is actually a *Triplophysa* species, and *Oreonectes liboensis* and *Nemacheilus liboensis* are all its synonyms. *Oreonectes liboensis* is an unpublished name and *N. liboensis* is a *nomen nudum*. Li et al. (2006) proposed three diagnoses of *Paracobitis posterodarsalus* from *P. longibarbatus*. The first one, relative position of dorsal and pelvic fins is actually not a prominent difference. The other two are fin ray counts of the dorsal and anal fins. From Fig. 1 of Li et al. (2006), the branched dorsal fin ray count is at least 7 vs. 6 in the diagnosis and the branched anal fin ray count is at least 6 vs. 4 in the diagnosis. Therefore, the accuracy of fin ray counts in this paper is problematic. We tentatively treat *Paracobitis posterodarsalus* as the junior synonym of *Paracobitis longibarbatus*. From its distribution range, we can see it is a relatively widely distributed species in the karst region of south Guizhou and north Guangxi.

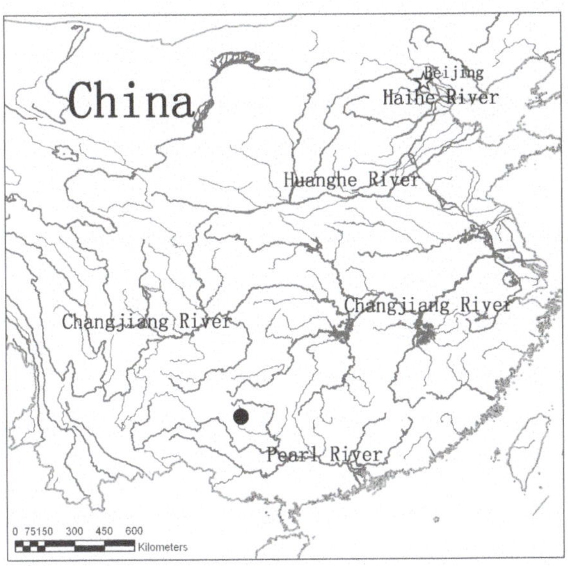

Fig. 143 Distribution of *Triplophysa longibarbatus*

72. *Triplophysa nandanensis* (Lan et al. 1995) (Fig. 144). **Common name:** Nandan blind plateau loach (E, translation from Chinese). **Etymology:** *nandanensis* (L) for the type locality, Nandan County: Guangxi. **Major synonyms:** None. **Ecological classification:** Troglophile. **History:** Type specimens were collected in 1991 (Lan et al. 1995). **Morphological characteristics:** D iv, 8 A iv, 5 P i, 10–11, V i, 7. Body elongated; naked; anterior and posterior nostrils situate closely together; nostril valve barbel-like; mouth inferior; eyes small; 3 pairs of barbels, well-developed; mouth ventral; lateral line present; dorsal fin insertion middle of snout tip to caudal fin base; caudal fin forked. Maximum standard length: 82.5 mm (Lan et al. 1995). Coloration of preserved specimens (in alcohol) is light yellowish, with many very small gray-black spots. **Troglomorphic characters:** Microphthalmic, scaleless. **Distribution:** China:

Fig. 144 *Triplophysa nandanensis*, paratype, ASIZB70034. (Photo by Zhao, Y.)

Guangxi Zhuang Autonomous Region: Nandan County: Liuzhai town (ca. 25°18′ N, 107°24′ E) (Fig. 145). **Habitat and ecology:** Exit of a subterranean stream. **Food and feeding:** Unknown. **Reproduction and development:** Unknown. **Other behavior:** Unknown. **Conservation status:** Undetermined. **Major threats:** No survey has been conducted to assess its possible threats. **Conservation plans:** None. **Phylogenetic relationships:** None known.

73. *Triplophysa nasobarbatula* Wang and Li, 2001 (Fig. 146). **Common name:** Nasal barbel plateau loach (E, translation from Chinese). **Etymology:** *nas* (L) nostril, nose; *barb* (L) barbel. **Major synonyms:** None. **Ecological classification:** Troglophile. **History:** Type specimens were collected in 1984 (Wang and Li 2001). **Morphological characteristics:** D iii, 8 A iii, 5 P i, 9, V i, 6. Body elongated; 3 pairs of barbels, well-developed; mouth inferior; nostril valve barbel-like; scales small, anterior ventral part naked; lateral line present; two well expanded chambers in anterior vesicle of gas bladder. Maximum standard length: 90 mm (Wang and Li 2001). Coloration of preserved specimens (in alcohol) is light brownish with six dark brown speckles on back and a few dots on dorsal and caudal fins. **Troglomorphic characters:** None known. **Distribution:** China: Guizhou Province: Libo County: Weng'ang town (ca. 25°14′ N, 107°54′ E), Dongtang town (ca. 25°17′ N, 108°01′ E) (Fig. 147). **Habitat and ecology:** In small waters

Fig. 146 *Triplophysa nasobarbatula*, KIZ2005001334. (Photo by Li, J.)

or pools connected to subterranean streams. **Food and feeding:** Unknown. **Reproduction and development:** Unknown. **Other behavior:** Unknown. **Conservation status:** Undetermined. **Major threats:** No survey has been conducted to assess its possible threats. **Conservation plans:** None. **Phylogenetic relationships:** None known.

74. *Triplophysa rosa* Chen and Yang, 2005 (Fig. 148). **Common name:** Rose blind plateau loach (E, translation from Chinese). **Etymology:** *rosa* (L) after the name of a caving club, Red Rose, which collected the holotype. **Major synonyms:** None. **Ecological classification:** Troglobite. **History:** The first specimen was collected on 26 November 2002. Two, one and two specimens were collected in 2004, 2005 and 2007, respectively by the Hongmeigui Caving Club and were deposited in KIZ (Kunming Institute of Zoology, Chinese Academy of Sciences).

Fig. 145 Distribution of *Triplophysa nandanensis*

Fig. 147 Distribution of *Triplophysa nasobarbatula*

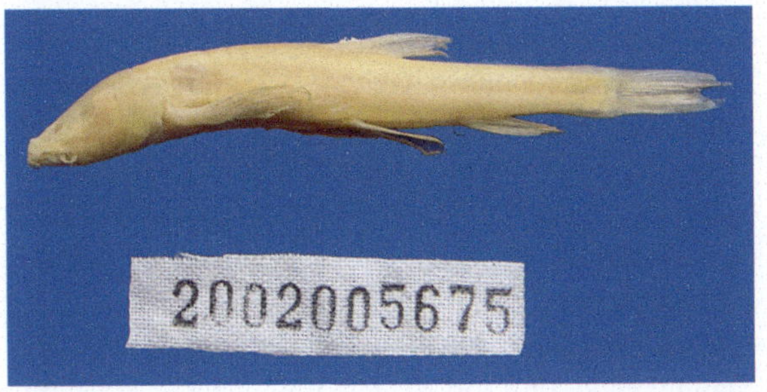

Fig. 148 *Triplophysa rosa*, holotype, KIZ2002005675, standard length 56 mm. (Photo by He, M.)

Morphological characteristics: D iii, 9 A iii, 6 P i, 12 V i, 7. Body elongated; naked; head slightly depressed; snout pointed; eyes vestigial; mouth ventral; distal margin of dorsal fin concave; tip of pelvic fin surpasses vertical level of anus; caudal fin deeply forked. Maximum standard length: 56.0 mm (Chen and Yang 2005). Coloration of preserved specimens (in alcohol) is pale, with no pigmentation and all fins transparent. **Troglomorphic characters:** Eyes degenerated, completely depigmented, scaleless. **Distribution:** China: Chongqing City: Wulong County: Jiangkou Town: Tianxing Town: Dongba Cave (29°13′40.44″ N, 107°55′16.46″ E) (Fig. 149). **Habitat and ecology:** A small pool 581 m underground (the pool is at 314 m above sea level), 10 m long and 0.5 m deep. Air temperature is 16–17°C, water temperature 13°C (November). About 12 individuals

Fig. 149 Distribution of *Triplophysa rosa*

were seen in another adjacent pool (Chen and Yang 2005). **Food and feeding:** Unknown. **Reproduction and development:** Unknown. **Other behavior:** Unknown. **Conservation status:** Undetermined. **Major threats:** No survey has been conducted to assess its possible threats. **Conservation plans:** None. **Phylogenetic relationships:** The species can be easily distinguished from other hypogean species of the genus by both body characters and distribution. Only one species, *T. bleekeri*, is distributed in the adjacent surface rivers. Although they share some common characters the relationships of both species still need to be further studied.

75. ***Triplophysa shilinensis*** Chen and Yang, 1992 (in Chen et al. 1992) (Fig. 150). **Common name:** Shilin blind plateau loach (E, translation from Chinese). **Etymology:** *shilinensis* (L) for the words Shi Lin which in Chinese means 'Stone Forest', the County where this species is found. **Major synonyms:** None. **Ecological classification:** Troglobite. **History:** The first two specimens were collected by the authors describing the species on 18 March 1991. **Morphological characteristics:** D iii, 7 A iii, 5 P i, 8–10 V i, 6. Body elongated; naked; blind; anterior and posterior nostrils situate closely together; nostril valve barbel-like; mouth inferior; pelvic fin origin opposite to dorsal fin insertion or slightly behind; caudal fin forked; posterior chamber of air bladder reduced, and anterior chamber enclosed in a bony capsule. Maximum standard length: 61.0 mm (Chen et al. 1992). Live coloration is semi-transparent. Coloration of preserved specimens (in alcohol) is whitish. **Troglomorphic characters:** Eyes completely degenerated, completely depigmented, scaleless. **Distribution:** China: Yunnan Province: Lunan (=Shilin) County: cave at Weiboyi village near Stone Forest (24°47′ N, 103°22′ E) (Fig. 151). **Habitat and**

Fig. 150 *Triplophysa shilinensis*, KIZ199100036 (Photo by He, M.)

Fig. 151 Distribution of *Triplophysa shilinensis*

ecology: The cave has a funnel-like opening and goes down vertically for 109 m before it reaches a horizontal stream where this fish can be found. **Food and feeding:** Insects have been found in their stomach contents (Chen et al. 1992). **Reproduction and development:** Unknown. **Other behavior:** Unknown. **Conservation status:** VU, in China's Species Red List (Wang and Xie 2004). **Major threats:** Excessive collecting (Wang and Xie 2004). **Conservation plans:** None. **Phylogenetic relationships:** More closely related to *T. gejiuensis* (Chen et al. 1992).

76. ***Triplophysa tianeensis*** (Chen et al. 2004) (Fig. 152). **Common name:** Tian'e blind plateau loach (E, translation from Chinese). **Etymology:** *tianeensis* (L) derived from the name of the type locality, Tian'e County, Guangxi. **Major synonyms:** None. **Ecological classification:** Troglobite. **History:** Type specimens were collected in January 2003 (Chen et al. 2004). **Morphological characteristics:** D iii, 7 A iii, 5 P i, 9 V i, 6. Body elongated; naked; eye degenerated resulting in a small concentration of eye tissue; mouth inferior; lateral line present, straight; caudal fin forked. Maximum standard length: 61.4 mm (Chen et al. 2004). Live coloration is light yellowish, semi-transparent, operculum transparent, sides with some very small dots. **Troglomorphic characters:** No externally visible eyes, depigmented, scaleless. **Distribution:** China: Guangxi Zhuang Autonomous Region: Tian'e County: Bala town (24° 58′ N, 107°02′ E) (Fig. 153). **Habitat and ecology:** Same cave where *Sinocyclocheilus furcodorsalis* was found. **Food and feeding:** Unknown. **Reproduction and development:** Unknown. **Other behavior:** Unknown. **Conservation status:** Undetermined. **Major threats:** No survey has been conducted to assess its possible threats. **Conservation plans:** None. **Phylogenetic relationships:** The species is most closely related and sympatric to *T. nandanensis* (Chen et al. 2004).

Fig. 152 *Sinocyclocheilus tianeensis*, cited from Lan and Zhang 2006. (Photo by Lan, J.)

77. ***Triplophysa xiangshuingensis*** Li, 2004 (Fig. 154). **Common name:** Xiangshuijing plateau loach (E, translation from Chinese). **Etymology:** *xiangshuingensis* (L) after the name of the type

Fig. 153 Distribution of *Sinocyclocheilus tianeensis*

Fig. 154 *Triplophysa xiangshuingensis*, cited from Li 2004

locality, Xiangshuiqing, Shilin County. **Major synonyms:** None. **Ecological classification:** Troglophile. **History:** Type specimen (only one) was collected in 2000 (Li 2004). **Morphological characteristics:** D iii, 6 A ii, 5 P i, 9, V i, 6. Body elongated; naked; eye small; anterior and posterior nostrils close to each other, nostril valve barbel-like; 3 pairs of barbels, well developed; mouth ventral; caudal fin forked; lateral line complete and straight. Maximum standard length: 87 mm (Li 2004). Live coloration is light yellowish with four black-brown saddle-backed speckles on back and three black-brown stripes on dorsal and caudal fins. **Troglomorphic characters:** Microphthalmic, scaleless. **Distribution:** China: Yunnan Province: Shilin County: Xiangshuijing village (ca. 24°52′ N, 103°21′ E) (Fig. 155). **Habitat and ecology:** Subterranean stream. **Food and feeding:** Unknown. **Reproduction and development:** Unknown. **Other behavior:** Unknown. **Conservation status:** Undetermined. **Major threats:** No survey has been conducted to assess its possible threats. **Conservation plans:** None. **Phylogenetic relationships:** None known.

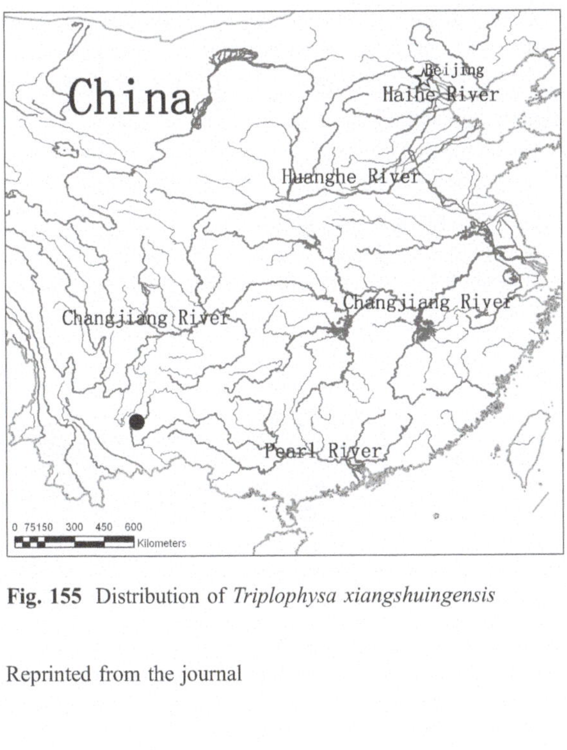

Fig. 155 Distribution of *Triplophysa xiangshuingensis*

Fig. 156 *Triplophysa xiangxiensis*, KIZ9705001. (Photo by Li, J.)

78. *Triplophysa xiangxiensis* (Yang et al. 1986) (Fig. 156). **Common name:** Xiangxi blind plateau loach (E, translation from Chinese). **Etymology:** *xiangxiensis* (L) because it is found in the area of West Hunan (Xiang is the abbreviation of Hunan Province in Chinese). **Major synonyms:** *Noemacheilus xiangxiensis* (Yang et al. 1986), *Schistura xiangxiensis* (in Zhu 1989). **Ecological classification:** Troglobite. **History:** First found was in 1984 (He et al. 2006). **Morphological characteristics:** D iii, 8 A iii, 6 P i, 11 V i, 16. Body elongated; naked; blind; nostril valve well-developed; 3 pairs of barbels; mouth ventral; pectoral fin very long, reaches anal fin base or even caudal fin base. Maximum standard length: 85.0 mm (Yang et al. 1986). Live coloration is slightly pinkish due to the circulatory system been visible through the depigmented skin. **Troglomorphic characters:** No externally visible eyes, depigmented, scaleless. **Distribution:** China: Xiangxi: Hunan Province: Longshan County: Huoyan Township: Feihu (=Flying tiger) Cave (ca. 27°27′ N, 109°26′ E) (Fig. 157). **Habitat and ecology:** Subterranean waters in karst cave (360 m above sea level), Yuanjiang River system, 20 m underground. **Food and feeding:** Could be fed in aquarium. **Reproduction and development:** External reproduction, non-guarder (Breder and Rosen 1966). **Other behavior:** Unknown. **Conservation status:** Undetermined. **Major threats:** No survey has been conducted to assess its possible threats. **Conservation plans:** None. **Phylogenetic relationships:** None known.

79. *Triplophysa yunnanensis* Yang, 1990 (Fig. 158). **Common name:** Yunnan plateau loach (E, translation from Chinese). **Etymology:** *yunnanensis* (L) after the name of the Province of Yunnan. **Major**

Fig. 157 Distribution of *Triplophysa xiangxiensis*

synonyms: None. **Ecological classification:** Troglophile. **History:** Type specimens were collected in 1987 (Yang 1990). **Morphological characteristics:** D iii, 7 A iii, 5 P i, 10–11 V i, 7. Body elongated; eye small; anterior and posterior nostrils close to each other, posterior edge of nostril slightly long; 3 pairs of barbels; mouth inferior; dorsal fin longer than body depth; caudal fin forked, upper lobe longer than lower one; scaled from dorsal fin base; lateral line complete. Maximum standard length: 62.8 mm (Yang 1990). Live coloration is light yellowish with a row of round brown speckles along midline, 8–10 round speckle on back, and one light strip on dorsal fin. **Troglomorphic characters:** Microphthalmic. **Distribution:** China: Yunnan Province: Yiliang: Jiuxiang (ca. 25°00′ N, 103°23′ E) (Fig. 159). **Habitat and ecology**: Found in an underground river. **Food and feeding:** Unknown. **Reproduction and development:** Unknown. **Other behavior:** Unknown. **Conservation status:** Undetermined. **Major threats:** No survey has been conducted to assess its possible threats. **Conservation plans:** None. **Phylogenetic relationships:** None known.

Fig. 158 *Triplophysa yunnanensis*, holotype, KIZ1987003994 (original number 874200), standard length 62.8 mm. (Photo by Ms. He, M. (KIZ))

Fig. 159 Distribution of *Triplophysa yunnanensis*

80. *Triplophysa zhenfengensis* Wang and Li, 2001 (Fig. 160). **Common name:** Zhenfeng plateau loach (E, translation from Chinese). **Etymology:** *zhenfengensis* (L) after the name of the type locality: Zhenfeng County. **Major synonyms:** None. **Ecological classification:** Troglophile. **History:** Type specimens were collected in 1990 (Wang and Li 2001). **Morphological characteristics:** D iii, 7 A iii, 5 P i, 9–11, V i, 5–7. Body elongated; 3 pairs of barbels, well-developed; mouth ventral; scales small, anterior ventral part naked; lateral line present; gill rakers on first gill arch 7–12; vertebrae 4+36. Maximum standard length: 83 mm (Wang and Li 2001). Coloration of preserved speci-

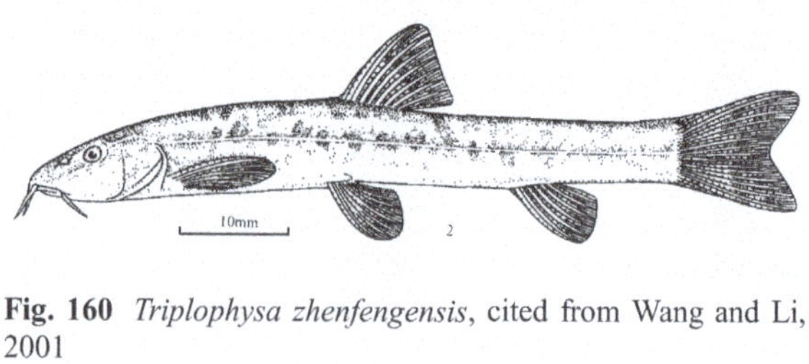

Fig. 160 *Triplophysa zhenfengensis*, cited from Wang and Li, 2001

mens (in alcohol) is brownish, back darker, with some round dark-brown speckles, 2 stripes on dorsal fin, and 2–3 stripes on caudal fin. **Troglomorphic characters:** Possibly small scales. **Distribution: China:** Guizhou Province: Zhengfeng County: Longchang town (ca. 25° 32′ N, 105°12′ E); Xingren County: Gaowu town (ca. 25 °28′ N, 105 °30′ E) (Fig. 161). **Habitat and ecology:** Exit of a subterranean stream. **Food and feeding:** Unknown. **Reproduction and development:** Unknown. **Other behavior:** Unknown. **Conservation status:** Undetermined. **Major threats:** No survey has been conducted to assess its possible threats. **Conservation plans:** None. **Phylogenetic relationships:** None known.

Genus *Yunnanilus* Nichols, 1925

About 30 species has been described for this genus. Most of species are only found in Yunnan Province. Other species are found in Guizhou Province of China, Myanmar (Burma), and Viet Nam. Hypogean species of this genus are all troglophile and almost all are found in Yunnan Province.

81. *Yunnanilus bajiangensis* Li, 2004 (Fig. 162). **Common name:** Bajiang Yunnan loach (E, translation from Chinese). **Etymology:** *bajiangensis* (L) after the type locality: Bajiang River. **Major synonyms:** None. **Ecological classification:** Troglophile. **History:** First collected in 1995 (Li 2004). **Morphological characteristics:** D iii, 9 A ii, 5 P i, 9–10, V i, 6. Body elongated and compressed; anterior nostrils tube-like, with a short distance to posterior one; mouth sub-ventral; dorsal fin insertion to snout tip longer than from it to caudal fin base; scales small; no lateral line. Maximum standard length: 52 mm (Li 2004). Live coloration is light yellowish, with brown horizontal stripes on sides. Coloration of preserved specimens (in alcohol) is gray-brownish. **Troglomorphic characters:** None known. **Distribution:** China: Yunnan Province: Shilin County: Heilongtan Reservoir (ca. 24°46′ N, 103°16′ E) (Fig. 163). **Habitat and ecology:** It is sympatric with *Sinocyclocheilus macrocephalmus* and *S. oxycephalus*. **Food and feeding:** Unknown. **Reproduction and development:** Unknown. **Other behavior:** Unknown. **Conservation status:** Undetermined. **Major threats:** Environment degradation due to modifications in the subterranean streams and containment into a small reservoir. **Conservation plans:** None. **Phylogenetic relationships:** None known.

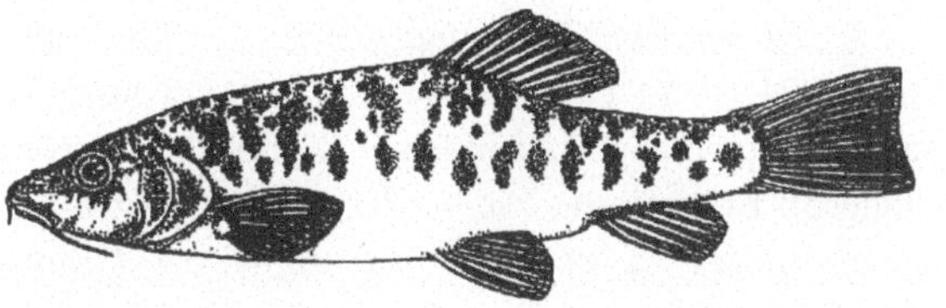

Fig. 162 *Yunnanilus bajiangensis*, cited from Li, 2004

Fig. 161 Distribuion of *Triplophysa zhenfengensis*

Fig. 163 Distribution of *Yunnanilus bajiangensis*

82. *Yunnanilus beipanjiangensis* Li, Mao and Sun, 1994 (in Li et al. 1994) (Fig. 164). **Common name:** Beipanjiang Yunnan loach (E, translation from Chinese). **Etymology:** *Yunnanilus* (L) meaning loach from Yunnan, a province on southwest China; *beipanjiangensis* (L) after the name of Beipanjiang River, where the species is found. **Major synonyms:** None. **Ecological classification:** Troglophile. **History:** Type specimens were collected in 1991 (Li et al. 1994). **Morphological characteristics:** D iii, 7–8 A ii, 5 P i, 10, V i, 6. Body elongated; naked; anterior nostrils tube-like, with a short distance to posterior one; mouth inferior; dorsal spine soft; caudal fin forked; lateral line incomplete, end above pectoral fin. Maximum standard length: 73 mm (Li et al. 1994). Coloration of preserved specimens (in alcohol) is light yellowish. Males have a black stripe along midline from posterior edge of operculum to caudal fin base. This stripe is not obvious in females. Both males and females have many black speckles on back and sides. **Troglomorphic characters:** Scaleless. **Distribution:** China: Yunnan Province: Zhanyi County: Xintun (26.02°N, 104.01°E) (Fig. 165). **Habitat and ecology:** Spring. **Food and feeding:** Unknown. **Reproduction and development:** Unknown. **Other behavior:** Unknown. **Conservation status:** Undetermined. **Major threats:** No survey has been conducted to assess its possible threats. **Conservation plans:** None. **Phylogenetic relationships:** None known.

83. *Yunnanilus discoloris* Zhou and He, 1989 (Fig. 166). **Common name:** Different color Yunnan loach (E, translation from Chinese). **Etymology:** *dis* (E) different; *color* (L) colorful. **Major synonyms:** None. **Ecological classification:** Troglophile. **History:** First collection was in 1983. **Morphological characteristics:** D iii, 8–9; A iii, 5; P i, 11–12; V i, 7. Nostrils well separated, the anterior tubular. Mouth inferior. Upper lip smooth, lower one wrinkle, interrupted at middle. Male with a black longitudinal stripe on both sides and a light blackish stripe of spots along back; female with black spots and blotches on back and lateral sides. Both sexes, dorsal, anal and ventral with one light blackish stripe, caudal with one or two stripes (Zhou and He 1989). **Troglomorphic**

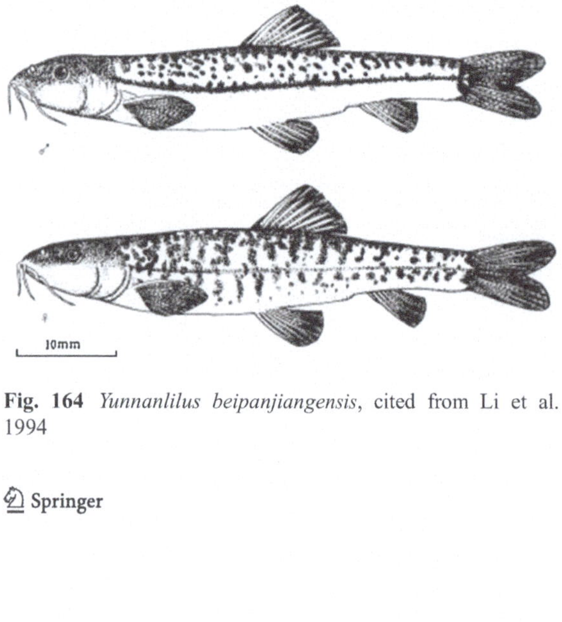

Fig. 164 *Yunnanlilus beipanjiangensis*, cited from Li et al. 1994

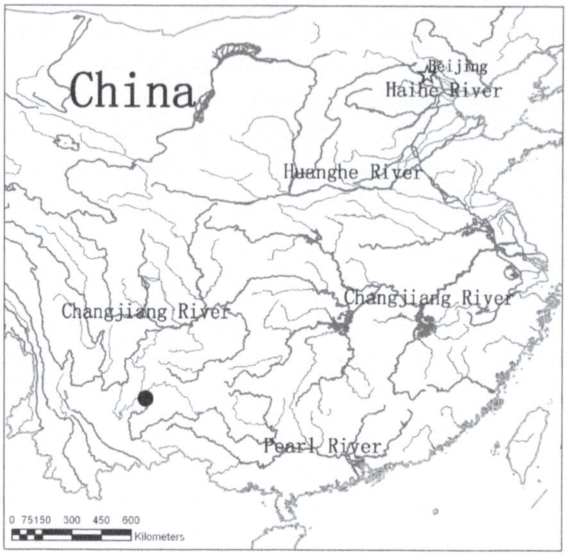

Fig. 165 Distribution of *Yunnanlilus beipanjiangensis*

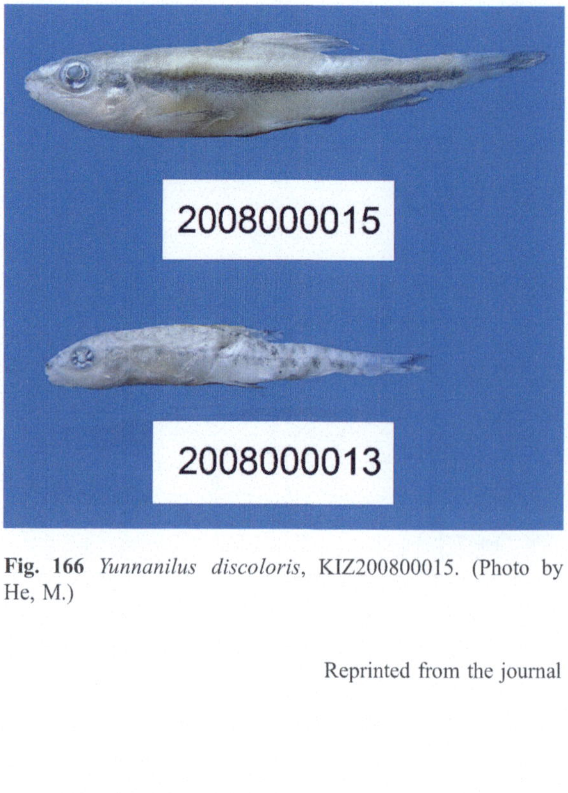

Fig. 166 *Yunnanilus discoloris*, KIZ200800015. (Photo by He, M.)

Fig. 167 Distribution of *Yunnanilus discoloris*

characters: None. **Distribution:** China: Yunnan: Chenggong County: White Dragon Spring (24 °54′ N, 102 °48′ E) (Fig. 167). **Habitat and ecology:** Lives in the spring water, the average water depth is about 80 cm, the bottom is made up of sand and gravel; the main aquatic macrophyte is *Hydrilla verticillata*. **Food and feeding:** Unknown. **Reproduction and development:** Unknown. **Other behavior:** Unknown. **Conservation status:** Protected by Buddhist temples. **Major threats:** Its loss from the lake likely due to introduced fish species, changes in water quality, loss of macrophytes. Current threats at the springs are introduced species, modification of spring structure. **Conservation plans:** None. **Phylogenetic Relationship:** None known.

84. ***Yunnanilus longidorsalis*** Li, Tao and Lu, 2000 (in Li et al. 2000a) (Fig. 168). **Common name:** Long back Yunnan loach (E, translation from Chinese). **Etymology:** *longi* (L) long; *dorsalis* (L) back, indicating typical characteristic of the species: a long dorsal fin. **Major synonyms:** None. **Ecological classification:** Troglophile. **History:** Type specimens were collected in 1994 (Li et al. 2000a). **Morphological characteristics:** D iii, 11 A ii, 6 P i, 10–11, V i, 6. Body elongated and compressed; eyes big; anterior nostrils tube-like, with a short distance to posterior one; mouth terminal; 3 pairs of barbels short; dorsal fin insertion to snout tip shorter than it to caudal fin base; dorsal fin long, longest ray opposite end of anal fin base; scales small; no lateral line. Maximum standard length: 37.6 mm (Li et al. 2000a). Coloration of preserved specimens (in alcohol) is gray-yellowish, with some brown spots on front of body and 7–8 brown worm-like speckles on sides. **Troglomorphic characters:** None known, except possibly small scales. **Distribution:** China: Yunnan Province: Luoping County: A'gang (ca. 25°05′ N, 104°08′ E) (Fig. 169). **Habitat and ecology:** Spring. **Food and feeding:** Unknown. **Reproduction and development:** Unknown. **Other behavior:** Unknown. **Conservation status:** Undetermined. **Major threats:** No survey has been conducted to assess its possible threats. **Conservation plans:** None. **Phylogenetic relationships:** None known.

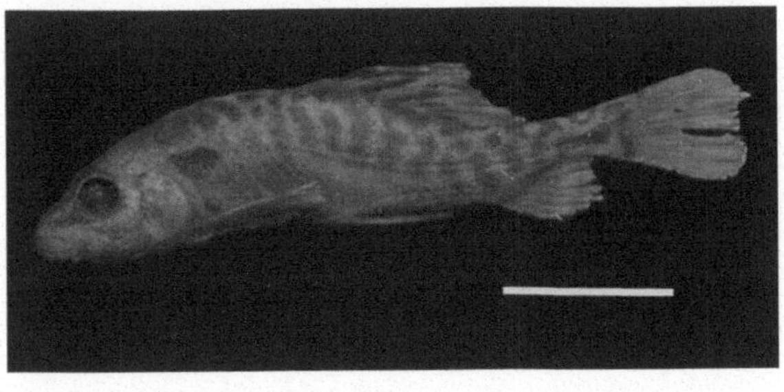

Fig. 168 *Yunnanilus longidorsalis*, paratype, ASIZB73118, standard length 36 mm. (Photo by Ye, E. and Zhao, Y.)

85. ***Yunnanilus macrogaster*** Kottelat and Chu, 1988 (Fig. 170). **Common name:** Big stomach

Fig. 169 Distribution of *Yunnanilus longidorsalis*

Fig. 170 *Yunnanilus macrogaster*, holotype, KIZ1980004273, standard length 70.5 mm. (Photo by He, M.)

Yunnan loach (E, translation from Chinese). **Etymology:** *macro* (Gr) large; *gaster* (Gr) stomach, an allusion to the large swollen stomach. **Major synonyms:** None. **Ecological classification:** Troglophile. **History:** First collected in 1980 (Kottelat and Chu 1988). **Morphological characteristics:** D iii, 8 A iii, 5 P i, 12–13, V i, 7. Body moderately elongated and compressed; lateral line short, with 6–10 pores; cephalic lateral line pores present; caudal peduncle 1.43–1.45 times longer than deep; eye diameter 17–19% dorsal HL. Maximum standard length: 63.2 mm (Yang 1990). Body yellowish brown with irregularly distributed dark brown spots on its upper two-thirds; top of head darker, a dark blotch on opercle behind eye; fins hyaline; black basal caudal bar vertical but not reaching upper and lower margins of caudal peduncle (Kottelat and Chu 1988). **Troglomorphic characters:** None known. **Distribution:** China: Yunan Province: Luoping city: Datangzi village (ca. 24°52′ N, 104°18′ E) (Fig. 171). **Habitat and ecology:** Small marsh, about 0.1 km^2 (Kottelat and Chu 1988), actually it is the exit of a subterranean stream. This species is sympatric with *Yunnanilus macrogaster*, *Y. paludosus* and *Sinocyclocheilus malacopterus*. **Food and feeding:** Insects and worms. **Reproduction and development:** External reproduction, non-guarder (Breder and Rosen 1966). **Other behavior:** Unknown. **Conservation status:** Undetermined. **Major threats:** Anthropogenic development. **Conservation plans:** None. **Phylogenetic relationships:** None known.

86. *Yunnanilus macrolepis* Li, Tao and Mao, 2000 (in Li et al. 2000a) (Fig. 172). **Common name:** Big scale Yunnan loach (E, translation from Chinese). **Etymology:** *macros* (Gr) long, big; *lepis* (Gr) scale. **Major synonyms:** None. **Ecological classification:** Troglophile. **History:** Type specimens were collected

Fig. 171 Distribution of *Yunnanilus macrogaster*

in 1993 (Li et al. 2000a). **Morphological characteristics:** D iii, 11 A ii, 6 P i, 10–11, V i, 6. Body elongated and compressed; eye small; anterior nostrils tube-like, with a short distance to posterior one; mouth small and inferior; pairs of barbels long; dorsal fin insertion to snout tip longer than it to caudal fin base; all fins short; scale relatively big; lateral line incomplete. Maximum standard length: 92 mm (Li et al. 2000a). Fixed specimens have many brown spots on back and sides, all fins dark brownish. **Troglomorphic characters:** Possibly the small size of the eyes. **Distribution:** China: Yunnan Province: Luoping County: Xuetian Village (24°53′ N, 104°18′ E) (Fig. 173). **Habitat and ecology:** Exit of a subterranean stream. **Food and feeding:** Unknown. **Reproduction and development:** Unknown. **Other behavior:** Unknown. **Conservation status:** Undetermined. **Major threats:** No survey has been con-

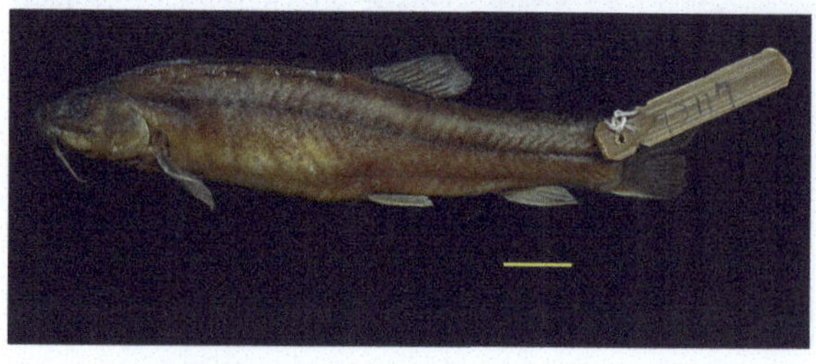

Fig. 172 *Yunnanilus macrolepis*, paratype, ASIZB73117, standard length 82 mm. (Photo by Ye, E. and Zhao, Y.)

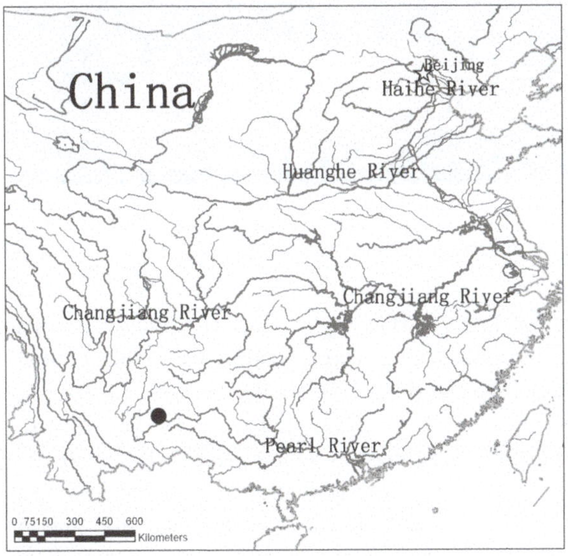

Fig. 173 Distribution of *Yunnanilus macrolepis*

ducted to assess its possible threats. **Conservation plans:** None. **Phylogenetic relationships:** None known.

87. *Yunnanilus nanpanjiangensis* Li, Mao and Lu, 1994 (in Li et al. 1994) (Fig. 174). **Common name:** Nanpanjiang Yunnan loach (E, translation from Chinese). **Etymology:** *nanpanjiangensis* (L) after the name of Nanpanjiang River, the type locality. **Major synonyms:** None. **Ecological classification:** Troglophile. **History:** Type specimens were collected in 1991. **Morphological characteristics:** D iii, 8–9 A ii, 6 P i, 10, V i, 6–7. Body elongated; naked or a few scales on caudal peduncle; anterior nostrils tube-like, with a short distance to posterior one; eye small; mouth inferior; dorsal spine soft; dorsal fin insertion to snout tip longer than it to caudal fin base; caudal fin forked; lateral line incomplete. Maximum standard length: 76 mm (Li et al. 1994). Male has a light black stripe along midline from posterior edge of operculum to caudal fin base. This stripe is not obvious in female. Both male and female have many black speckles on back and sides. **Troglomorphic characters:** Microphthalmic, almost entirely scaleless. **Distribution:** China: Yunnan Province: Luoping County: A'gang Town: She'Ao village (ca. 25°05′ N, 104°08′ E) (Fig. 175). **Habitat and ecology:** At the exit of subterranean waters. **Food and feeding:** Unknown. **Reproduction and development:** Unknown. **Other behavior:** Unknown. **Conservation status:** Undetermined. **Major threats:** No survey has been conducted to assess its possible threats. **Conservation plans:** None. **Phylogenetic relationships:** None known.

Fig. 174 *Yunnanilus nanpanjiangensis*, paratype, ASIZB78386, standard length 64 mm. (Photo by Ye, E.)

88. *Yunnanilus niger* Kottelat and Chu, 1988 (Fig. 176). **Common name:** Black body Yunnan loach (E, translation from Chinese). **Etymology:** *niger* (L) black. **Major synonyms:** None. **Ecological classification:** Troglophile. **History:** First collecting took place in 1980 (Kottelat and Chu 1988). **Morphological characteristics:** D iv, 8 A iii, 5 P i, 11, V i, 7. Body deep and compressed; no lateral line; no head pores; no scales along dorsal midline in from of dorsal fin; body depth 26% SL, 14 branched caudal rays. Maximum standard length: 62.5 mm (Kottelat and Chu 1988). Coloration of preserved specimens (in

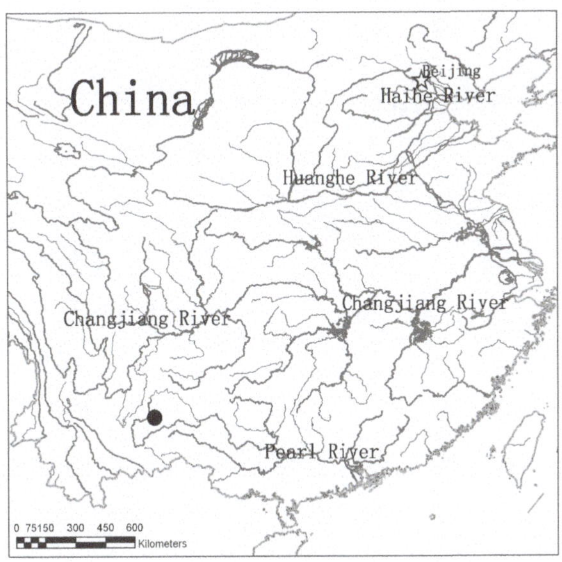

Fig. 175 Distribution of *Yunnanilus nanpanjiangensis*

Fig. 176 *Yunnanilus niger*, holotype, KIZ1980001275, standard length 62.5 mm. (Photo by He, M.)

Fig. 178 *Yunnanilus obtusirostris*, holotype KIZ1987004000, standard length 33.5 mm. (Photo by He, M.)

alcohol) is brownish black with dorsal, anal and pelvic fins black, caudal and pectoral is grayish, and a continuous black basal caudal bar (Kottelat and Chu 1988). **Troglomorphic characters:** None known; lack of scales may be one. **Distribution:** China: Yunan Province: Luoping city: Datangzi village (ca. 24°52′ N, 104°18′ E) (Fig. 177). **Habitat and ecology:** See above, *Yunnanilus macrogaster.* **Food and feeding:** Unknown. **Reproduction and development:** External, non-guarder (Breder and Rosen 1966). **Other behavior:** Unknown. **Conservation status:** Undetermined. **Major threats:** Environmental degradation. **Conservation plans:** None. **Phylogenetic relationships:** None known.

89. ***Yunnanilus obtusirostris*** Yang, 1995 (in Yang and Chen 1995) (Fig. 178). **Common name:** Obtuse snout loach. **Etymology:** *obtus* (L) obtuse; *rostrum* (L) mouth, snout. **Major synonyms**: None. **Ecological classification**: Troglophile. **History:** First record might be in 1958 (Yang and Chen 1995). **Morphological characteristics**: D iii,8; A iii,5; P i, 9–10; V i, 6–7. Snout obtuse and snout length is less than interorbital width. Nostrils well separated, the anterior tubular. Upper lip smooth, lower one wrinkled, interrupted at middle. Eyes present. Lateral line absent (Yang and Chen 1995). Maximum standard length: 41.5 mm (Yang and Chen 1995). Live coloration is grayish with a pearl blue longitudinal stripe on both sides (Yang and Chen 1995). **Troglomorphic characters:** None. **Distribution:** China: Yunnan: Chengjiang County: West Dragon Spring and Lake Fuxian (24 °30′ N, 102°53′ E) (Fig. 179). **Habitat and ecology**: Lives in the spring water, rivulets and an estuary around Lake Fuxian with an average water depth of 0.5–3.0 m. There are a lot of macrophytes in that habitat (Yang and Chen 1995). **Food and feeding:** *Chironomus* (85%), Coleoptera pupa (6%) and *Gammarus* (5%) (Yang and Chen 1995). **Reproduction and development:** The reproductive season is from June to July. **Other behavior:** Unknown. **Conservation status**: Undetermined. **Major threats:** No survey has been conducted to assess its possible threats. **Conservation plans:** None. **Phylogenetic Relationship:** None Known.

Fig. 177 Distribution of *Yunnanilus niger*

90. ***Yunnanilus paludosus*** Kottelat and Chu, 1988 (Fig. 180). **Common name:** Marsh Yunnan loach (E, translation from Chinese). **Etymology:** *palus, paludis* (L) marsh; after the habitat where the types were collected. **Major synonyms:** None. **Ecological classification:** Troglophile. **History:** First collection was in 1980 (Kottelat and Chu 1988). **Morphological characteristics:** D iii, 8 A iii, 5–6 P i, 9–11, V i, 6–7. Body elongated and slightly compressed, body depth 14–16% SL; caudal peduncle slender, 1.98–2.35 times longer than deep; body nearly naked, scales present only on dorsal and ventral margin of caudal

Fig. 179 Distribution of *Yunnanilus obtusirostris*

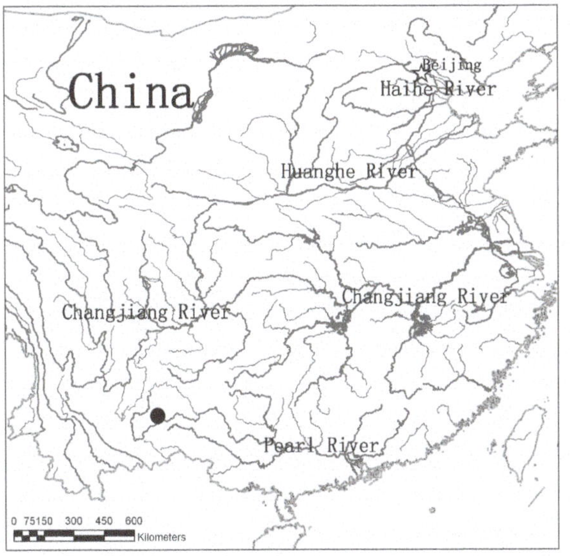

Fig. 181 Distribution of *Yunnanilus paludosus*

peduncle; lateral line pores 16–25; head pores present. Maximum standard length: 79.9 mm (Kottelat and Chu 1988). Body with 3 irregular longitudinal rows of spots; a sinuous black bar at base of caudal fin (Kottelat and Chu 1988). **Troglomorphic characters:** None except possibly for the reduction in the number of scales. **Distribution:** China: Yunan Province: Luoping City: Datangzi Village (ca. 24°52′ N, 104° 18′ E) (Fig. 181). **Habitat and ecology:** See above, *Yunnanilus macrogaster*. **Food and feeding:** Flies (Kottelat and Chu 1988). **Reproduction and development:** Egg diameter is about 0.8–1.0 mm (Kottelat and Chu 1988). **Other behavior:** Unknown. **Conservation status:** Undetermined. **Major threats:** Environmental change due to the local development. **Conservation plans:** None. **Phylogenetic relationships:** None known.

Fig. 180 *Yunnanilus paludosus*, holotype, KIZ1980001276, standard length 79.9 mm. (Photo by He, M.)

91. *Yunnanilus parvus* Kottelat and Chu, 1988 (Fig. 182). **Common name:** Small stripe Yunnan loach (E, translation from Chinese). **Etymology:** *parvus* (L) small. **Major synonyms:** None. **Ecological classification:** Troglophile. **History:** First collecting took place in 1984 (Kottelat and Chu 1988). **Morphological characteristics:** D iv, 8 A iii, 5 P i, 11–12, V i, 8. Body spindly and compressed; lateral line reaching, at most, to tip of pectoral fins and with only 7–16 pores; cephalic lateral line pores present; caudal peduncle 0.91–1.23 times longer than deep, its depth 11–13% SL; maximum size 37.6 mm. Maximum standard length: 38 mm (Kottelat and Chu 1988). Body with three longitudinal rows of spots, variable; head dark on top; a dark blotch on opercle; dorsal and anal fins have dark markings (Kottelat and Chu 1988). **Troglomorphic characters:** None

Fig. 182 *Yunnanilus parvus*, holotype, KIZ1984001244, standard length 26 mm. (Photo by He, M.)

Fig. 183 Distribution of *Yunnanilus parvus*

known. **Distribution:** China: Yunan Province: Kaiyuan City: Nandong Cave (ca. 23°39′ N, 103°17′ E) (Fig. 183). **Habitat and ecology:** Karst cave located a few kilometers south of Kaiyuan. A swift stream comes out from the cave all year round (Kottelat and Chu 1988). **Food and feeding:** Unidentifiable animal remains or detritus in stomach contents (Kottelat and Chu 1988). **Reproduction and development:** Egg diameter is around 1 mm (Kottelat and Chu 1988). **Other behavior:** Unknown. **Conservation status:** Undetermined. **Major threats:** Nandong Cave has been developed as a public park. Lights, signs, and some construction have taken place in the cave which has become a popular tourist spot. **Conservation plans:** None. **Phylogenetic relationships:** This species belongs to the *Y. pleurotaenia* group and seems to be most closely related to *Y. pulcherrimus* (Yang et al. 2004). **Remarks:** Original authors could not ascertain whether this fish occurs inside the cave since it cannot be penetrated by humans (Kottelat and Chu 1988).

92. *Yunnanilus pulcherrimus* (Yang et al. 2004) (Fig. 184). **Common name:** Beautiful barred Yunnan loach (E, translation from Chinese). **Etymology:** *pulcherrimus* (L) most beautiful. **Major synonyms:** None. **Ecological classification:** Troglophile. **History:** First collected in 1999 (Yang et al. 2004). **Morphological characteristics:** D iii, 8–9 A iii, 5–6 P i, 8–10, V i, 6–7. Body spindly and compressed;

Fig. 184 *Yunnanilus pulcherrimus*, holotype, KIZ1999001786, standard length 39.6 mm. (Photo by He, M.)

anterior nostrils tube-like, with a short distance to posterior one; eye big; mouth small, sub-ventral; dorsal fin insertion midway from snout tip to caudal fin base; scale small; lateral line incomplete. Maximum standard length: 49 mm (Lan and Zhang 2006). Coloration of preserved specimens is gray-brownish on back and upper part of sides, lower part light yellowish; 12–17 black-brown horizontal strips around the body; all fins light grayish or grayish. **Troglomorphic characters:** None known. **Distribution:** China: Guangxi Zhuang Autonomous Region: Du'an County (ca. 23°56′ N, 108°05′ E) (Fig. 185). **Habitat and ecology:** Subterranean stream. **Food and feeding:** Unknown. **Reproduction and development:** Unknown. **Other behavior:** Unknown.

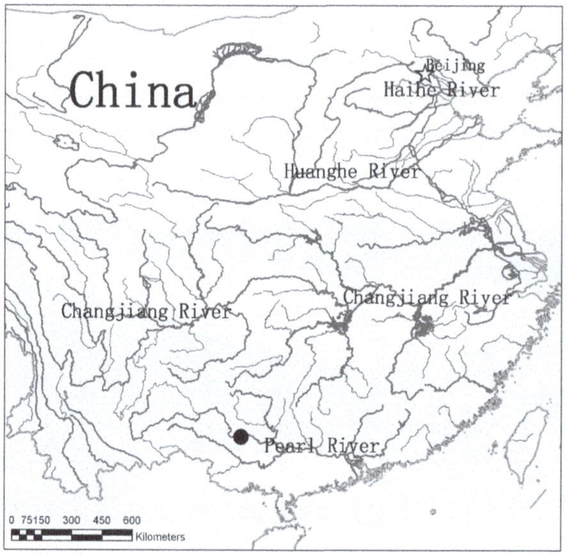

Fig. 185 Distribution of *Yunnanilus pulcherrimus*

Conservation status: Undetermined. **Major threats:** No survey has been conducted to assess its possible threats. **Conservation plans:** None. **Phylogenetic relationships:** It belongs to the *Y. pleurotaenia* group, and seems to be most closely related to *Y. parvus* (Yang et al. 2004).

Acknowledgments This study was funded by the projects of the National Natural Sciences Fund of China (NSFC-30470251 and NSFC-J0030092 to Y. Zhao, and NSFC-30730017 to X. Chen), and Basic Research Program of China (973 Program, 2007CB411600 to X. Chen). Many thanks go to Zhang C (IZB) for providing some suggestions and necessary stuffs. Appreciation to Ye E (IZB), Lan J (Du'an Fishery Bureau), He M (KIZ) and Li J (KIZ) for taking some pictures. Thanks to David L.G. Noakes for editorial suggestions.

References

Berra TM, Humphrey JD (2002) Gross anatomy and histology of the hook and skin of forehead brooding male nurseryfish, *Kurtus gulliveri*, from northern Australia. Environ Biol Fishes 65:263–270 doi:10.1023/A:1020523905635

Bleeker P (1871) Mémoire sur les cyprinoïdes de Chine. Verh Akad Amsterdam 1–91

Breder CM, Rosen DE (1966) Modes of reproduction in fishes. T.F.H., Neptune City, New Jersey

Brown RW (1956) Composition of scientific words. Smithsonian Institution, Washington, DC

Burr BM, Adams GL, Krejca J, Paul RJ, Warren ML Jr (2001) Troglomorphic sculpins of the *Cottus carolinae* species group in Perry County, Missouri: distribution, external morphology, and conservation status review. Environ Biol Fishes 62:279–296 doi:10.1023/A:1011819922403

Chen J, Lan J (1992) Description of a new genus and three new species of fishes from Guangxi, China (Cypriniformes: Cyprinidae, Cobitidae). Acta Zootax Sin 17:104–109

Chen X, Yang J (2005) *Triplophysa rosa* sp. nov.: a new blind loach from China. J Fish Biol 66:599–608 doi:10.1111/j.0022-1112.2005.00622.x

Chen J, Zhao Z, Zheng J, Li D (1988a) Description of three new barbine from Guizhou, China. Acta Acad Med Zunyi 11:1–4

Chen Y, Chu X, Luo Z, Wu J (1988b) A new blind cyprinid fish from Yunnan, China with a reference to the evolution of its characters. Acta Zool Sin 34:64–70

Chen Y, Yang J, Xu G (1992) A new blind loach of *Triplophysa* from Yunnan stone forest with comments on its phylogenetic relationship. Zool Res 13:17–23

Chen Y, Yang J, Zhu Z (1994) A new fish of the genus *Sinocyclocheilus* from Yunnan with comments on its Characteristic adaptation (Cypriniformes: Cyprinidae). Acta Zootax Sin 19:246–253

Chen Y, Yang J, Lan J (1997) One new species of blind cavefish from Guangxi with comments on its phylogenetic status (Cypriniformes: Cyprinidae: Barbinae). Acta Zootax Sin 22:219–223

Chen Y, Yang J, Sket B, Aljancic G (1998) A new blind caveloach of *Paracobitis* with comment on its characters evolution. Zool Res 19:59–63

Chen X, Cui G, Yang J (2004) A new cave-dwelling fish species of genus *Triplophysa* (Balitoridae) from Guangxi, China. Zool Res 25:227–231

Chen X, Yang J, Cui G (2006) A new fish species of genus *Sinocrossocheilus* (Cyprinidae) from Guangxi, China. Zool Res 27:81–85

Christiansen KA (1962) Proposition pour la classification des animaux cavernicoles. Spelunca Mem 2:76–78

Chu X, Chen Y (1979) A new blind cobitid fish (Pisces, Cypriniformes) from subterranean waters in Yunnan, China. Acta Zool Sin 25:285–287

Chu X, Chen Y (1982) A new genus and species of blind cyprinid fish from China with special reference to its relationships. Acta Zool Sin 28:383–388

Chu X, Cui G (1985) A revision of Chinese cyprinid genus *Sinocyclocheilus* with reference to the interspecific relationships. Acta Zootax Sin 10:435–441

Chu X, Cui G (1989) Barbinae. In: Chu X, Chen Y (eds) The fishes of Yunnan, China Part I Cyprinidae. Science, Beijing, pp 135–229

Dai D (1988) Un nouveau poisson cavernicole. pp. 88-89 Guizhou expe '86: Première expedition speleologique Franco-Chinoise dans le centre et le sud de la Province du Guizhou., Exped. Speleologiqu. Federation Franxaise de Speliologie, Paris

Du LN, Chen XY, Yang JX (2008) A review of the Nemacheilinae genus *Oreonectes* Günther with descriptions of two new species (Teleostei: Balitoridae). Zootaxa 1729:23–36

Ercolini A, Berti R, Chelazzi L, Messana G (1982) Researches on the phreatobic fishes of Somalia: achievements and prospects. Monit Zool Ital 17:219–241

Fang PW (1936) *Sinocyclocheilus tingi*, a new genus and species of Chinese barbid fishes from Yunnan. Sinensia 7:588–593

Fang S (1981) Genus *Crossocheilus*. In: F.I.o.G.Z.A. Region & C.A.o.S. Institute of Zoology (eds) Freshwater fishes of Guangxi, China (1st ed). Guangxi People's Publishing House, Nanning, pp 101–103

Gotch AF (1995) Latin names explained.. A guide to the scientific classification of reptiles, birds & mammals. Facts on File, New York

He L, Wang X, Chen Q, Xiang J (2006) Morphological description on *Triplophysa xiangxiensis*. Fresh Fish 36:56–58

Kottelat M (1988) Two species of cavefishes from northern Thailand in the genera *Nemacheilus* and Homaloptera (Osteichthyes: Homalopteridae). Rec Aust Mus 40:225–231

Kottelat M, Chu X (1988) Revision of *Yunnanilus* with descriptions of a miniature species flock and six new species from China (Cypriniformes: Homalopteridae). Environ Biol Fishes 23:65–94 doi:10.1007/BF00000739

Lan J, Zhang C (2006) Cobitidae. In: Zhou J, Zhang C (eds) Freshwater fishes of Guangxi, China (2nd edi). Guangxi People's Publishing House, Nanning, pp 80–122

Lan J, Yang J, Chen Y (1995) Two new species of the subfamily nemacheilinae from Guangxi, China (Cypriniformes: Cobitidae). Acta Zootax Sin 20:366–372

Lan J, Yang J, Chen Y (1996) One new species of cavefish from Guangxi. Zool Res 17:109–112

Lan J, Zhao Y, Zhang C (2004) A new species of the genus *Sinocyclocheilus* from China (Cypriniformes: Cyprinidae: Barbinae). Acta Zootax Sin 29:377–380

Li W (1985) Description on four species of *Sinocyclocheilus* from Yunnan, China (Pisces: Cyprinidae). Zool Res 6:423–429

Li G (1989) On a new fish of the genus *Sinocyclocheilus* from Guangxi China. Acta Zootax Sin 14:123–126

Li W (1992) Description on three species of *Sinocyclocheilus* from Yunnan, China. Acta Hydrobiol Sin 16:57–61

Li W (2004) The three new species of Cobitidae from Yunnan, China. J Jishou Univ 25:93–96 Nat Sci Ed

Li W, Lan J (1992) A new genus and three new species of Cyprinidae from Guangxi, China. J Zhanjiang Fish Coll 12:46–51

Li W, Tao J (1994) A new species of Cyprinidae from Yunnan—*Sinocyclocheilus rhinocerous* sp. nov. J Zhanjiang Ocean Univ 14:1–3 former J Zhanjiang Fish Coll

Li W, Zhu Z (2000) A new species of *Triplophysa* from cave Yunnan. J Yunnan Univ 22:396–398

Li W, Mao W (2007) A new species of the genus *Sinocyclocheilus* living in cave from Shilin, Yunnan, China (Cypriniformes, Cyprinidae). Acta Zootax Sin 32: 226–229

Li W, Mao W, Sun R, Lu Z (1994) Two new species of *Yunnanilus* from Yunnan province, China (Cypriniformes: Cobitidae). Acta Zootax Sin 19:370–374

Li W, Wu D, Chen A, Tao J (1997) Histological study on the horn-like projection of the head of *Sinocyclocheilus rhinocerous*. J Yunnan Univ 19:426–428

Li W, Wu D, Chen A (1998) Two new species of *Sinocyclocheilus* from Yunnan (Cypriniformes: Cyprinidae). J Zhanjiang Ocean Univ 18:1–5

Li W, Tao J, Mao W, Lu Z (2000a) Two new species of *Yunnanilus* from eastern Yunnan, China (Cypriniformes: Cobitidae). Acta Zootax Sin 25:349–353

Li W, Xiao H, Zan R, Luo Z, Li H (2000b) A new species of *Sinocyclocheilus* from Guangxi, China. Zool Res 21:155–157

Li W, Zong Z, Nong R, Zhao C (2000c) A new species of *Sinocyclocheilus* from Yunnan—*Sinocyclocheilus maculatus* Li, sp. nov. J Yunnan Univ 22:79–80

Li W, Liao Y, Yang H (2002a) Two new species of *Sinocyclocheilus* from Eastern Yunnan, China. J Yunnan Agric Univ 17:161–163

Li W, Mao W, Lu Z (2002b) A new species of *Sinocyclocheilus* from Yunnan China. J Zhanjiang Ocean Univ 22:1–3

Li W, Mao W, Lu Z, Tao J (2002c) Two new species of Cyprinidae from Yunnan. J Yunnan Univ 24:385–387

Li W, Lan J, Chen S (2003a) A new species of cave *Sinocyclocheilus* from Guangxi—*Sinocyclocheilus Jiuxuensis* Li *et* Lan, sp. nov. J Guangxi Norm Univ 21:83–85

Li W, Mao W, Lu Z, Yan W (2003b) The two new species of *Sinocyclocheilus* from Yunnan, China. J Jishou Univ 24:63–65 Natur Sci Ed

Li W, Ran J, Chen H (2003c) A new species of Cave *Sinocyclocheilus* in Guizhou and its adaptation comment. J Jishou Univ 24:61–63 Nat Sci Ed

Li W, Xiao H, Zan R, Luo Z, Ban C, Fen J (2003d) A new species of *Sinocyclocheilus* from caves in Guangxi. J Guangxi Normal Univ 21:80–81

Li W, Xiao H, Feng H, Zhao H (2005a) A new species of *Sinocyclocheilus* from Yunnan: *Sinocyclocheilus aluensis*. J Zhanjiang Ocean Univ 25:1–3

Li W, Xiao H, Jin X, Wu D (2005b) A new species of *Sinocyclocheilus* from Yunnan, China. Southw China J Agric Sci 18:90–91

Li W, Ran J, Chen H (2006) A new species of *Paracobitis* from Guizhou, China. J Zhanjiang Ocean Univ 26:1–2

Li W, Yang H, Han F, Tao C, Hong Y, Chen H (2007) A new species in cave of blind *Sinocyclocheilus* from Yunnan, China (Cypriniformes: Cyprinidae). J Guangdong Ocean Univ 27:1–3

Liang L, Liu C, Wu Q (1987) Description of a new species of subfamily Labeoinae (Cyprinidae) from Guangxi. J Guangxi Agri Coll 2:77–80

Liao J, Wang D, Luo Z (1997) A new species and a new subspecies of *Schistura* from Guangxi and Guizhou, China (Cypriniformes: Cobitidae: Noemacheilinae). Acta Acad Med Zunyi 20:4–7

Lin R, Luo Z (1986) A new blind barbid fish (Pisces, Cyprinidae) from subterranean water in Guangxi, China. Acta Hydrobiol Sin 10:380–382

Mao W, Lu Z, Li W, Ma H, Huang G (2003) A new species of *Sinocyclocheilus* (Cyprinidae) from cave of Yunnan, China. J Zhanjiang Ocean Univ 23:1–3

Nelson JS (2006) Fishes of the world. Wiley & Sons, New York

Pellegrin J (1931) Description de deus cyprinidés nouveaux de Chine appartenant au genre *Schizothorax* Heckle. Bull Soc Zool Fr 56:145–149

Ran J, Li W, Chen H (2006) A new species blind loach of *Paracobitis* from Guangxi, China (Cypriniformes: Cobitidae). J Guangxi Norm Univ 24:81–82

Regan CT (1904) On a collection of fishes made by Mr. John Graham at Yunnan Fu. Ann Mag Nat Hist 7:190–194

Riehl R, Baensch HA (1991) Aquarien Atlas. Band. 1. Melle: Mergus. Verlag für Natur- und Heimtierkunde, Germany

Romero A (2001) Scientists prefer them blind: the history of hypogean fish research. Environ Biol Fishes 62:43–71 doi:10.1023/A:1011830329016

Romero A, Paulson KM (2001) It's a wonderful hypogean life: a guide to the troglomorphic fishes of the world. Environ Biol Fishes 62:13–41 doi:10.1023/A:1011844404235

Romero A, Green SM (2005) The end of regressive evolution: examining and interpreting the evidence from cave fishes. J Fish Biol 67:3–32

Sbordoni M, De Matthaeis CE, Mattoccia M, Berti R, Sbordoni V (1996) Genetic variability and differentiation of hypogean cyprinid fishes from Somalia. J Zoological Syst Evol Res 34:75–84

Shan X, Lin R, Yue P, Chu X (2000) Barbinae. In: Yue P (ed) Fauna Sinica, Osteichthyes, Cypriniformes III. Science, Beijing, pp 3–170

Shi D, Zhou J, Liang H, He A (1994) Distribution and living environment of the *Sinocyclocheilus anatirostris*. Zool Res 15(supplement):196–197

Su R, Yang J, Cui G (2003) Taxonomic review of the genus *Sinocrossocheilus* Wu (Teleostei: Cyprinidae), with a description of four new species. Zool Stud 42:420–430

Trajano E, Mugue N, Krejca J, Vidthayanon C, Smart D, Borowsky R (2002) Habitat, distribution, ecology and behavior of cave balitorids from Thailand (Teleostei: Cypriniformes). Ichthyol Explor Freshwat 13:169–184

Wang D, Chen Y (1989) Descriptions of three new species of Cyprinidae from Guizhou province, China (Cypriniformes: Cyprinidae). Acta Acad Med Zunyi 12:29–34

Wang D, Liao J (1995) The karyotype of a cave-living fish *Ceratobarbus biangularis*. Acta Acad Med Zunyi 18:85–86

Wang D, Liao J (1997) A new species of *Sinocyclocheilus* from Guizhou, China (Cypriniformes: Cyprinidae: Barbinae). Acta Acad Med Zunyi 20:1–3

Wang D, Li D (2001) Two new species of the genus *Triplophysa* from Guizhou, China. Acta Zootax Sin 26:98–101

Wang S, Xie Y (2004) China species red list. Higher Education, Beijing

Wang D, Huang Y, Liao J, Zheng J (1995) Taxonomic revision of the genus *Gibbibarbus* Dai. Acta Acad Med Zunyi 18:166–168

Wu X (1963) Economic animals in China: Freshwater Fishes. Science Press, Beijing, p 159

Wu X, Lin R, Chen J, Chen X, He M (1977) Barbinae. In: Wu X (ed) Cyprinid fishes in China. Shanghai Scientific and Technical, Shanghai, pp 261–265

Wu HL, Shao KT, Lai CF (1999) Latin-Chinese dictionary of fishes names. The Sueichan, Keelung

Xiao H, Li W, Zan R (2004) The three new species of *Sinocyclocheilus* from Kunming, Yunnan. Southwest China J Agri Sci 17:521–524

Yang J (1990) Noemacheilinae. In: Chu X, Chen Y (eds) The fishes of Yunnan, China, Part II Cyprinidae. Science, Beijing, pp 12–63

Yang J, Chen Y (1995) The biology and resource utilization of the fishes of Fuxian Lake, Yunnan. Yunnan Science and Technology, Kunming

Yang G, Yuan F, Liao R (1986) A new blind cobitidae fish from the subterranean water in Xiangxi, China. J Huazhong Agric Univ 5:219–223

Yang J, Chen Y, Lan J (1994) *Protocobitis typhlops*, a new genus and species of cave loach from China (Cypriniformes: Cobitidae). Ichthyol Explor Freshwat 5:91–96

Yang J, Chen X, Lan J (2004) Occurrence of two new plateau-indicator loaches of Nemacheilinae (Balitoridae) in Guangxi with reference to zoogeographical significance. Zool Res 25:111–116

Yang J, Pan X, Li Z (2007) Preliminary report on the successful breeding of the endangered fish *Sinocyclocheilus grahami* endemic to Dianchi Lake. Zool Res 28:329–331

Yue P, Chen Y (1998) China red data book of endangered animals, pisces. Science, Beijing

Zhang C (1986) On the ecological adaptation and geographical distribution of the barbine fish *Varicorhinus* (*Scaphesthes*) *macrolepis* (Bleeker). Acta Zool Sin 32:266–272

Zhang C, Dai D (1992) A new species of *Sinocyclocheilus* from Guangxi, China (Cypriniformes: Barbinae). Acta Zootax Sin 17:377–380

Zhang E, Chen J (1997) A taxonomical revision of the genus *Crossocheilus* in China and a description of a new genus from China (Pisces: Cyprinidae). Acta Zootax Sin 22:321–326

Zhang C, Zhao Y (2001) A new fish of *Sinocyclocheilus* from Guangxi, China with a comment on its some biological adaptation. Acta Zootax Sin 26:102–107

Zhang E, Fang F (2005) *Linichthys*: a new genus of Chinese cyprinid fishes (Teleostei: Cypriniformes). Copeia 2005:61–67 doi:10.1643/CI-03-256R2

Zhang Z, Zhao Y, Zhang C (2006) A new blind loach, *Oreonectes translucens* (Teleostei: Cypriniformes: Nemacheilinae), from Guangxi, China. Zool Stud 45:611–615

Zhao Y (2006) An endemic cavefish genus *Sinocyclocheilus* in China—species diversity, systematics and zoogeography (Cypriniformes: Cyprinidae). Chinese Academy of Sciences, Beijing

Zhao Y, Zhang C (2006) Past research and future development on endemic Chinese cavefish of the genus *Sinocyclocheilus* (Cypriniformes, Cyprinidae). Acta Zootax Sin 31:769–777

Zhao Y, Zhang C, Lan J (2006a) *Sinocyclocheilus*. In: Zhou J, Zhang C (eds) Freshwater fishes of Guangxi, China (2nd edition). Guangxi People's Publishing House, Nanning, pp 259–282

Zhao Y, Watanabe K, Zhang C (2006b) *Sinocyclocheilus donglanensis*, a new cavefish (Teleostei: Cypriniformes) from Guangxi, China. Ichthyol Res 53:121–128

Zhao Y, Lan J, Zhang C (2009a) A new cavefish species, *Sinocyclocheilus brevibarbatus* (Teleostei: Cypriniformes: Cyprinidae), from Guangxi, China. Environ Biol Fishes (this issue)

Zhao Y, Zhang C, Zhou J (2009b) *Sinocyclocheilus guilinensis*, a new species from an endemic cavefish group (Cypriniformes: Cyprinidae) in China. Environ Biol Fishes (this issue)

Zheng B (1981) *Oreonectes anophthalmus* Zheng, sp. nov. In: F.I.o.G.Z.A. Region & C.A.o.S. Institute of Zoology (eds) Freshwater fishes of Guangxi, China (1st ed). Guangxi People's Publishing House, Nanning, pp 162–163

Zheng C, Xie J (1985) One new carp of the genus *Sinocyclocheilus* (Barbinae, Cyprinidae) from Guizhou Province, China. In: Liu J (ed) Transactions of the Chinese Ichthyological Society, No.4. Science, Beijing, pp 123–126

Zheng J, Wang J (1990) Description of a new species of the genus *Sinocyclocheilus* from China (Cypriniformes: Cyprinidae). Acta Zootax Sin 15:251–254

Zhou W, He J (1989) A new species of dwarfism in *Yunnanilus* (Cypriniformes: Cobitidae). Acta Zootax Sin 14:380–384

Zhou S, Li G (1998a) A new species of *Sinocyclocheilus* from Guangxi (Cypriniformes: Cyprinidae: Barbinae). Guangxi Sci 5:139–141,149

Zhou S, Li G (1998b) A new species of the genus *Sinocyclocheilus* from of cave Guangxi China (Cypriniformes, Cyprinidae, Barbinae). In: Sun J (ed) Contributions from Tianjin Natural History Museum, 15, Ocean Press, Beijing, pp 9–12

Zhou J, Zhang C (2006) Freshwater fishes of Guangxi, China, 2nd edn. Guangxi People's Publishing House, Nanning

Zhou J, Zhang C, He A (2004) A new species of the genus *Sinocyclocheilus* from Guangxi, China (Cypriniformes, Cyprinidae). Acta Zootax Sin 29:591–594

Zhu S (1983) A new genus and species of Nemachilinae (Pisces: Cobitidae) from China. Acta Zootax Sin 8:311–313

Zhu S (1989) The loaches of the subfamily Nemacheilinae in China (Cypriniformes: Cobitidae). Jiangsu Science and Technology Publishing House, Nanjing

Zhu Y (2006) Labeoninae. In: Zhou J, Zhang E (eds) Freshwater fishes of Guangxi, China. 2nd edn. Guangxi People's Publishing House, Nanning, pp 310–345

Zhu S, Cao W (1987) The nomeacheiline fishes from Guangdong and Guangxi with descriptions of a new genus and three new species. Acta Zootax Sin 12:323–331

Zhu Y, Lu Y-J, Yang J-X, Zhang S (2008) A new blind underground species of the genus *Protocobitis* (Cobitidae) from Guangxi, China. Zool Res 29:452–454 doi:10.3724/SP.J.1141.2008.00452

The manufacturer's authorised representative in the EU is Springer Nature Customer Service Centre GmbH, Europaplatz 3, 69115 Heidelberg, Germany. If you have any concerns regarding our products, please contact ProductSafety@springernature.com

Printed and bound by CPI Group (UK) Ltd, Croydon, CR0 4YY
26/09/2026
02222907-0001